Nontarget Effects of Agricultural Fungicides

Author

Subhash C. Vyas, M.Sc.(Ag), Ph.D.
Fellow of the Phytopathological
Society of India and
Professor of Plant Pathology
Department of Plant Pathology
J. N. Agricultural University
College of Agriculture
Indore, India

CRC Press, Inc.
Boca Raton, Florida

Library of Congress Cataloging-in-Publication Data
Vyas, Subhash C., 1937-
  Nontarget effects of agricultural fungicides.

  Bibliography: p.
  Includes index.
  1. Fungicides. 2. Fungicides--Physiological effect.
3. Plants, Effect of fungicides on. 4. Microorganisms--
Physiology. 5. Fungicides--Environmental aspects.
I. Title.
SB951.3.V93   1988   632'.952   87-15860
ISBN 0-8493-6889-8

PREFACE

The greatest challenge of our time is to produce sufficient food for our rapidly growing population. Of the various measures available, improvement in agricultural productivity is judged to be the ultimate means of augmenting food production and supply. Protection of crops from diseases is an important aspect of productivity. Primary reliance in the control of plant diseases has been placed on the use of fungicidal compounds of intrinsic fungitoxicity, as this is a very flexible method of plant disease control. Fungicides are both systemic and nonsystemic (surface active). They control disease in many ways. Fungi are also plants; therefore, when chemicals are used to inhibit or kill fungi, it is possible that these chemicals might affect the plants themselves which are meant to be protected through their use. Application of fungicides on plant foliage or other parts and seed ultimately reaches the soil. The soil harbors several beneficial organisms which are responsible for a variety of reactions which enrich the fertility of the soil and make micro- and macronutrients available to the plants. Because fungicides are also biocides, their effect extends to micro- and macroorganisms, which is not intended. Since fungicides are chemicals and many are systemic, they can enter the plant system and alter its physiology and metabolism. Therefore, it might be expected that such compounds would in some cases alter plant nutrition and hence the performance of the crop. In other cases, they may prove to be phytotoxic. These constitute the nontarget effects of agricultural fungicides.

A great deal of information has been accumulated on the nontarget effects of agricultural pesticides in the recent past. These developments are reflected in the proliferation of research papers and reports in scientific journals throughout the world. The progress of research on nontarget effects of pesticides has been summarized periodically in reviews or popular articles and in occasional brief chapters in books, monographs, or bulletins. Yet, to the best of the author's knowledge, no one has presented a thorough discussion of the nontarget effects of agricultural fungicides. Besides the pressing need for such a comprehensive work, the experience of the author in researching this field has prompted this compilation of information on the subject in order to present it in a useful form. This book presents a broader view of the subject than that generally included in bulletins and manuals. Most previous discussions on this development have been of a controversial nature and have been included to stimulate thinking, especially among young scientists and students. The information presented in this volume represents a careful synthesis of research papers and articles. It is not intended to condemn, condone, or propound arguments regarding particular fungicides according to their effects on nontarget microorganisms, since available evidence will not permit it. Instead, general statements have been made and trends identified. In some cases, the information obtained by various researchers conflicts. This merely reflects the inadequacy of our knowledge of this subject. However, the differences of opinion are often because of inconsistent methods in determining effects of fungicides on nontarget microorganisms. In most cases, original papers and articles are consulted. However, when access to original papers was not possible, particularly because of linguistic limitations, the help of *Review of Applied Plant Pathology, Field Crop Abstracts, Residue Reviews, Chemical Abstracts,* and *Biological Abstracts* has been sought in order to do justice to the numerous original publications in Russian, German, French, and Spanish.

The author has tried to compile all relevant and up-to-date data in this book, which will be of immense help not only to students, research workers, and teachers but also to extension workers and scientists engaged in related industries.

This book contains 16 chapters on the nontarget effects of agricultural fungicides on plants and soil. Each chapter has an introduction followed by a presentation of the effects of nonsystemic and systemic fungicides. In many cases a future outlook has been discussed.

Literature pertaining to the chapters has been cited so that the reader can obtain more detailed information on the topic of interest.

The first chapter of this book includes the introduction to the subject, mentioning briefly the development of fungicides, their chemical classification, method of application, and group of pathogens controlled. The mode of action of fungicides has also been discussed, because fungi are also plants and any biochemical effect of fungicides will also affect the plants themselves. Chapter 2 deals with the phylloplane microflora, which are responsible for the variety of reactions on the leaf surface. They are the first living microorganisms to be affected immediately after the application of fungicides on foliage.

Fungicides are chemicals and their application on plants and soil may result in "phytotonic" effects by their influence on plant nutrition, physiological and biochemical processes, growth, and other parameters responsible for yield. These effects are discussed in Chapters 3 and 4. Chapter 5 is devoted to the mechanisms of these phytotonic effects.

Some fungicides are specific and selective in their action and control a particular disease. Such fungicides aggravate other diseases for which they are not intended; these are termed "iatrogenic" diseases, which are discussed in Chapter 6. Plant virus and viral disease control is discussed in Chapter 7. Some systemic and nonsystemic fungicides showing high effectivity against air pollutants, which are becoming a serious menace in industrial countries, are discussed in Chapter 8. Indiscriminate use of fungicides results in the appearance of a variety of symptoms. This is also due to fungicide selectivity and specificity. Chapter 9 is devoted to this topic. Fungicides applied in the environment for the control of plant diseases sooner or later reach the soil and affect microflora and their activity and other biochemical processes responsible for maintenance of fertility of soil; this is discussed in Chapter 10.

The zone around the root in the soil is called the rhizosphere and the microorganisms associated with this zone are designated as rhizosphere microflora, which are responsible for the occurrence of soil-borne diseases and also make available the micronutrients. Thus, the rhizosphere is an area of extensive microbial processes. The application of fungicides may result in an array of effects on these microflora. The effects of fungicides on rhizosphere microflora are discussed in Chapter 11. Mycorrhiza, a fungus-root association, is becoming very important to the fertility of soil, plant disease control, and the ever-increasing demand for fertilizers and pesticides. The effects of fungicides on mycorrhizae are discussed in Chapter 12. In Chapter 13, the nontarget effects of agricultural fungicides on algae, a microorganism which is a potential source of biofertilizer, are discussed. In Chapter 14, the effects of fungicides on rhizobium, a symbiotic root-bacterial association, responsible for fixing nitrogen in plants, are discussed.

Chapter 15 is devoted to the nontarget role of fungicides on biocontrol agents, which is the solution for the control of soil-borne diseases. Scientists, politicians, and the public are concerned with fighting pollution. Plant pathologists are also concerned with this problem. Hence, fungicides with indirect effects on disease control of plants are discussed in Chapter 16.

I wish to express my sincere gratitude to Drs. K. V. V. Prasad, P. P. Shastry, M. Thomas, and A. Kotasthane for their assistance; to Mr. Mohd. Ahsan Quadri, for typing the manuscript; and to the authorities of J. N. Agricultural University, Jabalpur (India) for their encouragement in the completion of this book.

S. C. Vyas
Indore, India
March 1986

AUTHOR

Subhash Chandra Vyas is Professor of Plant Pathology in the College of Agriculture, J. N. Agricultural University, Indore, India.

He gained his Masters of Science in Agriculture with specialization in Plant Pathology from Vikram University, Ujjain, India, in 1961 and his Doctorate degree in Plant Pathology with a minor in Biochemistry and Crop Production and Plant Nutrition from G. B. Pant University of Agriculture and Technology, Pantnagar, India, in 1972. As Professor of Plant Pathology he has published over 100 papers, several bulletins on oilseeds, and bibliographies. He has authored a book called *Systemic Fungicides,* which was first available in Asia.

He is engaged in an investigation of the effects of fungicidal chemicals on host, pathogens, and disease control and also in the development of new and improved methods for the control of oilseed diseases. He has worked as Principal Investigator for various projects funded by the Indian Council of Agricultural Research (ICAR) and the Madhya Pradesh Council of Science and Technology, Bhopal, India. He is a member of several prestigious professional societies, such as the Indian Phytopathological Society, the Indian Society of Seed Technologists, the Society of Mycology of Plant Pathology, and the Plant Protection Association of India. He was elected as Councilor for the Indian Phytopathological Society in 1986.

In addition, Dr. Vyas teaches advanced courses on fungicides and guides graduate students in their research at J. N. Agricultural University, Jabalpur, and now at Indore, India.

ACKNOWLEDGMENTS

I wish to thank the following for their kind support in supplying reprints of their papers for the preparation of this manuscript: Drs. J. B. Sinclair, University of Illinois, H. D. Sisler, University of Maryland, R. W. Smiley, Cornell University, and J. W. Eckert, University of California (USA); R. J. W. Byrde and K. J. Bent, University of Bristol, and E. Griffiths, University of Wales (UK); L. V. Edgington, University of Guelph (Canada); S. G. Georgopolous, Athens (Greece); J. Dekker and G. J. Bollen, The Netherlands; H. Lyr and H. Domsch, West Germany; F. J. Schwinn, Basel (Switzerland); G. Simon-Sylverstre, Versailles (France); Drs. M. N. Khare, J. N. Agricultural University, Jabalpur, N. Sethunathan, Central Rice Research Institute, Cuttack, V. P. Agnihotri, Indian Sugarcane Research Institute, Lucknow, P. N. Thapliyal and A. P. Sinha, Govindvallab Pant University of Agriculture and Technology, Pantnagar, N. G. K. Karanth, Indian Institute of Sciences, Banglore, D. Lalithakumari, Madras University, Dharam Vir, Indian Agricultural Research Institute, New Delhi, and R. K. Grover, Hariyana Agricultural University, Hissar (India). Thanks are also due to the following for their supply of product profiles and other information on their products: Du Pont, Rohm & Haas, Eli-Lilly, Dow Chemical Co., Uniroyal (USA); Ciba-Geigy (Switzerland); ICI and May & Baker (UK); Celamerck, Hoechst, and BASF, (West Germany); Nippon Soda (Japan); and several other companies. Lastly, thanks to my several colleagues and students for their untiring efforts and help in the completion of this volume, and to my wife, Usha, for her inspiration and patience.

To my teacher, Dr. Y. L. Nene, for his inspiration

TABLE OF CONTENTS

Chapter 1

INTRODUCTION

I. OVERVIEW

Protection of crops from diseases is an important segment of productivity. Primary reliance in the control of plant diseases has been placed on fungicidal compounds of intrinsic fungitoxicity which became standard inputs of intensive crop production throughout the world. As this is a very flexible method of plant disease control, it can be scheduled well in advance (preventive treatment) for anticipated problems or adopted quickly in an emergency (specific treatment), and this will continue to be true in the foreseeable future. A great number of fungicides (systemic and nonsystemic) which specifically and effectively control the diseases caused by fungi have been developed and voluminous information has been generated on several aspects of plant disease control. The information will have relevance when the nontarget effects of agricultural fungicides are discussed in the following chapters of this book. It is difficult to provide all the relevant information available in this field; however, it is imperative to summarize briefly the important landmarks in relation to fungicides in plant disease control. The impact of fungicides on plant disease control and their nontarget effects are not mutually exclusive.

Fungicides control seed- and soil-borne crop diseases by seed and soil treatment, respectively, and air-borne diseases by foliar application. Over the past century, fungicide protectants have been developed from simple fungicides acting as surface protectants (Table 1) to systemic fungicides (Table 2) and finally to nonfungitoxic systemic compounds (Table 3) that suppress fungal pathogenicity or accentuate natural plant resistance systems. Nevertheless, most older fungicides (Bordeaux mixture and sulfur) are still in use, having been only partially displaced by newer compounds with more sophisticated mechanisms. Fungicides do the job either by killing (fungicidal) or by inhibiting growth and/or reproduction (fungistatic). These processes are mediated in fungi by inhibiting respiration; several metabolisms, and DNA, RNA, protein, and sterol biosynthesis; etc. (Table 4). Some fungicides act in a more subtle and sophisticated manner, such as by production of phytoalexins, by altering host defense mechanisms, and as antipenetrants. The various aspects of fungicides have been eloquently reviewed and presented in two classic volumes: *Antifungal Compounds*[1,2] and *Systemic Fungicides.*[3] A symposium of the British Mycological Society was held in 1984 and the proceedings were published in *Mode of Action of Antifungal Compounds.*[4]

The most commonly used fungicides have been classified according to their development from metallic compounds to the present-day nonfungitoxic and sophisticated compounds:

1. First-generation compounds:	This group includes fungicides based on metallic compounds which are broad spectrum and surface protectants, such as copper, zinc, mercury, sulfur, etc. In some instances they cause toxicity and in other instances they mitigate the deficiency symptoms.
2. Second-generation compounds:	This group contains organic fungicides such as dithiocarbamates, phthalamides, chlorothalonil, quinones, triphenyltins, etc. They are nonsite-specific and surface-active but rarely show phytotoxicity. They cause many nontarget effects on plants and in soil.

Table 1
COMMON SURFACE FUNGICIDES IN USE IN THE WORLD[1,2]

Chemical	Sensitive fungi	Method of application[a]
Elemental sulfur	Powdery mildew	2
Copper compounds		
Cuprous oxide	Many fungi, bacteria	2, 3
Copper oxychloride		
Bordeaux mixture		
Mercurials		
Phenylmercuric chloride	Cereal seed treatment, pruning wound fungi	1, 3
Mercurous chloride	Club root	1, 3
Dithiocarbamates		
Maneb	Many fungi/crops	2
Mancozeb		
Zineb		
Thiram	Many fungi	1, 3
Dinitrophenols		
Dinocap	Powdery mildew	2
Binapacryl		
Phthalimides		
Captan	Many fungi	1, 3
Captafol		1—3
Dichlofluanid		
Isophthalonitriles		
Chlorothalonil	Various fungi	1—3
Dicarboximides		
Iprodione	*Sclerotinia* and *Botrytis*	1—3
Vinclozolin		

[a] 1 = seed treatment, 2 = foliar application, 3 = soil treatment.

3. Third-generation compounds: The third-generation compounds are systemic and include carboximides, acylalanines, pyrimidines, piperazines, thiophanates, triazoles, etc. These compounds are synthesized according to need. They are site-specific and therefore development of resistance in fungi is one of their handicaps.

4. Fourth-generation compounds: This group includes compounds which are non-fungitoxic in vivo, but interfere specifically with fungal parasite systems and accentuate host defense mechanisms. Some common examples are probenazole, fosetyl-Al, and some melanin biosynthesis inhibitors (tricyclozole, pyroquilon, chlorbenthiozone, phthalide, etc.).

II. SELECTIVITY OF FUNGICIDES

Fungicides used throughout the world for the control of plant diseases are given in Tables 1 to 4. It is relevant, therefore, to review information on their selectivity. Recognition of the effects of pesticides on nontarget organisms has led to the need for identifying selective

Table 2
COMMON SYSTEMIC FUNGICIDES IN USE IN THE WORLD[1-3]

Chemical	Trade name	Sensitive fungi	Method of application[a]
Acetamides			
Chlorflurazole	Udonkor	Powdery mildew	2
Cymoxanil	Curzate	Oomycetes	
Acylalanines			
Metalaxyl	Ridomil	Oomycetes	1—3
Furalaxyl	Fongarid		
Benalaxyl			
Ofurace			
Cyprofuram			
Oxadixyl			
Aliphatics			
Prothiocarb	Dynone, Pervicur	Oomycetes	3
Propamocarb	Dynone-N		1—3
Benzimidazoles			
Benomyl	Benlate	Ascomycetes	1—3
Carbendazim	Bavistin	Basidiomycetes	
Cypendazole			1,2
Thiabendazole			1,2
Carboximides			
Carboxin	Vitavax	Smut, Rhizoctonia	1,3
Oxycarboxin	Plantvax	Rust	3
Dicarboximides			
Procymidone	Sumisclex	*Botrytis, Monilinia*	2,3
Imidazoles			
Imazalil	Fungiflor	Broad spectrum (not effective against oomycetes)	2
Morpholines			
Tridemorph	Calixin	Rust	2
Dodemorph	Meltatox		
Organophosphates			
IBP	Kitazin	Rice blast	1—3
Edifenfos	Hinosan		2
Phosphites			
Aluminum-tris	Aliette	Oomycetes	2,3
Piperazines			
Triforine	Saprol	Rust and powdery mildew	1—3
Pyrimidines and pyridines			
Ethirimol	Milgo	Powdery mildew (cereals)	2, 3
Dimethirimol	Milcurb	Powdery mildew (cucurbits)	
Triarimol	Milcurb	Powdery mildew	2
Bupiramate	Nimrod		2,3
Triazoles			
Triazbutyl	Indar	Brown rust	1,2,3
Triadimefon	Bayleton	Rust and powdery mildew	1—3
Triadimenol	Bayton	Smut	2
Bitertanol	Baycor	Powdery mildew	1—3

[a] 1 = seed treatment, 2 = foliar application, 3 = soil treatment.

pesticides having species-specific responses. Adverse effects of pesticides on nontarget organisms aroused public outcry, perhaps more than possible ill effects on human health might have. However, fungicides are seldom involved in such cases. The mercurials are the only disease-control chemicals that can definitely be classified as harmful to the environment and to humans. However, the quantity of mercury used in agriculture is only 1/20 of that

Table 3
FUNGICIDES WITH AN INDIRECT MODE OF ACTION[20,21]

Chemical	Trade name	Sensitive fungi	Method of application[a]
Acylalanines			
Metalaxyl	Ridomil	Oomycetes	1—3
Furalaxyl	Fongarid		
Phosphites			
Aluminum-tris	Aliette	Oomycetes	2,3
Pyrimidines			
Triarimol	Milcurb	Powdery mildew	2
Triazoles			
Tricyclozole	Beam	Rice blast	2, 3
Miscellaneous			
Probenazole	Oryzamate	Rice blast and bacterial blight	2
WL-28325		Rice blast	

[a] 1 = seed treatment, 2 = foliar application, 3 = soil application.

Table 4
MODE OF ACTION OF FUNGICIDES USED IN PLANT DISEASE CONTROL[1-3]

Fungicides	Metabolic process	Biochemical pathway	Molecular site
Captan, thiram, folpet, chlorothalonil, zineb, mancozeb, and all metal-based fungicides	Multisite inhibitors	Tricarboxylic acid cycle	Electron transport chain
Dodine, guazatine	Membrane structure and function	—	Cell membrane
Kitazin, edifenfos, isoprothiolane	Cell wall synthesis	Chitin incorporation via membrane transport	Greenberg pathway
Dimethirimol, ethirimol, bupirimate	Nucleic acid synthesis	Inhibit adenine incorporation in RNA	Nucleus
Metalaxyl, furalaxyl, benalaxyl, ofurace, cyprofuram, metoxazon	Nucleic acid synthesis	Inhibit uredine incorporation in RNA	Nucleus
Benomyl, carbendazim, thiophanates, thiabendazole, chloroneb	Nuclear function	Mitotis and microtubule function	Tubulin
Carboxin	Respiration	Succinic dehydrogenase	Electron transport
Fenaminosulf	Respiration	NADH oxidase	Electron transport
Pyrazophos, tridemorph, terrazole	Respiration	Oxidation	Electron transport
Iprodione, vinclozolin, procymidone	Nuclear function, cell wall synthesis	Mitotic instability	Tubulin
Triforine, tridemorph, fenpropimorph, imazalil, fenarimol, triadimefon, triadimenol, bitertanol, diclobutrazole, propiconazole, etaconazole, euthiobate	Membrane synthesis and function	Ergosterol biosynthesis	14-Demethylation
Prothiocarb, propamocarb	Cell membrane	ND[a]	ND
Tricyclozole, pyroquilon, chloroenthiozone	Cell membrane	Melanin synthesis inhibitors	ND

[a] ND = no data.

used in industry[5] and even that amount is being reduced drastically around the world. The current trend of utilizing relatively specific-acting, nonpersistent pesticides in crop production is desirable in principle because such agents are less likely to adversely affect nontarget species.

Selectivity was known to exist among traditional fungicides for many years. However, existence of fungicide selectivity became better known after the introduction of systemic fungicides starting in the mid-1960s. Systemic fungicides are selectively toxic to processes unique to fungi. This selectivity is so specific that among the hundreds of different fungi, only certain taxonomic groups of fungi are sensitive to each particular systemic fungicide. A systemic fungicide is so selective because it is toxic to a single site within the fungus.[6] This can cause problems since a single change in the fungus can result in fungicide resistance. The threat of resistance is so acute that some of the fungicides developed became obsolete and were discontinued, resulting in the loss of millions of dollars; however, this resistance can be reduced by using systemic fungicides, such as ergosterol biosynthesis inhibitors (EBI) and/or topical fungicides in combination or alternatively in a disease-management program. The chances for development of resistance in EBIs are very limited.

A. Surface Fungicides

Very few traditional fungicides exhibit selectivity, and it is not as clear-cut as that found in systemic fungicides. Inorganic sulfur fungicides are more commonly effective against powdery mildews than other fungal pathogens. Dithiocarbamates are relatively less effective against powdery mildews, whereas copper-based fungicides are less effective against rust and powdery mildews. Captan is relatively more effective against *Pythium* and *Phytophthora* than against *Fusarium* and *Rhizoctonia*. Pentachloronitrobenzene (PCNB) is more effective against sclerotial fungi (e.g., *Sclerotinia, Sclerotium, Botrytis*) than it is against others. Dicarboximides such as vinclozolin, iprodione, and procymidone also indicate selectivity toward *Botrytis* and *Sclerotinia* fungal pathogens (Table 1).

B. Systemic Fungicides

Most systemic fungicides are selective (Table 2). Research during the last decade led to the discovery of systemic fungicides effective against most of the major fungal pathogens of world crops. These fungicides have been found to be effective when used as seed treatments, soil drenches, or foliar sprays. Most systemic fungicides move upwards, and some show downward movement, but few indicate ambimobile (two-directional) movement.[7] Control of plant diseases with systemic fungicides in general requires lesser amounts than topical systemic fungicides. Some fungicides such as metalaxyl and carbendazim have longer residual action (more than 30 days) than many others.

III. FUNGICIDES AND PLANTS

Studies conducted using radiolabeled compounds indicated that the fungicides are observed either in traces (protectants) or in sufficient amounts (systemic). On their absorption the chemicals are altered within the plant. Many chemicals of the third and fourth generations are not really fungicidal. A broader and more meaningful term is "xenobiotics",[8] which describes a chemical foreign to the plant. These compounds are systemic xenobiotics. The xenobiotics may enter a plant and move in the direction of evapotranspiration streams in the apoplast (i.e., the entire nonliving cell wall continuum of the plant) or they may move with the photosynthates to "sinks" in the symplast where the translocation is in the living parts of the cells. When the xenobiotic moves in both directions it is termed an "ambimobile". The symplast is enclosed by membranes, protoplasts, plasmadesmata, including the phloem, and sieve cells. The apoplast includes the cell wall and the cuticle including xylem vessels and tracheids. Long-distance transport is passive whereas symplast transport is active and requires expenditure of metabolic energy. Symplast transport takes place in the phloem.[9] Apoplastic compounds may enter plant roots through the cell walls of the root hairs and then pass to the intercellular spaces of the cortex in a manner similar to the ion-exchange phenomenon. This free space communicates freely with the soil environment. The cell wall pathway between the cortex and the stele is blocked at the endodermis by the impervious band of Casparian strips in the radial walls. Compounds which move entirely in the apoplast are unable to pass this barrier. Thus lanthanum accumulates in cell walls and on the outside of the plasmalemma of epidermal, cortical, and endodermal cells up to the Casparian strip but is unable to pass from the cortex to the stele. Systemic fungicides, because they are translocated in the plant, are often stable and specific in action; they may alter the morphology of the growing plant and its metabolism.[10]

Fungicides are biocides. Their introduction into the environment and soil for the control of plant diseases can be expected to result in a diverse array of effects on target as well as on nontarget organisms. Because of this it is often not immediately apparent whether the effects are target or nontarget. Only when the nontarget effects become apparent are the inadequacy of fungicides for generalized field use and the environmental and economic importance of nontarget effects realized. Most of the fungicides have a wide range of biological activities extending beyond the intended target function. Some synonyms of nontarget effects are "not-so-obvious effects", "side effects", or "unintended effects".[11] Analysis of nontarget effects of fungicides indicated that they can be classified as either direct or indirect. The fungicides are developed and designed for the control of plant diseases. They act on fungi through a variety of mechanisms (Table 4). However, to the farmer, disease control per se is not as important as is increase in yield, or, more precisely, profitability of crop production.

Since fungi are plants and basically all their metabolic processes are almost identical to those of plants, the mode of action of fungicides in fungal systems could bear a relationship to plant metabolic processes. Those concerned with promotion of fungicidal treatments are well aware of this and rightly stress the yield response of treatments rather than the disease control that they achieve. The increase in yield is normally accounted for by the control of well-defined diseases, but there are occasions when the explanation seems inadequate. For example, the extent of the disease may seem minimal or its control insufficient to have produced the observed effects. In these circumstances, it is difficult to believe that fungicides do not have some stimulative "phytotonic" effect on growth which is indepedent of normal disease control.[12,13] Perhaps the most important and least understood nontarget effects are those upon the host plant, which result in a change of the basic physiology of the plant. As discussed above, the presence of fungicides in plant systems profoundly influences the performance of the crop plant, its physiology, and hence its nutrition. It is not often possible to decide whether changes in fungicide-induced nutrient uptake that seem to have been brought about indicate changes in the nutrient requirement or result in more or less efficient uptake and utilization.[14] These changes result in an increase in yield which is not directly related to disease control but is due to a change in the developmental pattern of the crop. This difference in the developmental pattern might be increased tillering, earlier initiation of tiller buds or changes in the apical dominance allowing more rapid lateral growth. These effects closely resemble the effects of kinetin applications in crops. Some fungicides such as carbendazim (methyl-2-yl-benzimidazole carbamate) and carbendazim-generating compounds have been shown to produce cytokinin-like effects in plants.[15-17] These effects increase the physiological life of crops such as cereals and fruits, thus delaying senescence and increasing yield. The application of fungicides results in unexpected (nontarget) effects on yield when they are applied at maturity time or at the flag-leaf stage. This is due to control of weak pathogens at earhead formations. Several general effects of fungicides on host metabolism, which may interfere with yield without producing visible symptoms, are discussed in subsequent chapters. Sometimes late application of fungicides results in the control of saprophytes which would otherwise damage the epidermis and reduce the yield. These many examples of direct effects of fungicides on plants indicate that they should not be overlooked when drawing conclusions about increased yields after fungicidal application.[3] The subject is so important that in 1981 the European Plant Protection Organization (EPPO) devoted one complete issue of their prestigious journal to discussing the nontarget effects of agricultural fungicides.

The fungicide-induced physiological stress upon plants can affect root and foliage exudates and the microflora which interact with pathogens at the soil-root or environment-foliage interface. These "rhizosphere effects" or "phyllosphere effects" include changes in the activity of mycorrhizal fungi, rhizobia and other root microbes, associated organisms, and phylloplane microorganisms, resulting in a change in the basic physiology of the plant and influencing the yield of crop plants.[18,19]

The knowledge acquired from studies of the physiology and biochemistry of plant-pathogen-fungicide relations speeds substantially the current progress of chemical disease-control technology. There is evidence that chemicals act by modulating the susceptibility of the host, rather than by killing the fungus directly; there is hope that this mode of action will lead to a technically advanced fungicide that can overcome the problems of present-day fungicides.[20] The fourth-generation fungicides illustrate this trend as they show very little or no fungitoxicity in vitro but decrease infection or reduce disease intensity when applied to plants. This indicates that the fungicides are metabolized to fungitoxicants in the host plant or that they affect the pathogenicity of the fungus or the resistence of the host. A compound of this type may interfere with factors critical for pathogenesis, such as the production or activity of fungal toxins and enzymes or elicitors and repressors of host resistance. On the other hand, the primary action may be on the host rather than the pathogen. A fungicide may increase passive resistance of the host by affecting cell wall composition, or it might accentuate active resistance mechanisms that are triggered by a challenge of the pathogen. This again is a classic case of exploiting the nontarget effects of agricultural fungicides for plant disease control.[20,21]

Cereals are important crops throughout the world and any reduction in the yield of these crops can result in famine. Fungicides are being used extensively for the control of several diseases of cereals at various critical growth stages. The flag leaf is a single organ important for supplying the developing grain with assimilate. The prolongation of the active life of that leaf, particularly during the latter stages of grain filling, largely accounts for yield increases. Apart from the pathogenic fungi, several weakly parasitic and saprophytic fungi on the aerial parts of the plants grow on the green tissues and affect the development of the plant. As the tissues reach maturity they may grow even more actively and accelerate senescence, resulting in deterioration of the grain quality. This can, in extreme instances, lead to the head-blackening syndrome. In addition, phyllosphere and glume microflora are responsible for similar effects. Late application of fungicide around anthesis or grain-filling stage helps to overcome all these effects. However, it has been cautioned that broad-spectrum fungicides should be avoided, as they eliminate both beneficial and harmful microflora.

Some fungicides show selectivity and specificity when they are applied to plants, but others are broad spectrum. Indiscriminate use or application of more than the recommended dosage results in the production of phytotoxicity in plants. Manifestation of phytotoxic symptoms depends on several factors such as atmospheric conditions, cultivar, type of cultivations, and crops. Although avian toxicity due to fungicide use in the field has been noticed, any mammalian toxicity that was observed was due solely to indiscriminate usage. However, this warrants thorough monitoring in the future.

Plant viruses and viral diseases are of real concern to mankind because of their effect on crop quality and quantity loss. These plant enemies cause damage to crops such as tobacco, tomatoes, rice, wheat, cucurbits, cardamom, sugar cane, groundnut, potatoes, etc. A conservative estimate of loss would be several million dollars. Plant pathologists, horticulturists, breeders, and agronomists are all devoting much of their time to discovering control measures. One of the potential control measures is the use of a resistant variety. However, this measure has its own inherent drawbacks which are beyond the scope of this chapter. While attempting to control plant diseases, Tomlinson[22] observed a unique phenomenon: the systemic fungicide carbendazim can mitigate the symptoms of viral diseases in tobacco plants after soil and foliar applications. Reports concerning several systemic fungicides and crops soon followed. However, systemic fungicides per se are not toxic to viruses, but result in modification of metabolism of the plant in ways not favourable to viral multiplication.[23] Another mechanism suggested by the researchers is that systemic fungicides possess cytokinin-like effects which enhance senescence by antagonizing the activity of the chlorophylase enzyme and thus reducing viral multiplication in plants (nontarget effect). Some claims by the researchers are so promising that they merit further exploration when the control measures are not in effect.

Excessive use of petroleum products in industry, transport, and other areas polluted the environment and resulted in the appearance of several plant diseases due to air pollutants such as ozone (O_3), paracetyl nitrate (PAN), sulfur dioxide, etc. These diseases cause loss in yield. Although plant pathologists and breeders are engaged in evolving a variety potentially resistant to air pollutants, progress is very slow. Meanwhile, several systemic fungicides were effective in reducing the losses in yield by seed, soil, and foliar application. The effectiveness of systemic fungicides is so high that they are used at field scale in some countries to minimize the phytotoxicity of air pollutants.

IV. FUNGICIDES AND SOILS

Soil is a living system which supports several root microbe associations, both beneficial and harmful. The microbes are saprophytic for much of their life but can invade living root tissue in certain circumstances, for example: (1) when soil-inhabiting pathogens and certain mycorrhizal fungi are present, (2) when saprophytic microbes exist in the soil around the roots or on the root surface, or (3) when the microbes are active mainly or only within living roots. These microbe associations are very delicate and any physical or chemical disturbances or alteration in the cropping pattern drastically alters microbial balance, which in turn affects the association. The bacterium *Rhizobium* symbiotically fixes nitrogen to legume roots by nodulation; this can be accomplished by seed inoculation. Similarly, the mycorrhizal fungi have a large role in the protection of plants from fungal infection and also augment phosphate uptake from the soil. The mycorrhizal association is also accomplished by seed and soil inoculation. Another member of soil biota is the blue-green algae, composed of chlorophyllous microbes whose role in the maintenance of soil fertility has recently been realized. These associations are often specific between one microbial species and one plant species. Because of great qualitative and quantitative diversity and variability of microorganisms and the physiological requirements of individual species, any fungicide that comes in contact with field soil is likely to induce at least a temporary change in microbial balance and behavior. Such application also results in ecological change around the seed and the seedlings; of primary importance are the effects of fungicides on inoculum density and the environmental factors that determine the capacity of a pathogen to induce disease. The action of fungicides on soil microbes indicates that many fungicides, even at the recommended rates, produce some form of change, since even stimulation can be regarded as undersirable in certain cases. In addition, the effect is not limited to the target microbes but extends to nontarget organisms, including plants. The observed effect is usually temporary, but in other instances extended periods of effects have been observed. For example, the inhibition of nitrification for 1 month may be unimportant in agronomy, but the inhibition of the natural antagonist of a plant pathogen for a week may be sufficient for the pathogen to gain a hold on the crop. The quantitative nature of the soil population may be altered for varying periods of time and chemical changes may occur within the soil. The application of fungicide results in the enhancement of some bacterial species, providing organic matter and thus increasing the antibiotic activity. The inactivation of the fungal species by the fungicide reduces competition for available nutrients, oxygen, space, and water. The effects of selective, specific, or broad-spectrum fungicide application at the normal rates to control plant pathogens are disturbing because of the elimination of natural antagonists and for other reasons. Griffiths[13] quoted several examples of the increase in plant pathogenic infection of crops following the application of fungicides (iatrogenic diseases). A number of possible interactions between fungicides and crop plants and between plant and soil biophase leading to effects on microorganisms have been discussed.[24,25] Hence, it is imperative to investigate and review rhizobial, mycorrhizal, and nitrogen-fixing bacteria, along with mineral-transforming bacteria, and their roles and interaction with fungicides, since the fungicides are not applied for these microbes. The significant and undesirable changes in soil biota in fungicide-amended soils indicate urgent necessity for microbial ecological-monitoring programs.

The concept of integrated pest management systems (IPMS) envisions minimum use of fungicides and pesticides to reduce the cost of plant disease control and at the same time to avoid indiscriminate use of pesticides which pollute the environment. A number of pesticides indirectly induce an increase in antagonistic activity of mycoflora on plant and soil interfaces.[24,25] Therefore, it is considered integrated control and has been called a blending technique.[24-27] A prerequisite for its use is either that the antagonist be less sensitive or specific to fungicide use than the pathogen, or that the antagonist be more successful in colonizing the treated surface than the pathogen. The possibility that plant diseases can be controlled by manipulating fungicides that trigger the defense mechanisms of plants has been suggested several times. Recently, several such compounds have been developed, such as WL-28325, probenazole, aluminum-tris, and tricyclozole. The compound WL-28325 is unique in that it does not stimulate phytoalexin production itself, but rather increases the capacity of the plant to synthesize phytoalexins in response to infection, whereas a fungicide like aluminum-tris stimulates phytoalexin production directly, and still others modify host defense mechanisms which stimulate resistant-variety reactions. Chemicals that specifically interfere with pathogenesis open up the possibility of a more sophisticated method of plant disease control. The ecological advantages are obvious since saprophytic fungi and other organisms are not affected. It should be realized that highly selective agents may be economically less attractive, especially when different pathogens on the same host plant have to be eradicated. It is pertinent to study the fundamental aspects of their interference with pathogenesis, however, since increased knowledge in this area may open new perspectives for the design of fungicides.

Chemophobia is the fear that widespread use of chemicals will destroy our environment and adversely affect our personal health. Conservationists point out the damage to wildlife caused by pesticides and fungicides,[28] and the public has developed a very justifiable fear of agrochemicals. This has prompted many pathologists and environmentalists to review and develop alternative methods of plant disease control. Few can doubt that, with the growing world population and the need for greater food production, fungicides will play an increasingly important role in the future. Though developments in the nonchemical methods of plant disease control are welcomed — including the use of genetic engineering to make pathogens less viable, hyperparasites, antagonists, disease-resistant cultivars, sanitation, and biological control — it is unlikely that these methods will be applied universally. Although these alternative methods do reduce dependency upon fungicides and hazard due to pollution, these standby nonchemical approaches to disease control will not replace or dispense with the need for fungicides, but will serve as supplements to the effective use of fungicides. Therefore, it is imperative to derive the greatest benefits and least damage from the fungicides that are used. At present, plant pathologists engaged in fungicidal control of plant diseases are simultaneously evaluating the nontarget impact of agricultural fungicides in various components of the terrestrial ecosystem where a network of interactions keeps everything in a state of dynamic equilibrium.

There are now more than 300 systemic and nonsystemic fungicides being used throughout the world for the control of plant diseases. Many are broad spectrum whereas others are specific. Their application results in an array of effects which may be intended or unintended. Most researchers and farmers are not aware of these nontarget effects. As a result they are not recognized at an early stage, and when recognized they are often not described accurately. Further, lack of information on how and why nontarget effects develop or appear makes it difficult to devise counter measures to cope with the problem or — if the nontarget effects are favorable — how to utilize them.

There is no doubt that fear of nontarget effects of fungicides is totally justified; however, at the same time, there is also no doubt that without chemicals, our control over plant diseases would be greatly, and perhaps irreparably, handicapped. Chemical control measures and other methods of plant disease control are not mutually exclusive. On this issue the pathologists are not alone. The support of the Environmental Protection Agency (EPA) is critical. The agrochemical industry must also improve public awareness of the safety and benefits of fungicides. The so-called "four Rs" of proper fungicide usage are

1. Right chemical
2. Right dosage
3. Right coverage
4. Right timing

and these must be reinforced by judicious use of chemicals. Along with this, the principles laid down for IPMS must be adhered to strictly. With these controls, the short–sighted attitudes of those who fear the nontarget effects of fungicides will be replaced by long-term goals for proper fungicide use.[28]

This work sets out to show, therefore, how knowledge obtained by research and experience is made available to the field of practical agriculture, and how a better understanding of the relationships of plants and microorganisms to the whole ecology of our environment may be realized. It will be of value to plant pathologists, microbiologists, environmentalists, conservationists, agricultural extension workers, students, and industrialists, for few of these have first-hand information on the nontarget effects of agricultural fungicides. However, it will be particularly valuable to biologists who intend to understand the role of microorganisms in this ecological problem, which is of great economic importance in relation to future problems of food production. This is one of the reasons behind the preparation and presentation of this work.

REFERENCES

1. **Siegel, M. R. and Sisler, H. D., Eds.,** *Antifungal Compounds,* Vol. 1, Marcel Dekker, New York, 1977, 600.
2. **Siegel, M. R. and Sisler, H. D., Eds.,** *Antifungal Compounds,* Vol. 2, Marcel Dekker, New York, 1977, 647.

3. **Vyas, S. C.,** *Systemic Fungicides,* Tata McGraw-Hill, New Delhi, 1984, 360.
4. **Trinci, A. P. J. and Ryley, J. F.,** *Mode of Action of Antifungal Compounds,* Cambridge University Press, Cambridge, 1984, 405.
5. **Anon.,** *Contemporary Control of Plant Diseases with Chemicals: Present Status, Future Prospects and Proposal for Action,* The American Phytopathological Society, St. Paul, Minn., 1979, 170.
6. **Edgington, L. V., Martin, R. A., Bruin, G. C., and Parsons, J. M.,** Systemic fungicides: a perspective after 10 years, *Plant Dis.,* 64, 19, 1980.
7. **Rowe, R. C.,** Translocation of metalaxyl and RE 26745 in potato and comparison of foliar and soil application for control of *Phytophthora infestans, Plant Dis.,* 66, 989, 1982.
8. **Edgington, L. V.,** Structural requirements of systemic fungicides, *Annu. Rev. Phytopathol.,* 19, 107, 1981.
9. **Crowdy, S. H.,** Translocation, in *Systemic Fungicides,* Marsh, R. W., Ed., Longman, London, 1977, 92.
10. **Dimond, A. E. and Rich, S.,** Effects on physiology of the host and on host/pathogen interaction, in *Systemic Fungicides,* Marsh, R. W., Ed., Longman, London, 1977, 115.
11. **Rodriguez-Kabana, R. and Curl, E. A.,** Nontarget effects of pesticides on soil borne pathogens and disease, *Annu. Rev. Phytopathol.,* 18, 311, 1980.
12. **Griffiths, E. and Scott, S. W.,** Possible phytotonic efffects of fungicides on barley, in *Crop Protection Agents — Their Biological Evaluation,* McFarlane, N. R., Ed., Academic Press, London, 1977, 465.
13. **Griffiths, E.,** Role of fungicides in maximizing grain yield of barley, *EPPO Bull.,* 11, 347, 1981.
14. **Hance, R. J.,** Effects of pesticides on plant nutrition, *Residue Rev.,* 78, 13, 1981.
15. **Skene, K. G. M.,** Cytokinin-like properties of the systemic fungicide benomyl, *J. Hortic. Sci.,* 47, 179, 1972.
16. **Tripathi, R. K. and Schlosser, E.,** Effect of fungicides on the physiology of plants. I. Effect of carbendazim on biochemical changes and membrane permeability of senescing cabbage and cucumber leaves, *Meded. Fac. Landbouwwet. Rijksuniv. Gent.,* 42, 1073, 1977.
17. **Mukhopadhyaya, A. N. and Bandopadhyaya, R.,** Cytokinin-like activity of carbendazim, *Pesticides,* 11(7), 24, 1977.
18. **Ketznelson, H.,** Nature and importance of the rhizosphere, in *Ecology of Soilborne Plant Pathogens,* Baker, K. F. and Snyder, W. C., Eds., University of California Press, Berkeley, 1965, 187.
19. **Blakeman, J. P. and Fokkema, N. J.,** Potential for biological control of plant diseases on the phylloplane, *Annu. Rev. Phytopathol.,* 20, 167, 1982.
20. **Wade, M.,** Antifungal agents with an indirect mode of action, in *Mode of Action of Antifungal Compounds,* Trinci, A. P. J. and Ryley, J. F., Cambridge University Press, Cambridge, 1984, 283.
21. **Dekker, J.,** The development of resistance to fungicides, *Prog. Pestic. Biochem.,* 4, 165, 1985.
22. **Tomlinson, J. A.,** Chemotherapy of plant virus diseases, *Proc. 1977 Br. Crop Prot. Conf. Pests and Diseases,* 1, 807, 1977.
23. **Fraser, R. S. S. and Whenham, R. J.,** Chemotherapy of plant virus diseases with methyl-benzimidazole-2yl-carbamate: effects on plant growth and multiplication of tobacco mosaic virus, *Physiol. Plant Pathol.,* 13, 51, 1978.
24. **Bollen, G. J.,** Side effects of pesticides on microbial interaction, in *Soil-Borne Plant Pathogens,* Schippers, B. and Gams, W., Eds., Academic Press, New York, 1979, 451.
25. **Bollen, G. J.,** Nontarget effects of pesticides on soil-borne pathogens, *Proc. FAO Eur. Coop. Network Pesticides with Special Reference to their Impact on the Environment, Versailles, France,* June 4 to 8, 1984, 11.
26. **Papavizas, G. C.,** Soil-borne plant pathogens: new opportunities for biological control, *Proc. 1984 Br. Crop Prot. Conf. Pests and Diseases,* 1, 371, 1984.
27. **Baker, K. F. and Cook, R. J.,** *Biological Control of Plant Pathogens,* Freeman and Company, San Francisco, 1974, 433.
28. **Thomas, A. Z.,** Effect of citizen chemophobia on plant pathology, *Plant Dis.,* 68, 655, 1984.

Chapter 2

PHYLLOPLANE MICROFLORA

I. INTRODUCTION

Much attention has been focused during the last two decades on better understanding the microbial communities that exist on leaf surfaces. They are often large and complex. The ecological niche of the leaf surface has been termed the "phyllosphere" to characterize saprophytic microflora adopted more or less simultaneously.[1,2] This immediately suggests a parallel between epiphytes of leaves and organisms of the rhizosphere; it might therefore be helpful, and perhaps more correct, to refer to the leaf habitat as the "phylloplane". This term does not overemphasize the comparison between the root and leaf habitats, but only stresses the fact that the microorganisms grow on the leaf surface and not in the zones around them, as might be suspected from a comparison with the rhizosphere.[3] The microflora are also referred to as "phyllophora", "epiphytic microflora", and "leaf surface microflora". An aspect of central importance in assessing the microbial ecology of aerial plant surfaces, and one which has received virtually no attention, concerns the effect of fungicides on phylloplane mycoflora. *A priori* there are reasons to expect that epiphytic microbial distribution could be extremely nonuniform in time and space. These reasons include variation due to microclimate; to anatomical features such as wax, pubescence, and epidermal contours; and to physiology, as reflected, for example, in leachates. Further, more fungi show diurnal periodicity which may conceivably contribute to their preferential colonization of certain leaves, for example, those covered by dew.

The aerial surfaces of plants are inhabited by a range of microflora which do not usually produce any symptoms of disease. They are divided into residents and transients, even though some components may not always be so easily classified.[4,5] The resident microflora grow on or within the leaf surface but the transient microflora are those which have been deposited on plants from the atmosphere or by insects, and which do not develop on the leaf but may be recovered as viable units when the flora are assayed.[4]

The beneficial qualities of phyllosphere saprophytes mainly consist of their capacity to interact with or antagonize a number of plant-pathogenic fungi. The phyllosphere saprophytes easily antagonize necrotrophic pathogens, which develop after spore germination a branched mycelium network from which penetration originates. Therefore, the antagonistic action, which seems to be based on nutrient competition,[6] will reduce spore germination and affect growth.[7] Although *Sporobolomyces roseus* can reduce the growth of *Colletotrichum graminicola* on maize, it produces, even then, limited superficial growth and forms appressoria.[8] An intriguing question is whether common saprophytic phyllosphere fungi can antagonize biotrophic pathogens. It has been observed that *Cladosporium herbarum* antagonizes *Puccinia graminis* but *C. cladosporioides* stimulates the infectivity of *P. recondita*.[9,10]

There are now more than 300 fungicides available for plant disease control. Copper and sulfur fungicides are exempted from tolerance limits in food crops and high tolerances of most organic fungicides are permitted. Mercury has zero tolerance in some countries and is now considered a major pollutant of certain seas and estuaries, although quantitatively fungicides contribute little to this problem. The persistence of fungicides in plants and soil is affected by many factors, e.g., hydrolysis, photolysis, volatility, and leaching. Most of the fungicides do not accumulate in soil, although copper is an exception, as some orchard soils have been so contaminated as to be almost devoid of animal life, but surprisingly can still support good plant growth. The effects of such chemicals are also likely to be enhanced by the current trend toward using formulations containing mixtures of fungicides which increase their spectrum of activity against pathogens as well as nonpathogens. Furthermore, many fungicides have been used from the germination of the crop to a week before harvest. This results in large disturbances in the epiphytic microflora at any time during the season.

Crop-protection fungicides with a broad-spectrum activity are likely to have a considerable effect on the elements of microflora associated with the plant, although not on the largest pathogens. These interactions may restrict pathogen development, a form of biological control, or enhance pathogen success. Senescence of leaves, flowers, and fruits is accompanied, and possibly influenced, by prolific microbial activity. These effects can obviously be influenced by broad-spectrum fungicides. These might compete with and suppress the target pathogens. In the simplest case, other pathogens than those directly controlled may be allowed to develop. This may be straightforward opportunism on the part of a pathogen not within the activity spectrum of fungicides. Some of the foliar fungicides approved for use on cereals are very specific and do not interfere with the growth of the nonparasitic microflora constituting phylloplane mycoflora. In contrast, other chemicals including captafol, carbendazim, and dithiocarbamates have a broad spectrum of activity against both pathogenic and nonpathogenic organisms.[11-16] Broad-spectrum fungicides such as captan and dithiocarbamates strongly inhibit growth of saprophytic fungi, whereas bacteria are mainly unaffected; benzimidazoles inhibit *Sporobolomyces*, *Cladosporium*, and *Aureobasidium* spp. and to a lesser extent *Cryptococcus* spp. This may be the cause of the increase in *Helminthosporium* disease in cereals[17] and brown rust in wheat[18] following the use of the specific fungicides benomyl and ethirimol, respectively, for the control of these diseases.

Interaction of fungicides and leaf surface microorganisms has been reviewed.[6,11] Hislop[5] and more recently Khare and Vyas[19] provided a comprehensive overview of the entire subject of agrochemicals and the microbiology of the aerial surfaces of plants. Accordingly this chapter emphasizes recent studies.

II. NONSYSTEMIC FUNGICIDES

Nonsystemic fungicides usually comprise metal-based fungicides, metal-based organic fungicides, and organic fungicides; they are broad spectrum and nonspecific in their action on the phylloplane microflora. Research indicates that the nonsystemic fungicides have adversely affected diseases and phylloplane microflora.

Metal-based fungicides such as copper, mercury, and sulfur, and their combinations with several organic fungicides have been used for the control of plant diseases without producing nontarget effects on the microbiology of the leaf surfaces of plants. Application of copper-based fungicides resulted in a severe outbreak of disease caused by *Colletotricum coffeanum*. The explanation of this unexpected phenomenon was that the fungicides disturbed the microbial balance among the nonpathogenic strains of *C. coffeanum*.[20-23]

Gibbs[22] reported that spraying copper on coffee increases the sporulating capacity (SC) of *Phoma* sp. more than *Phomopsis* sp. sprayed with tuzet but both increased SC relatively more than captafol. It was suggested that copper and tuzet increased the SC of coffee berry disease strains more than captafol. The fungicides selectively encouraged the development of the saprophytic strain of the fungus and if the pathogenic strain was not reinfected, the increase remained unaltered.[23] Copper fungicides stimulated colonization of the woody tissues of the coffee plant by another *Colletotrichum* sp. which apparently is not overgrown by the pathogenic strain; however, they are antagonistic in vitro.[23] An unexplained significant positive correlation between copper spraying and the production of spores of *Colletotrichum* sp. (pathogenic strain) has been established.[20] Mercury fungicide changes the protecting effect of *Trichoderma viride* applied to plum trees to prevent subsequent infection by *Stereum purpureum*.[24]

Copper fungicides at the rate of 0.3% effectively controlled *Clasterosporium carpophilum*, but they may cause an increase in the severity of apricot die-back disease by decreasing surface antagonists.[25] Similarly, copper aggravated rust of *Antirrhinum* because *Fusarium roseum* infection in uredial pustules was reduced while *Puccinia antirrhini* was unaffected.[26]

The possibility of changing from sulfur and lime sulfur as scab fungicides to organic materials such as captan, which have a wider spectrum of activity, has been examined.[27] The study showed that populations of molds, filamentous fungi and other yeasts were greatly affected by the fungicide. While captan controlled scab, it worsened powdery mildew.[28] However, except for the findings of Moore,[29] there is little evidence of a direct cause-and-effect relationship between the use of captan and the incidence of apple mildew. Captan applied six times at the rate of 1.8 kg ha^{-1} on apple had a marked effect in reducing epiphytic fungal flora on buds and leaves. However, when spraying was stopped, the number and type of nonbacterial organisms returned to near normal within a few months.[30]

The nonsystemic fungicides dichlone, ferbam, and thiram at the strength of 0.25% foliar spray greatly reduced the growth of saprophytic mycoflora on petals of tomato flowers when these fungicides were used for the control of *Botrytis cinerea*.[31] The control of disease was obtained at the expense of losing a considerable amount of saprophytes which may act as antagonists to the pathogens. Initially this does not matter, but biological control improves with time whereas the opposite is true for fungicides. Both the nontarget effect of fungicides and the control of gray mold infections by spraying spores of *Cladosporium herbarum* or *Penicillium* sp. have been demonstrated.[31] The spores colonized dead petals and prevented infection by *B. cinerea*. The recovery of *Cladosporium* spp. early in the season[30] agrees with the suggestion of Dickinson[32] that they can colonize living leaves, although Holloman[33] did not find colonies on potato leaves. Spore suspension of *C. herbarum* sprayed on flowering strawberries in the field increased the yield of marketable fruit, attributing the similar effect found by Newhook[31] to the prevention of blossom blight.[34] *Penicillium* spp. were also involved in antagonizing the development of pathogens.[30] Simard et al.[35] found that out of 34 organisms antagonistic to *Venturia inaequalis* in vitro, 29 were *Penicillium* spp. and that inhibition was probably due to fungistatic action. Inhibition of these fungi has been observed on apples after the application of captan at 0.25% on the foliage.[30] Carbendazim and captafol at 0.05 and 0.25% respectively, markedly reduced the activity and viability of several saprophytic fungi on potato leaves, in contrast to captan and captafol (0.25%); however, zineb did not appear to inhibit the sporulation of *Sporobolomyces* sp.[36]

Stott[37] showed that the spores of *Alternaria chartarum*, *Aureobasidium pullulans*, and *Cladosporium cucumerinum* were inhibited at low concentrations of fungicides: funginex at 50 ppm completely inhibited the spores, and partial inhibition was observed at concentrations as low as 1 ppm. Captan was less effective against *Alternaria chartarum* and inhibited spore germination at 100 ppm, but was extremely effective against *Aureobasidium pullulans* and *C. cucumerinum*. Funginex and captan were more inhibitory to spore germination than to mycelial growth.[37] Similar results were reported by Kuthubutheen and Pugh.[38] The response of *C. cucumerinum* in growth studies was somewhat contrary to its patterns of incidence on fungicide-treated leaves; its number on the leaf surface was suppressed by thiram and Verdasan immediately after application but was identical to that of the control on the leaf surface. The growth studies, however, showed *C. cucumerinum* to be a slow-growing fungus with relatively low tolerance for fungicides and weak cellulolytic activity. Increasing the concentration of fungicides either reduced or prevented its growth and physiology. Studying the effects of captan applied to apple leaves,[30] a profound but short-lived inhibitory effect capable of controlling pathogens on saprophytes has been observed. Similarly, quantitative but not qualitative reductions in the microflora of sugar beet leaves treated with captan and funginex (0.3%) were observed, but within a few weeks of fungicidal applications the leaf flora had returned to normal.[37]

III. SYSTEMIC FUNGICIDES

Systemic fungicides on application to foliage may enter a plant and move in the direction of the evapotranspiration stream, with the photosynthates to sinks, or in both directions.[39] These three patterns of movements are termed "apoplastic" (upward), "symplastic" (downward), and "ambimobile," respectively. The symplast is the living part of the plant that is enclosed by membranes, i.e., protoplasts and plasmadesmata, including the phloem and sieve cells. Thus long-distance transport in the phloem is symplastic. The apoplast is the nonliving part of the plant, i.e., cell walls and cuticle, including xylem vessels and tracheids. Thus long-distance movement on the transpiration stream is apoplastic. Systemic fungicides after absorption into the plants remain in the phloem or xylem and thus affect the metabolic processes of the treated plant profoundly. One of the important properties of systemic fungicides is that they enhance leaf exudation. The leaf exudates contain a variety of organic compounds, and thus favor colonization of mycoflora on the leaf and selectively eliminate others.[48] The physiological and biochemical effects of systemic fungicides have also been reviewed.[40,41]

The harmful effect of broad-spectrum fungicides on the buffering capacity of saprophytic fungi merits discussion, since it has been observed that fungicides affect saprophytic colonization on the leaf surface.[11,16,17,43,44] Systemic fungicides reduce antagonists which can suppress disease organisms, thus allowing the disease to develop. This effect is perhaps even more noticeable where the use of broad-spectrum fungicides such as benzimidazoles is concerned. Fehrmann[45] reviewed several reports concerning increased infection after benomyl application by pathogens tolerant to this fungicide. One would expect that a tolerant pathogen would not be affected or stimulated by benomyl. However, the investigation by Fokkema et al.[46] confirmed, through several years of study, that the benomyl-sprayed leaf showed more severe infection than the untreated control, due to reduction of the saprophytic mycoflora to a level at which they were no longer antagonistic. While using fungicides, it is necessary to be aware of the susceptibility of the naturally occurring biological control mechanism which is evident only after a disturbance of the biological balance.

The nontarget effect of systemic fungicides in vitro on the saprophytic phylloplane mycoflora has been studied by several workers using poisoned food techniques. The objective of such studies is to anticipate the potential effect of foliar fungicides on the composition of phylloplane mycoflora in the field. The fungitoxic spectrum of the systemic benzimidazoles (benomyl, carbendazim, and thiophanates) has been observed to be very extensive; however, they are not effective on phycomycetes (except *Olpidium brassicae*), *Alternaria, Drechslera, Curvularia, Stemphylium* or certain saprophytic yeasts common on plant surfaces.[47] The benzimidazoles did not affect *Cryptococcus;* thiophanate-methyl was not effective on *Sporobolomyces;* tridemorph was somewhat effective on all the saprophytes[48] but triadimefon, triforine, ethirimol, and oxycarboxin were ineffective in reducing the growth of saprophytes in vitro isolated from wheat phylloplane. Benomyl at 40 ppm under in vitro conditions was tolerated by 9 species of *Cochliobolus,* 14 species of *Drechslera,* 4 species of *Helminthosporium,* and 5 species of *Pyrenophora.*[49] Benzimidazoles, particularly carbendazim, were reported to be highly effective against *Cryptococcus* sp. and *Sporobolomyces* sp. at very low concentrations,[48] while thiophanate-methyl at low concentrations was ineffective against *Sporobolomyces* spp. Similarly, triforine, oxycarboxin, and ethirimol were not fungitoxic to phylloplane mycoflora; however, tridemorph was observed to be highly toxic for all mycoflora used in the study. Similar results were also reported earlier.[50]

Fokkema and de Nooij studied the effect of systemic fungicides on the colonization by saprophytic mycoflora on phylloplane of flag leaf of wheat. The flag leaf is the most important single organ for supplying the developing grain photosynthate assimilate.[51] The prolongation of the active life of the leaf, particularly during the latter stages of grain filling, as was achieved by fungicide, can be expected to largely account for the yield increases recorded.[52] Late application (before or after anthesis) of zineb (0.25%), carbendazim (0.15%), carbendazim + zineb (1:1 ratio at the rate of 0.2%), and tridemorph (0.05%) to winter wheat reduced the saprophytic mycoflora and increased yield up to 21%. The increase was mainly attributed to prolonged photosynthetic activity in flag leaf rather than to control of specific diseases. Fokkema et al.[46] and Prew and McIntosh[53] found increased levels of *Cochliobolus* sp. on rye and sharp eyespot of rye *(Rhizoctonia cerealis)* following the application of benomyl for the control of these diseases. These increases in the diseases have been shown to be the result of extensive use of benzimidazole fungicides.[54] According to Fokkema and de Nooij,[48] the fungicidal application reduces antagonists which can suppress low levels of disease organisms; however, removal of the antagonists allows the diseases to develop. Again, this effect is perhaps more noticeable where use of broad-spectrum fungicides is concerned. Direct suppression of the buffering capacity of the leaves may allow fungi normally regarded as saprophytes, such as *Alternaria* spp., to become weak pathogens at the end of the season. These possibilities have been discussed.[55] Fokkema and de Nooij[48] conducted an elegant experiment to demonstrate the interaction of phylloplane mycoflora and fungicides. Four days before fungicide spraying, the leaves were inoculated with a mixture of saprophytes. On the fourth day, half of the leaves were sprayed with fungicide and the remaining half with water; colonization was determined by culturing leaf washings.[56] Surprisingly, carbendazim and other benzimidazoles did reduce the growth of *Cryptococcus* spp., and thiophanate-methyl did affect *Sporobolomyces,* contrary to its effect on potato dextrose agar (PDA). This may be due to the differences in the concentration of fungicides in different tests. On the other hand, when benzimidazoles were sprayed in the field, *Cryptococcus* appeared to be less sensitive than *Sporobolomyces* spp.[46] It is not clear why *Cladosporium* is hardly reduced; it is possible that the dead yeast cells form a food base for overcoming the fungicidal effect. The effect of tridemorph on saprophytic colonization was in accordance with the results of the in vitro test.[48] Foliar application of benomyl on wheat at 0.05%, one or two times, was reported to affect the population of *Aureobasidium pullulans* more than that of *Alternaria* spp., which were insensitive to benomyl.[57] *Aureobasidium* sp. was highly affected by benomyl application on oak, more than were *Cladosporium* sp. and *Colletotrichum coffeanum.*[58]

Field studies give an indication of the duration of the effect of fungicides on saprophytic mycoflora.[17,48] Studies of the saprophytic colonization of wheat flag leaves in a large field experiment indicated that pink yeast (*Sporobolomyces* sp.) and white yeast (*Cryptococcus* sp.) formed the most important groups throughout the season; *Cladosporium* spp. increased on the older leaves and *A. pullulans* was not important. The treatment with triadimefon at 3 kg + carbendazim at 2 kg ha^{-1} had a considerable effect on the population of yeasts and almost neutralized the adverse effect on *Cladosporium* spp. for a short period; if the field had been sprayed regularly for another month (two sprays of 5 kg sulfur + mancozeb in 1:1 ratio per hectare and three sprays of 0.3% of benomyl + maneb + triadimefon on in the ratio of 1:1:1), the effect on the white yeast particularly would have been extended. Since mixtures of fungicides are used, it is not possible to determine the contribution of the individual fungicide, but the dithiocarbamates alone could have been responsible for the decrease in the saprophytic population.[48] One or two applications of benomyl to a wheat crop disturbed the normal fungal flora of the leaf surface for several weeks.[17] This is true for leaves which had already emerged at treatment and for those which were still enclosed in the leaf sheath. They observed an average fall of 30% in the fungal population: *A. pullulans* was most affected (almost a complete disappearance) and *Epicoccum* sp. became less prevalent. *Alternaria* spp., which are insensitive to carbendazim, remained unaffected. These effects are due to carbendazim in cuticular exudates. According to these workers the disturbances in fungal antagonism in the phyllosphere may be associated with current records of an increase of *Helminthosporium* diseases, particularly in extensive cereal cultivation. Any increase in the number of treatments will tend to further reduce the number of fungal antagonists in the phyllosphere. Foliar application of benomyl (0.1%) to rye grass and white clover decreased the population of pathogenic fungi such as *Cereospora zabrina, Pithomyces chartarum, Cladosporium herbarum,* and *Cymadothea trifolii;* increased *Alternaria alteranata, Stemphylium botryosum,* and *Helminthosporium* spp.; but produced no change in *Puccinia graminis, Uromyces sp., Epicoccum purpurascens,* and *Colletotrichum coccoides.*[44] In all these cases, the role of phylloplane mycoflora, over and above that of fungicides, in reduction of foliar pathogens on the leaf surface has been demonstrated. Application of benomyl and thiophanate-methyl on flag leaves of winter wheat caused a reduction of *Sporobolomyces* sp. and *Cladosporium* sp. when applied at field rates for the control of plant diseases.[50] When the microflora of rye leaves were selectively reduced by benomyl sprays by 10 to 20% on control leaves, susceptibility to the *Drechslera* state of *Cochliobolus sativus* increased; this effect was not observed if the number of microflora on control leaves was high, about 500 propagules per square centimeter.[46] A comparison of the effects of benomyl and captan on the component of microflora of oak leaves indicated that, although benomyl caused less reduction than did captan in the total numbers of fungi on sprayed leaves, there was a shift in the nonfilamentous white yeasts, which increased the microflora from 33 to 68%. The effect was observed for more than 2 months.[58]

The phylloplane organisms are potential antagonists or competitors of pathogens on plant surfaces and, because of their exposed position, are especially vulnerable to fungicidal gases, sprays, and drenches. Consequently, disease-control treatments can lead to major reductions in their activity or even in the numbers of viable cells.[59]

A very important aspect of the effects of fungicides on soil fungi concerns the phylloplane microflora. Most of the saprophytic phylloplane fungi originate in the soil. The inhibition/ stimulation effect of a foliar-applied fungicide on an organism may be highly significant in subsequent events when the leaf eventually becomes incorporated into the soil. Ethirimol (1.5 kg ha^{-1}) did not affect the population of saprophytic filamentous fungi or ballistosporic yeasts; however, zineb (0.35 kg ha^{-1}) slightly reduced fungal population but had no effect on *Sporobolomyces* spp.[11] Triforine and tridemorph applied at the rate of 0.05% on flag leaves of winter wheat had little effect on fungi; however, tridemorph (0.1%) with sticker considerably reduced the incidence of *Cladosporium* sp. and ethirimol had no effect on fungi.[60] Similarly, tridemorph at 0.5 kg active ingredient (a.i.) ha^{-1} was less effective as a general fungicide and on saprophytic mycoflora than was fentin acetate (0.1 kg ha^{-1}) + maneb (0.9 kg ha^{-1}) which reduced several yeasts on leaves.[48] Application of metalaxyl at 0.1% against blue mold (*Peronospora tobacina*) reduced the phylloplane mycoflora by 20.7 to 33.6%.[61] Carboxin (0.15%) and benomyl (0.05%) controlled leaf blotch of wheat (*Septoria nodorum*) and produced no effect on phylloplane mycoflora.[62] These studies suggest that a significant proportion of leaf microbes are susceptible to fungicides, especially those known to have a wide spectrum of activity against pathogens. Systemic fungicides, it is postulated, could have less effect. Whereas Dickinson[11] notes that the consequences of fungicide-reduced population on leaves could be twofold, i.e., (1) it could kill saprophytes which compete with pathogens leading to the establishment of spray-tolerant pathogens or (2) it could extend leaf life by delaying fungal-induced senescence, it is also highly probable that alterations to the phylloplane population could significantly affect microbial decomposition of the leaf tissues of soil.

The mycoflora of the culm base seem to be more susceptible to the effects of fungicide than do those of roots.[63] When benomyl was applied at dosages ranging from 0.24 to 4.80 kg ha^{-1} and samples were taken three times during the growing season, *Pseudocercosporella herpotrichoides, Alternaria* sp., and *Gerlachia nivalis* were predominant on the untreated plants, and those from the benomyl-treated plants (*Alternaria* spp. and *Fusarium culmorum*) were isolated. In later stages of growth, *G. nivalis* sharply declined and *Alternaria* spp. and *F. culmorum* increased; at the end of the season, *Periconia macrospinosa* and *Typhula incarnata* appeared in treated plants. A drastic shift in the fungal population has been observed in the culm bases of rye and wheat,[64] in contrast to the work of Bollen et al.[63] In the fields investigated by Reinecke,[64] *Alternaria* spp., pycnidial fungi and *R. cerealis* were significantly stimulated. Reasons for different responses under different situations, especially in relation to *R. cerealis,* have been discussed.[65] The relation to *T. incarnata* and *Drechslera* spp.[63] and *D. sorokiniana*[46] in rye and *Typhula* sp.[66] in wheat in fungicide-treated plots may be due to their resistance to chemicals. The dynamics of the mycoflora of the culms show an increase in *Alternaria* spp. and *P. macrospinosa* towards the end of the season in rye[63] and wheat[67] in fungicide-treated plots. The occurrence of *P. macrospinosa* is consipicuous as this fungus was recently thought to cause "crater disease" in wheat in South Africa.[68] *G. nivalis* was only prevalent on culm bases during the earlier stages of the crop,[63] although this does not appear to be a general rule. *F. culmorum* and *F. avenaceum* became predominant after tillering and *G. nivalis* remained present at a low level[69] but did not decline.[63]

Fungicide treatment also modified the phylloplane microflora of glumes of wheat and barley.[11-13,70,71] *Cladosporium* and *Sporobolomyces* were reduced more on wheat flag leaves treated with benomyl and thiophanate-methyl at the operational rate[12] than with captafol. On the other hand, Hill and Lacey[71] reported that captafol caused a considerably greater reduction in the flora than benomyl and tridemorph, both of which reduced the number of microorganisms to approximately a third of those on grain in controls. However, with respect to particular taxa, benomyl was more effective against *Cladosporium* but less effective against *Alternaria* then was captafol. The latter fungus was more abundant on treated grain than on the controls. Yeasts were little affected by any fungicide except captafol.[70,71] These changes could be particularly important if the seeds are used to determine relative resistance to pathogens, since it has been shown that susceptibility to *Helminthosporium sativum* was increased by reducing superficial microflora by chemical treatment.[72] Two to six preharvest foliar sprays of benomyl in wheat showed that it increased the number of *Alternaria* and *Bipolaris* spp. on seed and that colonies of pathogenic fungi (*Septoria* sp. and *Epicoccum* sp.) showed a decrease.[73] These results could be used for the control of pathogens such as *Phoma betae* on *Beta vulgaris,*[74] *D. sorokiniana (H. sativum)* on rye,[75] and *Alternaria brassicicola* on *Brassica* spp.[76]

The effect of foliar application of carbendazim (0.1%) on the mycoflora of phylloplane of groundnut plants was studied.[77] The fungicide markedly increased the growth of shoot and root systems of groundnut and fungal number on the leaves. The fungi comprise *Penicillium islandicum, P. rubrum, Aspergillus terreus,* and *Cladosporium herbarum.* However, the numbers of pathogenic fungi decreased. These saprophytic fungi are known to be antagonistic to several foliar pathogenic fungi.[8]

Phyllospheric nitrogen-fixing bacteria play a significant role in the nitrogen economy of plants.[78-80] The phyllospheric nitrogen-fixing bacteria comprise *Corynebacterium* sp. from wheat, potato, and tomato; *Azotobacter chroococcum* from rice; and *Bacillus* sp. from jute.[80] Phyllospheric nitrogen fixers have been sprayed on the wheat as substitutes for nitrogenous fertilizers and showed marked improvement in yield and growth of the plants: an average increase in yield by 70% was obtained, which was very near to that obtained by fertilizer treatment. Since the spraying of bacteria and fungicidal application coincide, it is pertinent to study the effects of various fungicides on these bacteria. Application of captan at 0.3% on apples for the control of scab reduced the bacterial flora and produced variable effects.[30] The effect of Bordeaux mixture on bacteria colonizing cherry leaves was similarly shown by Crosse[81] to be marked, but in this instance the effect persisted longer, so that populations on sprayed leaves never attained the same level as on the control leaves. The in vitro effect of technically pure-grade ziram and edifenfos (the fungicides used for the control of rice diseases) on the phylloplane nitrogen-fixing bacteria was studied, and it was found that at a concentration of 700 $\mu\ell$ mℓ^{-1} they completely inhibited the growth of the bacteria.[82] These results indicate that the side effects of fungicides used for the control of plant pathogens should be studied before final recommendation is made for their use.

Application of preharvest fungicides during bud initiation and flowering time also influences the fruit microflora. Application of dichlofluanid and benomyl commercial formulations on soft fruit reduced the numbers of *Botrytis cinerea, Cladosporium* spp., *Penicillium* spp., and *Aureobasidium pullulans* on fruit picked early in the season.[83] The latter three fungi were reported to be antagonists to several pathogenic fungi. Benomyl, thiophanate-methyl, and carbendazim fungicides are being used for the control of infection of *B. cinerea* on soft fruit rot.[84] These fungicides are ineffective against phycomycetes. Continuous use of these fungicides results in the disappearance of the antagonistic fungi and development of resistance in *B. cinerea*, which aggravates the disease.

Late application of fungicides after heading (or preharvest) often suppresses markedly the development of black molds on the glumes. These observations have revealed that *Cladosporium* spp. are pathogenic and thereby lower grain weight. Previous research on this aspect is incomplete. O'Donnell and Dickinson[85] reported that *Alternaria alternata, C. herbarum,* and *C. cladosporoides* are able to penetrate bean leaves through their stomata. Most cereal pathologists are of the opinion that fungicidal spray for the control of black molds on the ears does not contribute to yield. Sinclair[86] reviewed the subject and concluded that preharvest sprays of fungicides improved seed quality with or without an effect on yield in rice and soybean. Most of the seed-borne pathogens infect the seed in the field itself and preharvest spray reduces the phyllosphere pathogens which become seedborne under adverse storage conditions. Phylloplane mycoflora are of greater importance than was previously thought; hence, in some seasons, their control by fungicides may lead to useful yield increases. In addition, if the effects of fungicides persist after anthesis, there will be less head blackening. This syndrome is usually attributed to sooty molds but is, in fact, mainly caused by the sporulation of *Cladosporium* sp. and *Alternaria* sp. on the real flower parts. Reduction of the colonization of the glumes, and hence the amount of sporulation, brings about an added bonus in that the quality of grain is higher, especially desirable if it is intended for flour milling.[55]

Systemic fungicides applied to seed or soil are absorbed by the seeds and subsequent emerging seedlings. When a particular concentration of fungicides is reached in leaves, it can then be excreted in potent form from the surfaces of the treated leaves.[88] *Sporobolomyces roseus, Leucosporidium scottii,* and *Aureobasidium pullulans* were inhibited by foliar application of benomyl and thiabendazole at 0.15% in vitro as well as in vivo; however, these fungi were not inhibted in vitro by carboxin (0.15%), but were inhibited in vivo on pea and barley seedlings.[87] This suggests that only the fungicides benomyl and thiabendazole induced the excretion from the leaves of substances toxic to sensitive yeasts and yeast-like test fungi. The authors concluded that the inhibition observed in carboxin-treated pea and barley seedlings may be the result of hormonal imbalance since treated plants were greener than the controls, no guttation occurred, and the stomata tended to remain open. It is believed that carboxin affects the epiphytic microflora by rendering the leaf surface incapable of sustaining them, possibly through dryness. Its effect was reversed by spraying plants with the cytokinin antagonist, abscissic acid.[87] The sprayed plants behaved as did plants without carboxin. The initiation of some fungal leaf diseases may be prevented even before the organisms penetrate if the concentration of the translocated fungicide is high enough to prevent germination. Carboxin was shown to have an unfavorable effect on some plants and sensitivity to drought, but in dry weather its use might prevent germination on leaves of spores of pathogens to which it is not toxic, such as *Alternaria* and *Helminthosporium* spp. It seems likely that there are fungicides that could modify all the epiphytic flora by soil application no less than by foliar application. Such changes may have wide repercussions. van den Heuvel,[88] for instance, has shown that *Aureobasidium pullulans,* common on leaves, may reduce the number of lesions caused by *Alternaria zinniae* on dwarf beans. The development of *S. roseus* was inhibited on leaves of pea and barley treated with carboxin, benomyl, and thiabendazole. Bioassay of leaf pieces showed that these effects were due to fungicide residue on leaves from treated seed. They concluded that some systemic fungicides applied to seed might modify all the epiphytic microflora. The epiphytic microflora of strawberry plants raised after soil application of systemic fungicides were studied.[89] Some of the fungicides accumulated in or on leaves,[90] but no consistent changes in the epiphytic microflora of the treated leaves were detected in glasshouse or field experiments. However, even if systemic fungicides do affect phylloplane organisms, the microbiology of other plant parts, such as some fruits and petals which do not possess stomata and transpire very little, thus accumulating no residues,[39] will remain unaltered.[5,91]

The application of fungicides affected the physiological activity of phylloplane mycoflora and thus their survival. The leaf surface is a complex habitat in which, under normal conditions, microbial populations are held in dynamic balance by interactions between the phylloplane microorganisms and the host plant, and the delicate balance of the occurrence of microbial numbers, species composition, and activities on the leaf surface can be significantly modified following the application of fungicides and other fungitoxic substances.[91-93] The fungicides on natural substrates such as the leaf surface create a microbial vacuum which is open to recolonization either from the air flora or from spores that have survived the fungicide treatment. Under these circumstances, for a fungus to successfully recolonize the fungicide-treated leaf, it must possess among other things some degree of tolerance to at least low concentrations of the fungicide. Kuthubutheen and Pugh[38] studied the effects of fungicides on the physiology of phylloplane fungi and reported that the radial growth rates were reduced with increasing concentration of the fungicides. Captan and dicloran were less inhibitory than thiram and Verdasan. The fungi could tolerate lower concentrations of fungicides in liquid media than in solid media; 0.05 ppm of Verdasan prevented the mycelial growth of *Drechslera biseptata* and *Trichoderma viride.* Spore germination and germ tube elongation were arrested at lower concentrations of fungicides than was mycelial growth. Concentrations of captan and dicloran up to 50 ppm were only fungistatic to the spores of *A. alternata* and *Cladosporium cladosporoides.* Thiram and Verdasan at 50 and 10 ppm, respectively, were fungicidal after exposure of the spores to the fungicides for 24 hr. *T. viride* and *D. biseptata* were capable of growth on cellulose agar with Verdasan at 0.1 and 1 ppm, respectively; these species did not show growth in the liquid medium with 0.05 ppm of Verdasan. This could be explained by two factors: (1) the fungal mycelia are constantly being bathed by fungicide in the liquid medium or (2) in calcium the mercury-chelating thiols of the yeast extract and the amino groups of asparagine might have bound the fungicide.[49]

According to Fokkema et al.,[46] the application of fungicides drastically alters the phylloplane mycoflora which are partly responsible for maintaining microbial balance on the leaves. Gachichi[96] attempted to isolate mycoflora on fungicide-sprayed or unsprayed leaves in *Kharif* (July through October) and *Rabi* (October through March) of groundnut *(Arachis hypogaea* L.). The variety of fungi isolated indicate that the group supports several saprophytic fungi in both seasons. Dickinson[55] showed that these fungi are antagonistic and can reduce the occurrence of several pathogenic fungi. Further, a study was made assaying the sensitivity of the fungal leaf saprophytes in vitro to determine to what extent these saprophytes are affected. The fungicides were dispensed in PDA[96,97] utilizing the technique discussed by Fokkema and de Nooij.[48] Fungicides (Table 1) such as dithiocarbamates, chlorothalonil, tridemorph, and benzimidazoles reduce the colony diameter of saprophytic fungi, whereas specific fungicides such as triforine and biloxazol have little or no effect.

Table 1
SENSITIVITY OF THE SAPROPHYTIC FUNGI ISOLATED FROM GROUNDNUT LEAVES TO VARIOUS FUNGICIDES (500 PPM) ON PDA

Fungicide	DS	AA	TK	AF	FE	AO	CC
Systemic							
Thiophanate-methyl	+	+	+	+	+	+	+
Carbendazim	+	+	+	+	+	+	+
Triforine	+	0	+	+	−	−	−
Biloxazol	+	+	+	+	+	+	+
Tridemorph	0	0	0	0	0	0	0
Nonsystemic							
Zineb	0	0	−	−	−	0	−
Mancozeb	−	−	−	−	−	−	−
Chlorothalonil	±	±	0	+	±	±	+

Note: DS = *Drechslera specifera*, AA = *Alternaria alternata*, TK = *Trichoderma koningii*, FE = *Fusarium equisetii*, AF = *Aspergillus flavus*, AO = *Aspergillus ochraceus*, CC = *Choanophora circinans*. Colony size on PDA with the fungicide has been expressed as a percentage of that on PDA (100%) using the following symbols: 0 = no growth, − = 40% growth, ± = 40 to 60% growth, + = 60% growth.

The same study was extended in the field with the objective of determining the extent of nontarget effects of systemic and nonsystemic fungicides on the microbial balance on the phyllosphere of groundnut (Tables 2 to 4), and secondly the possible consequences of a reduction of the saprophytic population for the development of fungal leaf diseases. The observations in Table 2 indicate that the mycoflora possess different sensitivities to the fungicides. The fungicides are toxic to several saprophytes immediately but in 2 weeks several fungi recolonize, which has the potential for biological control of foliar fungal pathogens. Further, *Trichoderma* sp., a potential antagonist, was isolated from all the treated leaves after the first and second sprays of fungicides, except from those treated with thiophanate-methyl, in which it was isolated only after the second spray. The effect of fungicides on qualitative and quantitative occurrence of phylloplane mycoflora after seed and foliar application was also studied (Tables 3 and 4). The data indicate that the seed treatment and foliar application of fungicides support saprophytic mycoflora differently: the broad-spectrum fungicides are more toxic than the systemic; however, the systemic fungicides showed specificity in the supporting mycoflora.[47] The results, when compared with in vitro test results, showed that some fungicides highly toxic under in vitro conditions were not equally effective in vivo; this may be because artificial conditions could not be simulated under natural conditions, and vice versa. Similar observations were reported by Fokkema and de Nooij.[48]

Abaxial and adaxial leaf surfaces possess different physiologies because of variation in stomatal numbers, size, shape, and exudations. Fungicidal application can drastically change stomatal size, shape, and the pattern of leaf exudations.[46,48] Application of fungicides on seed and foliage was observed to support a number of mycoflora (Table 5). Some fungicides were active on seed whereas others were active on foliage.[96] This may be due to the differential uptake of these fungicides by roots and foliage.[98] Recently similar observations have been reported for dithiocarbamates and benzimidazole fungicides.[97,99] Changes due to benomyl application in the size and shape of stomata in sunflowers, cowpeas, and cotton have also been reported.[99]

The effects of fungicides such as carbendazim, thiophanate-methyl, and zineb on disease intensity, delay in senescence, and yield are correlated in Table 6. However, a deviation has been observed in mancozeb. Therefore, the growth of phylloplane mycoflora and in-

Table 2

PHYLLOPLANE MYCOFLORA ISOLATED IMMEDIATELY AND AFTER 15 DAYS OF FOLIAR APPLICATION OF FUNGICIDES AT 1500 PPM

Treatments

Mycoflora	Chloro-thalonil				Thiophanate-methyl				Carbendazim				Zineb			
	I		II		I		II		I		II		I		II	
	1	2	1	2	1	2	1	2	1	2	1	2	1	2	1	2
Acremonium sp.	−	−	−	+	−	+	−	+	−	+	−	+	−	+	+	+
Alternaria sp.	+	+	+	+	−	+	−	+	−	−	−	−	+	−	+	−
Aspergillus sp.	+	+	+	+	−	+	+	+	−	+	−	+	−	+	−	+
Cladosporium sp.	−	+	−	+	−	−	−	−	−	+	−	+	−	−	−	−
Drechslera sp.	−	+	−	+	−	+	−	+	−	+	−	+	−	−	−	−
Fusarium sp	−	+	−	+	−	+	−	+	−	+	−	+	−	−	−	−
Mucor sp.	−	−	−	−	−	+	−	+	−	−	−	−	−	−	−	−
Myrothecium sp.	−	+	−	−	−	−	−	−	−	−	−	+	−	−	−	−
Penicillium sp.	−	+	−	+	−	−	−	+	−	+	+	+	−	+	−	−
Pithomyces sp.	−	+	−	−	−	+	−	+	−	+	−	+	−	+	−	+
Rhizopus sp.	−	+	−	+	−	+	−	+	−	+	+	+	−	+	−	+
Trichoderma sp.	−	+	−	+	−	−	−	+	−	+	−	+	−	+	−	+
Nonsporulating	+	+	−	+	+	+	−	+	−	+	−	+	−	+	−	+

Treatments

Mancozeb				Triforine				Tridemorph				Biloxazol				Control			
I		II		I		II		I		II		I		II		I		II	
1	2	1	2	1	2	1	2	1	2	1	2	1	2	1	2	1	2	1	2
−	+	−	+	−	+	−	+	−	+	−	+	−	+	−	+	+	+	+	+
+	−	−	−	−	−	−	+	−	+	−	+	−	+	−	−	+	+	+	+
−	−	−	+	−	+	−	+	−	+	−	+	−	+	−	+	+	+	+	+
−	−	+	−	−	−	+	−	−	−	−	−	−	+	−	−	+	+	+	+
−	−	+	−	−	−	+	−	−	−	−	−	−	+	−	+	+	+	+	+

Table 2 (continued)
PHYLLOPLANE MYCOFLORA ISOLATED IMMEDIATELY AND AFTER 15 DAYS OF FOLIAR APPLICATION OF FUNGICIDES AT 1500 PPM

<table>
<thead>
<tr><th colspan="20">Treatments</th></tr>
<tr>
<th colspan="4">Mancozeb</th>
<th colspan="4">Triforine</th>
<th colspan="4">Tridemorph</th>
<th colspan="4">Biloxazol</th>
<th colspan="4">Control</th>
</tr>
<tr>
<th colspan="2">I</th><th colspan="2">II</th>
<th colspan="2">I</th><th colspan="2">II</th>
<th colspan="2">I</th><th colspan="2">II</th>
<th colspan="2">I</th><th colspan="2">II</th>
<th colspan="2">I</th><th colspan="2">II</th>
</tr>
<tr>
<th>1</th><th>2</th><th>1</th><th>2</th>
<th>1</th><th>2</th><th>1</th><th>2</th>
<th>1</th><th>2</th><th>1</th><th>2</th>
<th>1</th><th>2</th><th>1</th><th>2</th>
<th>1</th><th>2</th><th>1</th><th>2</th>
</tr>
</thead>
<tbody>
<tr><td>−</td><td>−</td><td>−</td><td>+</td><td>−</td><td>−</td><td>−</td><td>+</td><td>−</td><td>−</td><td>−</td><td>+</td><td>−</td><td>+</td><td>−</td><td>+</td><td>+</td><td>+</td><td>+</td><td>+</td></tr>
<tr><td>−</td><td>−</td><td>−</td><td>+</td><td>−</td><td>−</td><td>−</td><td>+</td><td>−</td><td>−</td><td>−</td><td>+</td><td>−</td><td>+</td><td>−</td><td>+</td><td>+</td><td>+</td><td>+</td><td>+</td></tr>
<tr><td>−</td><td>−</td><td>−</td><td>−</td><td>−</td><td>+</td><td>−</td><td>−</td><td>−</td><td>+</td><td>−</td><td>−</td><td>+</td><td>−</td><td>−</td><td>−</td><td>+</td><td>+</td><td>+</td><td>+</td></tr>
<tr><td>−</td><td>+</td><td>−</td><td>−</td><td>−</td><td>−</td><td>−</td><td>−</td><td>−</td><td>+</td><td>−</td><td>−</td><td>−</td><td>+</td><td>−</td><td>−</td><td>+</td><td>+</td><td>+</td><td>+</td></tr>
<tr><td>−</td><td>+</td><td>−</td><td>+</td><td>−</td><td>+</td><td>−</td><td>−</td><td>−</td><td>+</td><td>−</td><td>+</td><td>−</td><td>−</td><td>+</td><td>+</td><td>+</td><td>+</td><td>+</td><td>+</td></tr>
<tr><td>−</td><td>−</td><td>−</td><td>−</td><td>−</td><td>−</td><td>−</td><td>−</td><td>−</td><td>−</td><td>+</td><td>−</td><td>−</td><td>−</td><td>−</td><td>+</td><td>+</td><td>+</td><td>+</td><td>+</td></tr>
<tr><td>−</td><td>+</td><td>−</td><td>+</td><td>−</td><td>+</td><td>−</td><td>+</td><td>−</td><td>+</td><td>−</td><td>+</td><td>−</td><td>+</td><td>−</td><td>+</td><td>+</td><td>+</td><td>+</td><td>+</td></tr>
<tr><td>−</td><td>+</td><td>−</td><td>+</td><td>+</td><td>+</td><td>+</td><td>+</td><td>+</td><td>+</td><td>+</td><td>−</td><td>−</td><td>+</td><td>−</td><td>+</td><td>+</td><td>+</td><td>+</td><td>+</td></tr>
</tbody>
</table>

Note: I and II = fungicidal sprays, 1 and 2 = silations, + = presence, − = absence.

Table 3

EFFECT OF FUNGICIDES ON QUALITATIVE AND QUANTITATIVE OCCURRENCE OF PHYLLOPLANE MYCOFLORA AFTER SEED TREATMENT AT 1000 PPM IN GROUNDNUT

Mycoflora	Treatments							
	Triforine	Metalaxyl	Triadimefon	Captafol	Thiram	Delton	Carbendazim	Control
Aspergillus sp.	18	11	51	3	6	5	10	10
Alternaria sp.	6	1	0	0	1	0	3	4
Drechslera sp.	5	2	0	0	0	0	1	3
Fusarium sp.	3	3	0	1	3	0	2	1
Mucor sp.	1	0	0	1	0	0	0	0
Pithomyces sp.	0	1	1	2	1	0	0	1
Penicillium sp.	0	2	2	3	3	2	0	2
Rhizopus sp.	2	0	0	0	1	0	4	0
Trichoderma sp.	4	6	3	0	2	1	4	5
Total	39	27	11	10	17	8	24	26

Note: Data are derived from mean of three replications; the fungicides used are recommended for the control of seed-borne diseases.

Table 4

QUALITATIVE AND QUANTITATIVE ESTIMATION OF PHYLLOPLANE MYCOFLORA OF GROUNDNUT ON FOLIAR APPLICATION OF FUNGICIDES AT 1500 PPM

Mycoflora				Treatments					
	Chlorothalonil	Thiophanate-methyl	Carbendazim	Zineb	Mancozeb	Biloxazol	Tridemorph	Triforine	Control
Alternaria sp.	1	0	0	0	0	0	0	0	10
Acremonium sp.	0	0	0	0	0	1	0	0	0
Aspergillus sp.	6	7	4	10	2	7	1	2	8
Cladosporium sp.	1	0	0	0	0	10	0	0	0
Drechslera sp.	1	0	0	0	0	0	0	0	2
Mucor sp.	0	0	0	3	0	0	0	0	1
Myrothecium sp.	1	0	0	0	0	0	0	0	0
Pithomyces sp.	1	0	0	0	0	0	0	0	0
Penicillium sp.	1	1	0	0	1	2	0	0	0
Rhizopus sp.	1	0	0	0	0	2	0	0	1
Trichoderma sp.	1	2	1	3	4	0	0	1	3
Nonsporulating	2	3	60	1	0	4	0	0	2
Totals	18	14	65	20	7	26	2	4	21

Note: Data are derived from mean of three replications; the fungicides used are recommended for the control of foliar disease of groundnut.

Table 5

ESTIMATION OF PHYLLOPLANE MYCOFLORA ON UPPER AND LOWER SURFACES OF GROUNDNUT LEAF ON SEED TREATMENT (1000 PPM) AND FOLIAR SPRAY (1500 PPM) WITH DIFFERENT FUNGICIDES

	Seed treatment		Foliar treatment	
Fungicides	**Upper surface**	**Lower surface**	**Upper surface**	**Lower surface**
Chlorothalonil	—	—	15	4
Thiophanate-methyl	—	—	6	5
Carbendazim	—	—	3	3
Zineb	—	—	13	4
Biloxazol	—	—	13	9
Mancozeb	—	—	5	3
Tridemorph	8	5	2	2
Triforine	25	19	3	4
Metalaxyl	16	15	—	—
Captafol	6	6	—	—
Thiram	11	5	—	—
Delton	3	5	—	—
Rhizobium-inoculated plant only	8	7	—	—
Control	11	12	10	9

Note: The data are derived from the mean of three replications.

fection by foliar pathogens on pretreated leaves reflects a response to the change of stimulators and inhibitors.[100] A different effect was observed on the biloxazol-treated leaves. The fungicide supported a large number of mycoflora (see Table 5), and the low disease intensity of 17.34 (Table 6) indicated ineffectivity in the control of phylloplane mycoflora. These differing effects on the control of foliar diseases and the inhibition of mycoflora by the fungicides may be due to the fact that biloxazol showed specific fungitoxicity while zineb showed broad-spectrum activity. Still further, tridemorph, with a high disease intensity of 37.34 (Table 6) — over half that of control, supported a low number of phylloplane mycoflora (see Table 5), which allowed an increase of groundnut leaf spot diseases. This detrimental effect is undesirable since it promotes disease severity, possibly by the removal of mycoflora potentially antagonistic to leaf spot pathogens. Failure to control foliar diseases by fungicide applications may be due to alteration of the microbial balance of the leaf and thus elimination of possible biocontrol agents.

Phylloplane mycoflora were isolated from leaves of wheat (*Triticum aestivum* L.), rice (*Oryzae sativa* L.), potato (*Solanum tuberosum* L.), peas (*Pisum sativum*), sorghum (*Sorghum vulgare*), and sesamum (*Sesamum orientale*), both unsprayed and sprayed with systemic and nonsystemic fungicides. Some of the fungi isolated from the crop leaves are potential antagonists to foliar pathogens.[55] This will be discussed in Chapter 15.

IV. FUTURE OUTLOOK

The composition of the microflora of plants is determined by the interplay of many factors, including fungicides. It is clear that fungicides applied to plants have more than one role; they may alter the microflora directly, by affecting particular components, or indirectly, by altering the physiology of the host.[5] Fungicides may increase the pattern of pathogenesis positively.[20,22,45,48,55] Secondly, they may affect the balance of phylloplane mycoflora so that the antagonistic action of these fungi may be inhibited, and potential biocontrol may be eliminated.[45,48] Last, minor pathogens may assume major roles in pathogenesis and may

Table 6
**EFFECT OF FOLIAR-APPLIED FUNGICIDES (1500
PPM) ON TIKKA[a] DISEASE OF GROUNDNUT
DISEASE INTENSITY (DI), DELAY OF SENESCENCE
(DS) (%), AND YIELD (kg/ha)**

Treatments	DI[b]	DS (%)[c]	Yield (kg/ha)
Triforine	16.00	75.00	216.67
Thiophanate-methyl	17.34	100.00	300.00
Biloxazol	17.34	100.00	238.90
Carbendazim	18.67	83.34	255.57
Mancozeb	20.00	68.64	227.78
Chlorothalonil	21.34	100.00	233.34
Wettable sulfur	24.00	—	316.67
Zineb	27.34	66.67	288.90
Tridemorph	37.34	83.34	247.23
Control	70.66	58.34	105.57
Standard error of measure ±	8.45	5.96	2.72
Critical difference at 5%	26.57	17.69	57.22

Note: Mean of three replications.

[a] Tikka disease (leaf spots) caused by *Cercospora arachidicola* and *Phaeo-isariopsis personata*.
[b] Disease intensity was accorded as per Cobb's Scale.
[c] Delay of senescence was low (50%), moderate (75%), and high (100%) and showed presence of high levels of chlorophyll in leaves.

lead to an aggravated disease situation.[48,55] Nonspecific (nonsystemic) fungicides result in major shifts in the microbiological equilibrium and induce resistance in pathogens more frequently than do selective or systemic fungicides, which cause minimal disturbance in the component of microflora on the leaves. Thus, it is clear that indiscriminate use of fungicides may have indirect and direct effects on plants and diseases, but their application has not had as many undesirable biological side effects as has the application of insecticides.[91] Fungicide use in relation to the balance of saprophytic and parasitic microorganisms of the phylloplane was also discussed. A reported increase in coffee berry disease was an interesting example of the effect of fungicide; however, the mechanism of replacement by severe strains of the pathogen remains unknown. This acts as a warning against the possible side effects of fungicides and other agrochemicals.

The many examples of the direct effects of fungicides on plants indicate that they should be considered when evaluating increased yield after fungicide treatment. The effect of saprophytes on plants can be more readily studied when their numbers are artificially increased and then evaluated. Understanding the interacting factors that affect the microflora of plants in the presence of fungicides might lead to a way to foster indigenous biological control agents and eventually facilitate an integrated approach to disease control with resulting reductions in cost and pollution of the environment by fungicides. The empirical approach to disease control may prove most successful today. Such information, however is scarce. It can be gained in part from more precise assessments of fungicide-treated fields and also from more experiments under controlled conditions. In particular, the range of cultivars in common use should be examined, as there will be different responses to these fungi, based in part on host-resistance factors and also on the pattern of senescence in each cultivar. This information may lead to an ability to predict losses due to disease and exceptionally low grain yield. Evidence suggests that such losses should be predictable just after anthesis, as treatment with broad-spectrum protectant fungicides at this point can have positive results.[5,91,94] A more fundamental approach, utilizing information with detailed epidemiological data, may produce a more effective system of crop disease control. It is pertinent that such fungicides persist for several weeks, thus inhibiting microbial activity throughout the whole period of senescence. Broad-spectrum protectant fungicides are not now readily available, and it will be advantageous for renewed emphasis to be placed on developing these compounds, which has, in recent years, been postponed in favor of the development of more popular fungicides.

REFERENCES

1. **Last, F. T.,** Seasonal incidence of sporobolomyces on cereal leaves, *Trans. Br. Mycol. Soc.,* 38, 221, 1955.
2. **Ruinen, J.,** Occurrence of *Beijerinckia* species in the phyllosphere, *Nature (London),* 177, 220, 1956.
3. **Kerling, L. C. P.,** De Microflora op het Blad van *Beta vulgaris, Tijdschr. Plantenziekten,* 64, 402, 1958.
4. **Hislop, E. C.,** Side effects of fungicides: the effects of fungicides on the epiphytic microflora, *Sci. Hortic.,* 23, 143, 1971.
5. **Hislop, E. C.,** Some effects of fungicides and other agrochemicals on the microbiology of the aerial surfaces of plants, in *Microbiology of Aerial Plant Surfaces,* Dickinson, C. H. and Preece, T. F., Eds., Academic Press, New York, 1976, 41.
6. **Blakeman, J. P.,** Microbial competition for nutrients and germination of fungal spores, *Ann. Appl. Biol.,* 89, 151, 1978.
7. **Fokkema, N. J.,** Antagonism between fungal saprophytes and pathogens on serial plant surfaces, in *Microbiology of Aerial Plant Surfaces,* Dickinson, C. H. and Preece, T. F., Eds., Academic Press, New York, 1976, 477.
8. **Blakeman, J. P. and Parberry, D. G.,** Stimulation of appressorium formation in *Colletotrichum acutatum* by phylloplane bacteria, *Physiol. Plant Pathol.,* 11, 313, 1977.
9. **Mishra, R. R. and Tewari, R. P.,** Studies on biological control of *Puccinia graminis tritici,* in *Microbiology of Aerial Plant Surfaces,* Dickinson, C. H. and Preece, T. F., Eds., Acadmic Press, New York, 1976, 559.
10. **Doherty, M. A. and Preece, T. F.,** *Bacillus cereus* prevents germination of uredospores of *Puccinia allii* and the development of rust diseases of leek, *Allium portum,* in controlled environments, *Physiol. Plant Pathol.,* 12, 123, 1978.
11. **Dickinson, C. H.,** Interaction of fungicides and leaf saprophytes, *Pestic. Sci.,* 4, 563, 1973.
12. **Jenkyn, J. F. and Prew, R. D.,** The effect of fungicides on the incidence of *Sporobolomyces* spp. and *Cladosporium* spp. on flag leaves of winter wheat, *Ann. Appl. Biol.,* 75, 253, 1973.
13. **Dickinson, C. H. and Wallace, B.,** Effect of late applications of foliar fungicides on activity of microorganism on winter wheat flag leaves, *Trans. Br. Mycol. Soc.,* 67, 103, 1976.
14. **Mappes, C. J. and Hampel, M.,** Yield responses of winter barley to late fungicide treatments, *Proc. 1977 Br. Crop Prot. Conf. Pests and Diseases,* 1, 49, 1977.
15. **Andrews, J. H. and Kenerley, C. M.,** The effects of pesticide programs on non-target epiphytic microbial populations on apple leaves, *Can. J. Microbiol.,* 24, 1058, 1978.
16. **Andrews, J. H. and Kenerley, C. M.,** The effects of pesticide programs on microbial populations from apple litter, *Can. J. Microbiol.,* 25, 1331, 1979.
17. **Skajennikoff, M. and Rapilly, F.,** Variations de la phylloflora du bléliees des traitements fongicides à base de benomyl, *EPPO Bull.,* 11, 295, 1981.
18. **Little, R. and Doodson, T. K.,** The comparison of yield of some spring barley varieties in the presence and absence of mildew and when treated with fungicide, *Proc. 6th Br. Insectic. Fungic. Conf.,* 1, 91, 1971.
19. **Khare, M. N. and Vyas, S. C.,** Effect of systemic fungicides on phylloplane mycoflora, in *Progress in Microbial Ecology,* Mukerji, K. G., Agnihotri, V. P., and Singh, R. P., Eds., Print House, Lucknow, India, 1984, 134.
20. **Furtode, I.,** Effect of copper fungicides on the occurrence of the pathogenic form of *Colletotrichum coffeanum, Trans. Br. Mycol. Soc.,* 53, 325, 1969.
21. **Firhman, J.,** Possible side effects of fungicides on banana and coffee diseases, *Nature (London),* 225, 1161, 1970.

22. **Gibbs, J. N.,** Effects of fungicides on the population of *Colletotrichum* and other fungi in berry of coffee, *Ann. Appl. Biol.,* 70, 35, 1972.

23. **Nutman, F. J. and Roberts, F. M.,** The stimulating effects of some fungicides on *Glomerella cingulata* in relation to the control of coffee berry disease, *Ann. Appl. Biol.,* 64, 335, 1969.

24. **Grosclude, C.,** Premiers essais de protection biologique des blessures de taille vis-à-vis du *Stereum purpureum* Pers., *Ann. Phytopathol.,* 2, 507, 1970.

25. **Carter, M. V.,** Biological control of *Eutypa armeniacae, Aust. J. Exp. Agric. Anim. Husb.,* 11, 687, 1971.

26. **Dimock, A. W. and Baker, K. F.,** Effect of climate on disease development, injuriousness and fungicidal control, as exemplified by snapdragon rust, *Phytopathology,* 41, 536, 1951.

27. **Byrde, R. J. W.,** The vulnerability of fungus spores to fungicides, in *The Fungus Spore,* Colston Paper No. 18, Butterworths, London, 1966, 289.

28. **Horsfall, J. G. and Lukens, R. J.,** Selectivity of fungicides, *Conn. Agric. Exp. Stn. New Haven Bull.,* 676, 1, 1966.

29. **Moore, M. H.,** Glasshouse experiments on apple scab. II. Direct and translocated fungicidal activity, *Ann. Appl. Biol.,* 57, 451, 1966.

30. **Hislop, E. C. and Cox, T. W.,** Effects of captan on the non-parasitic microflora of apple leaves, *Trans. Br. Mycol. Soc.,* 52, 223, 1969.

31. **Newhook, F. J.,** The relationship of saprophytic antagonism to control of *Botrytis cinerea* Pers. on tomatoes, *N.Z. J. Sci. Technol.,* 38, 473, 1957.

32. **Dickinson, C. H.,** The mycoflora associated with *Hilimione portulacoides.* III. Fungi on green and moribund leaves, *Trans. Br. Mycol. Soc.,* 48, 603, 1965.

33. **Holloman, D. W.,** Observations on the phylloplane flora of potatoes, *Eur. Potato. J.,* 10, 53, 1967.

34. **Bhatt, D. D. and Vaughan, E. K.,** Inter-relationships among fungi associated with strawberries in Oregon, *Phytopathology,* 53, 217, 1963.

35. **Simard, J., Pelletier, R. L., and Coulson, J. G.,** Screening of microorganisms inhibiting apple leaf for their antibiotic properties against *Venturia inaequalis* (Cke), *Wint. Rep. Que. Soc. Prot. Plants,* 39, 59, 1957.

36. **Bainbridge, A. and Dickinson, C. H.,** Effects of fungicides on microflora of potato leaves, *Trans. Br. Mycol. Soc.,* 59, 31, 1972.

37. **Stott, M. A.,** Studies on the physiology of some leaf saprophytes, in *Ecology of Leaf Surface Microorganisms,* Preece, T. F. and Dickinson, C. H., Eds., Academic Press, New York, 1971, 203.

38. **Kuthubutheen, A. J. and Pugh, G. J. F.,** Effects of fungicides on physiology of phylloplane fungi, *Trans. Br. Mycol. Soc.,* 71, 261, 1978.

39. **Edgington, L. V.,** Structural requirements of systemic fungicides, *Annu. Rev. Phytopathol.,* 19, 107, 1981.

40. **Vyas, S. C. and Nene, Y. L.,** Influence of some fungicides on plants, *Pesticides,* 8(2), 41, 1974.

41. **Vyas, S. C.,** *Systemic Fungicides,* Tata McGraw-Hill, New Delhi, 1984, 360.

42. **Dickinson, C. H.,** Effects of ethirimol and zineb on the phylloplane microflora of barley, *Trans. Br. Mycol. Soc.,* 60, 423, 1973.

43. **Jenkins, J. E. E., Melville, S. C., and Jemmett, J. L.,** The effect of fungicides on leaf diseases and yield in spring barley in southwest England, *Plant Pathol.,* 21, 49, 1972.

44. **McKenzie, E. H. C.,** Seasonal changes in fungal spore number in rye grass, white clover, and pasture and effect of benomyl on parasitic fungi, *N.Z. J. Agric. Res.,* 14, 379, 1977.

45. **Fehrmann, H.,** Modern developments in fungicide use on cereals, *EPPO Bull.,* 11, 259, 1981.

46. **Fokkema, N. J., van der Laar, J. A. J., Nelis-Blomberg, A. L., and Schippers, B.,** The buffering capacity of the natural mycoflora of rye leaves to infection by *Cochliobolus sativus* and its susceptibility to benomyl, *Neth. J. Plant Pathol.,* 81, 176, 1975.

47. **Edgington, L. V., Khew, K. L., and Barron, G. L.,** Fungitoxic spectrum of benzimidazole compounds, *Phytopathology,* 61, 42, 1971.

48. **Fokkema, N. J. and de Nooij, M. P.,** The effect of fungicides on microbial balance in the phyllosphere, *EPPO Bull.,* 11, 303, 1981.

49. **Greenway, W.,** *In vitro* tests of the toxicity of mercury compounds to *Pyrenophora avenae, Trans. Br. Mycol. Soc.,* 60, 87, 1973.

50. **Jenkyn, J. F. and Prew, R. D.,** The effect of fungicides on the incidence of *Sporobolomyces* spp. and *Cladosporium* spp. on flag leaves of winter wheat, *Ann. Appl. Biol.,* 75, 253, 1973.

51. **Thorne, G. N.,** Physiological aspects of grain yield in cereal, in *The Growth of Cereals and Grasses,* Milthorpe, F. L. and Ivins, J. D., Eds., Butterworths, London, 1966, 88.

52. **Walpole, P. R. and Morgan, D. G.,** A quantitative study of grain filling in *Triticum aestivum* L., cultivar Maris Widgeon, *Ann. Bot. (London),* 34, 309, 1970.

53. **Prew, R. D. and McIntosh, A. H.,** Effect of benomyl and other fungicides on take-all, eyespot and sharp eyespot diseases of winter wheat, *Plant Pathol.,* 24, 67, 1975.

54. **van der Hoeven, E. P. and Bollen, G. L.,** Effect of benomyl on soil fungi associated with rye. I. Effect on incidence of sharp eyespot caused by *Rhizoctonia cerealis, Neth. J. Plant Pathol.,* 86, 163, 1980.
55. **Dickinson, C. H.,** Interactions of fungicides with minor pathogens on cereals, *EPPO Bull.,* 11, 311, 1981.
56. **Bashi, E. and Fokkema, N. J.,** Environmental factors limiting growth of *Sporobolomyces roseus,* an antagonist of *Cochliobolus sativus* on wheat leaves, *Trans. Br. Mycol. Soc.,* 68, 17, 1977.
57. **Skidmore, A. M. and Dickinson, C. H.,** Effect of phylloplane fungi on the senescence of excised barley leaves, *Trans. Br. Mycol. Soc.,* 61, 49, 1973.
58. **Klarren, R. C.,** Differential effects of fungicides on phylloplane fungi isolated from oak, *Trans. Br. Mycol. Soc.,* 62, 215, 1977.
59. **Dickinson, C. H. and Wallace, B.,** Effects of late applications of foliar fungicides on activity of microorganisms on winter wheat flag leaves, *Trans. Br. Mycol. Soc.,* 67, 103, 1976.
60. **Jenkyn, J. F. and Prew, R. D.,** The effect of fungicides on the incidence of *Sporobolomyces* spp. and *Cladosporium* spp. on flag leaves of winter wheat, *Ann. App. Biol.,* 75, 253, 1973.
61. **Duccoman, P. and Corbaz, R.,** Influence of blue mold fungicides on the tobacco leaf mycoflora, *Phytopathol. Z.,* 104, 19, 1982.
62. **Luke, H. H., Barnett, R. D., and Morey, S. A.,** Effects of foliar fungicides on the mycoflora of wheat seed using a new technique to assess seed infestations, *Plant Dis. Rep.,* 61, 773, 1977.
63. **Bollen, G. J., van der Hoeven, E. P., Lamers, J. G., and Schoonnen, M. P. M.,** Effect of benomyl on soil fungi associated with rye. II. Effect on fungi of culm bases and roots, *Neth. J. Plant Pathol.,* 89, 55, 1983.
64. **Reinecke, P.,** Untersuchungen zum Erregerspektrum des Fusskrankheitskomplexes an Getreide unter besonderer Berucksichtigung von *Rhizoctonia solani* Kuhn, Thesis, University of Göttingen, West Germany, 1977, 141.
65. **van der Hoeven, E. P. and Bollen, G. J.,** Effect of benomyl on soil fungi associated with rye. I. Effect on the incidence of sharp ryespot caused by *Rhizoctonia cerealis, Neth. J. Plant Pathol.,* 86, 163, 1980.
66. **Hossfeld, R.,** Forderung der *Typhula* -Faule an Wintergerste durch Einsatz von Fungiziden zur Halmbruchbekampfung, *Nachrichtenbl. Dtsch. Pflanzenschutzdienst (Braunschweig),* 26, 19, 1974.
67. **Hoes, J. A.,** Dynamics of the mycoflora of subterranean parts of winter wheat in dryland areas of Washington, *Phytopathology,* 52, 736, 1962.
68. **Scott, D. B., Visser, C. P. N., and Rufenacht, E. M. C.,** Crater disease of summer wheat in African dryland, *Plant Dis. Rep.,* 63, 836, 1979.
69. **Duben, J. and Fehrmann, H.,** Vorkommen und Pathogenitat von *Fusarium* -Arten an Winterweizen in der Bundesrepublik Deutschland. I. Artenspektrum und Jähreszeitliche Sukzession an der Halmbasis, *Z. Pflanzenkr. Pflanzenschutz.,* 86, 638, 1979.
70. **Hill, R. A. and Lacey, J.,** Biodeterioration. Incidence of molds and toxins in barley grains, *Rothamsted Exp. Stn. Rep. 1973,* p. 120, 1974.
71. **Hill, R. A. and Lacey, J.,** The microflora of ripening barley grain and the effects of pre-harvest fungicide application, *Ann. Appl. Biol.,* 102, 455, 1983.
72. **Ledingham, R. J., Sallans, B. J., and Simmonds, P. M.,** The significance of the normal flora on wheat seed in inoculation studies with *Helminthosporium sativum, Proc. Can. Phytopathol. Soc.,* 16, 10, 1949.
73. **Luke, H. H. and Barnett, R. D.,** Effect of fungicides on disease development, seed contaminator and yield of wheat, *Plant Dis. Rep.,* 60, 117, 1976.
74. **Warren, R. C.,** Interference by common leaf saprophytic fungi with the development of *Phoma betae* lesions on sugar beet leaves, *Ann. Appl. Biol.,* 72, 137, 1972.
75. **Fokkema, N. J.,** The role of saprophytic fungi in antagonism against *Drechslera sorokiniana (Helminthosporium sativum)* on agar plates and on rye leaves with pollen, *Physiol. Plant Pathol.,* 3, 195, 1973.
76. **Pace, M. A. and Campbell, R.,** The effect of saprophytes on infection of leaves of *Brassica* spp. by *Alternaria brassicicola, Trans. Br. Mycol. Soc.,* 63, 193, 1974.
77. **Khavi, M. A. and Ramarao, M.,** Effect of foliar application of Bavistin on the mycoflora of phylloplane and rhizosphere and nonrhizosphere soils of groundnut, *Proc. Natl. Acad. Sci. India Sect. B,* 51, II, 129, 1981.
78. **Ruinen, J.,** Nitrogen fixation in the phyllosphere, *Plant Soil,* 22, 375, 1965.
79. **Sen, A.,** Possibilities of utilization of an organism from the phyllosphere of *Eichhornia crassipes* for improvement of yields of wheat and paddy, *Indian J. Genet. Plant Breed.,* 35, 315, 1975.
80. **Pati, B. R. and Chandra, A. K.,** Effect of spraying nitrogen fixing phyllospheric bacterial isolates on wheat plants, *Plant Soil,* 61, 419, 1981.
81. **Crosse, J. E.,** Control of bacterial canker of cherry by eradicant sprays, *Proc. 4th Br. Insectic. Fungic. Conf.,* p. 528, 1967.
82. **Pati, B. R., Chandra, A. K., and Gupta, S.,** The *in vitro* effect of some pesticides on the nitrogen fixing bacteria isolated from the phyllosphere of some crop plants, *Plant Soil,* 80, 215, 1984.
83. **Dennis, C.,** Effect of pre-harvest fungicides on the spoilage of soft fruit after harvest, *Ann. Appl. Biol.,* 81, 227, 1975.

84. **Cohen, E. and Dennis, C.,** Effect of fungicides on the mycelial growth of soft fruit spoilage fungi. *Ann. Appl. Biol.,* 80, 237, 1975.

85. **O'Donnell, J. and Dickinson, C. H.,** Pathogenicity of *Alternaria* and *Cladosporium* isolates on *Phaseolus, Trans. Br. Mycol. Soc.,* 74, 335, 1980.

86. **Sinclair, J. B.,** Fungicide spray for the control of seed-borne pathogens of rice, soybean and wheat, *Seed Sci. Technol.,* 9, 697, 1981.

87. **Gross, Y. and Kenneth, R.,** The fate of carboxin, benomyl and thiabendazole in seed and soil treated plants as shown by *in vitro* and *in vivo* bioassay on some epiphytic yeast, *Ann. Appl. Biol.,* 73, 307, 1973.

88. **van den Heuvel, D.,** Effect of *Aureobasidium pullulans* on numbers of lesions on dwarf bean leaves by *Alternaria zinniae, Neth. J. Plant Pathol.,* 75, 300, 1979.

89. **Hislop, E. C. and Barnaby, V. M.,** The host microflora-strawberry leaf, *Long Ashton Res. Stn. Univ. Bristol Rep., for 1973,* p. 135, 1974.

90. **Nicholson, J. F., Sinclair, J. B., White, J. C., and Kirpatrick, B. L.,** Upward and lateral translocation of benomyl in strawberry, *Phytopathology,* 62, 1183, 1972.

91. **Hislop, E. C., Barnaby, V. M., and Fox, K.,** Effects of sufactants on plants and their associated microflora, *Long Ashton Res. Stn. Univ. Bristol Rep. for 1974,* p. 135, 1975.

92. **Saunders, P. J. W.,** Modification of the leaf surface and its environment by pollution, in *Ecology of Leaf Surface Microorganisms,* Preece, T. F. and Dickinson, C. H., Eds., Academic Press, New York, 1971, 81.

93. **Kuthubutheen, A. J.,** The effects of Fungicides on Soil and Leaf Surface Fungi, Ph.D. thesis, University of Ashton in Birmingham, 1977, 87.

94. **Dickinson, C. H. and Walpole, P. R.,** The effect of late application of fungicides on the yield of winter wheat, *Exp. Husb.,* 29, 23, 1975.

95. **Skidmore, A. M. and Dickinson, C. H.,** Effect of phylloplane fungi on the senescence of excised barley leaves, *Trans. Br. Mycol. Soc.,* 60, 107, 1973.

96. **Gachichi, J. G.,** Studies on Effects of Fungicides on Phylloplane Mycoflora of Groundnut, M. Sc. (Ag) thesis, J. N. Agricultural University, Jabalpur, India, 1983, 106.

97. **Anon.,** Studies on the Effects of Benzimidazoles and Dithiocarbamates Fungicides on the Nutrient Status of Groundnut besides Control of Tikka and Rust Diseases, Indian Council of Agricultural Research (*Ad hoc* scheme), J. N. Agricultural University, Jabalpur, India, 1986, 206.

98. **Nene, Y. L. and Thapliyal, P. N.,** *Fungicides in Plant Disease Control,* Oxford & IBH Publishing, New Delhi, 1979, 507.

99. **Thomas, M.,** Studies on the Effects of Benzimidazoles and Dithiocarbamates Fungicides on Nutritional Status of Groundnut, Ph.D. thesis, J. N. Agricultural University, Jabalpur, India, 1986, 156.

100. **Blakeman, J. P.,** The chemical environment of the leaf surface in relation to growth of pathogenic fungi, in *Ecology of Leaf Surface Microorganisms,* Preece, T. F. and Dickinson, C. H., Eds., Academic Press, New York, 1971, 255.

Chapter 3

PLANT NUTRITION AND PHYSIOLOGICAL AND BIOCHEMICAL PROCESSES

I. INTRODUCTION

In the last decade fungicides have come into standard usage for extensive crop production throughout the world. During this period there has been debate between those who prefer routine prophylactic treatment and those who prefer specific protective applications of fungicides in response to an identified risk of imminent, severe disease development. Support for specific treatment is essentially based on a combination of economic and conservationist arguments. This is related to a threshold level of disease in the crop above which economic loss is considered likely. The arguments in favor of prophylactic treatment are more complex but paradoxically more attractive. The treatment is well established in medical and veterinary science. The supporters of prophylactic treatment also point to results of fungicide trials which show qualitative and quantitative (yield) benefits (phytotonic effects) in the absence of apparently damaging disease levels. These increases over the untreated plots are frequently not statistically significant, but mean increase if consistently recorded does represent an economic advantage. Those concerned with promotion of fungicidal treatments are well aware of this and rightly stress the yield response of treatments rather than the disease control which they achieve. The effects of fungicides on plant nutrition and physiological and biochemical processes are evaluated below.

II. PLANT NUTRITION

Application of fungicides, especially systemic, changes the performance of the crop and hence its nutritional requirements. It is not clear whether this is the result of more or less efficient uptake and utilization of nutrients or changes in requirement. The interpretation, therefore, requires more evidence than is currently available. The subject has been reviewed by Hance.[1] Apart from this, the effects of organic fungicides on soil properties may also affect the availability of nutrients to plants and consequently their nutritional requirements. Although the effect of herbicides on plant nutrition has been comprehensively reported on, the effects of fungicides have not been explored.[2]

With the growing awareness of the use of fertilizers for extensive cropping, it is and will be difficult to cope with the demand for fertilizers in the near future. Many of the organic and systemic fungicides which possess the potential to correct nutritional deficiencies should be used. In addition, they also possess growth stimulatory effects which result in increased chlorophyll content and higher grain and fruit yields.

A. Nutrient Status

Metal-based organic fungicides may correct micronutrient deficiencies or cause a trace element imbalance in treated plants, depending upon the nature of metal constituents. Cox and Winfall[3] reported that zinc accumulation was as much as ten times higher in plants sprayed with zineb than in controls. Swelling of root tips and growth of lemon seedlings caused by excess of copper were prevented through the use of zinc sulfate. An increase in the zinc content of the leaves and roots where zineb had been applied was accompanied by a reduction in copper content. Emge and Linn[4] reported that application of zineb and zinc sulfate on tomato leaf increased the zinc content of leaf tissues. The increase was sufficient to overcome or prevent the occurrence of zinc deficiency symptoms on plants growing under zinc-deficient nutrient culture or soils.

By administering metal-based dithiocarbamates, it could be possible to reduce the inhibition of root formation, growth retardation, and chlorosis of plants caused by nutritional imbalance, besides correcting nutrient deficiencies. However, manganese aggravated the harmful effects of antibiotics used for plant disease control. Symptoms of manganese and zinc deficiencies in apples were substantially alleviated by three foliar sprays of mancozeb (0.25%).[5] Other workers[6,7] have also found that spray materials affected the mineral content of leaves. Ross and Longley[8] and Ross et al.[9] established that the performance of apple trees is affected by the dithiocarbamate fungicides used for disease control; in a series of trials on mature McIntosh apple trees, the highest yields were obtained with captan followed by dodine, phenyl-mercury acetate and dichlone in descending order at the operational dose. This was followed by another long-term experiment in which lower dosages of dodine (0.05%) resulted in yields comparable to those of captan. While investigating the effect of dithiocarbamates on the performance of apple trees, Ross et al.[9] obtained a favorable yield response with seasonal sprays of zineb which was associated with increased zinc content of foliage; the foliage of McIntosh and Cortland trees not sprayed with zineb contained 9.4 and 10.4 ppm zinc, whereas trees sprayed with zineb (0.91 kg ha^{-1}) contained 63.0 and 87.0 ppm, respectively.[9] Ross and Webster[10] observed large increases in foliar concentration of zinc from the fungicides metiram, zineb, and dikar which contain about 14, 18, and 2% zinc, respectively. According to Heeny et al.,[11] apple foliage containing less than 10 ppm zinc would probably be deficient for this element and 10 to 15 ppm would be a threshold level. The fungicidal mixture ''Tricarbamix Zn'' was developed for dual-purpose control of plant disease and correction of mineral deficiencies.[12] The mixture contains 45% ziram, 15% ferbam, and 15% maneb; tests of the usefulness of this material as a foliar spray for apple nutrition indicated that zinc, iron, and manganese contents were significantly higher than control varieties. It was concluded that these fungicides increase the micronutrient status of leaf tissues well above the critical level established and are safer to use than inorganic compounds which can be phytotoxic when applied as foliar sprays. Dikar contains about 12% manganese and it markedly increased the foliar concentration of this nutrient. Leaves sprayed with benomyl and metiram had an increased level of manganese. Pesticides not containing nutrients have been reported to affect the mineral content of apple foliage;[6,7] however, the application of fungicides did not significantly affect the foliar concentrations of nitrogen, phosphorus, potassium, magnesium, calcium, and copper or their deficiencies, and no interactions were observed.[10]

The retention of zinc in cuticular disks from leaves was significant after 24-hr immersion in propioneb, captan + zineb, and metiram at 0.2% w/v. The distribution coefficient (zinc retained: zinc in immersion fluid) varied from 25 to 88.2% for the fungicides but was 5.7% for zinc chloride. Laser probemass spectrography indicated that zinc supplied with fungicides remained superficially localized for 24 hr leaf application on maize; zinc was detectable only in the surface zone corresponding to a maximum depth of 30 μm from the adaxial leaf surface.[13]

Cole et al.[14] studied the effect of a number of fungicides on nutrient uptake from soil that had been steam-treated. They reported that disulfoton at 10 and 100 ppm, dibromochloropropane (DBCP) at 3 and 30 ppm, captan at 100 and 200 ppm, thiram at 200 ppm, quintozone at 200 ppm, fenaminosulf at 200 ppm, and maneb at 200 ppm decreased manganese and increased zinc uptake. In addition, DBCP decreased nitrogen, calcium, magnesium, and strontium accumulation and increased potassium, iron, and copper, whereas captan (100 ppm), thiram, fenaminosulf, and maneb (at the dosage mentioned above) increased the fresh weight of maize harvested after 5 to 8 weeks of treatment.

DBCP did not affect broad bean development whether applied at planting or 28 days later.[15] Chlorothalonil increased calcium in peanut leaves, which may have been due to disease control rather than to a direct effect on plant metabolism.[16] Thiram and Ceresan can inhibit rhizobia in culture but, in pot experiments, Elik and Kecskes[17] found that both compounds improved yield and nodule formation in vetch (*Vicia sativa*) plants grown from seeds inoculated with rhizobia in an acid forest soil and were without effect in a calcareous soil. However, Taha et al.[18] observed that Ceresan can inhibit broad bean and lentil nodulation and growth. Laurence et al.[19] reported that peanut appears to be tolerant of chlorothalonil, which can increase oil content and kernel yield by delaying maturity.

B. Uptake and Efflux of Ions

Besides controlling plant diseases, application of systemic fungicides results in a more significant increase in the yield of treated plants than in untreated controls, and at the same time may help to produce a higher quality grain. These results suggest that such biologically active compounds may affect the mineral nutrition of plants as well. No explanation has been reported, however, as to why these compounds also influence the nutrient metabolism of cultivated plants. Sometimes adverse effects, such as leakage of ions, have been observed.

Soil application of benzimidazole was found to increase the accumulation of potassium, sodium, and calcium ions in excised barley roots *(Hordeum vulgare)* and potassium uptake by intact barley plants.[20] It was effective in decreasing the loss of previously absorbed potassium ions when roots were subjected to low pH. A similar effect of benzimidazole on potassium ion uptake in tobacco has also been observed.[21] McClurg and Bergman[22] reported that soil application of benomyl at the rate of 20 ppm on a soil weight basis resulted in a significant increase in the calcium content of bean leaves.

Zsoldos[23] studied the effect of Kitazin, a systemic fungicide for rice blast, on the ion uptake and leakage in rice roots; it was observed that potassium ion uptake decreases as compared to the control. At higher concentrations, almost no potassium ion uptake takes place; at lower concentrations (10^{-5} and $10^{-6}M$), the ion uptake was stimulated, though to a smaller degree. It was further observed that there is no difference in the initial ion uptake trends between the excised roots and intact plants. Although the rates of uptake by the roots and plants exhibit numerical differences, the latter always being higher, the effects of Kitazin treatment are qualitatively the same for each type of treatment: either both are stimulated or both are inhibited. They observed a similar pattern for phosphate. At 10^{-3} M Kitazin concentration no ion uptake could be detected, showing the inhibiting effect of this compound on the active ion flux processes. Zsoldos[23] observed that fungicides exert a considerable influence on the permeability of cell membranes. Therefore, he conducted ion efflux experiments to study the behavior of membranes during Kitazin treatment. The results indicate that the rate of efflux tends to increase with the Kitazin concentration, with a definite increase of rate for 10^{-4} M Kitazin and a much more marked one for 10^{-3} M Kitazin. However, at lower rates, the ion leakage is not significantly different from that of a control; this is not surprising, since in a normal healthy tissue the membrane acts as a barrier to diffusion and exchange of ions.[24] When the membranes responsible for this retention are injured, as was found at high Kitazin concentrations, there is a rapid leakage of ions out of the tissue, following their own diffusion gradients. This result suggests that the biologically active compounds in high concentrations result in an extensive disorganization of the cell membranes. Root tissues become leaky and thus direct membrane damage results.

Dilz and Schepers[25] compared spring wheat response to nitrogen with and without benomyl. The highest yield in its absence was obtained with 8.0 kg ha^{-1} of fertilizer nitrogen, whereas on benomyl-treated replicates the maximum occurred with 120 kg ha^{-1} of fertilizer. This is not necessarily a direct stimulation by benomyl as, in a subsequent year, high nitrogen levels led to yield reductions because of increased disease susceptibility, which was counteracted by treatment with benomyl.[26] Therefore, they suggested that nitrogen fertilizer levels should be adjusted each year. However, benomyl may have a direct effect on wheat, as Ellen and Spiertz[27] found that it can increase the percentage of nitrogen, phosphorus, and potassium in the grain. Dilz and Schepers[25] reported that thiophanate improved spring wheat responses to nitrogen in a similar fashion, but Moyer and Paulsen[28] found analizine was without effect on winter wheat production. The possibility of including a fungicide in a combined spray application with nitrogen and herbicide has been considered. Penny and Jenkyn[29] added tridemorph to a spray of liquid nitrogen fertilizer, but fertilizer effectiveness was reduced as compared to nitrochalk applied as a solid, probably because it caused leaf scorch. Benomyl increased calcium in bean and did not have any other side effects.[22] Benomyl also increased calcium in peanut leaves.[16]

III. EFFECT OF NONSYSTEMIC FUNGICIDES ON RESPIRATION AND TRANSPIRATION

A. Copper
Sometimes copper fungicide sprays induce chlorosis, which may be due to a mineral deficiency resulting from synergistic or antagonistic interactions among nutrients. Copper fungicides such Bordeaux mixture (BM) and Burgundy mixture can induce iron and manganese deficiencies.[30] Southwick and Childers[31] reported that copper-based sprays, especially BM, reduced photosynthetic and transpiration activity of treated apples, but increased the rate of transpiration of tomatoes. BM increased the respiratory activity of grapevine after 7 days of application as compared to control and zineb application.[32] It was concluded that the influence of BM and other copper fungicides was primarily physiological rather than mechanical. Many injurious effects of fungicides on leaf color have been reported, particularly in apples and pears. Llewelyn[33] described discoloration and abscission of apple leaves as the results of copper-sulfate sprays.

B. Lime-Sulfur
The effect of lime-sulfur and other forms of sulfur on respiration, transpiration, and photosynthesis accounts for their interference with growth and yield of plants. A marked reduction in photosynthesis in apple leaves has been reported by many workers.[34] Hoffman[35] found that lime-sulfur increased respiration, but decreased photosynthesis to a greater extent than was principally responsible for the chlorisis that lime-sulfur produces in apple leaves. Agnew and Childers[34] found that lime-sulfur spray had a more inhibitory effect on carbon dioxide assimilation than did colloidal sulfur or sulfur dusts. They further observed that reduction in the relative rate of photosynthesis because of sulfur compounds is greater at higher temperatures, low humidities, and continuous sunshine. According to Berry,[36] the gaseous or volatile compounds produced after the spray of lime-sulfur, rather than dessication, are the major causes of lime-sulfur injury. It is possible that these gaseous materials are responsible for the reduction in photosynthesis.[35] The formation of hydrogen sulfide from lime-sulfur and colloidal sulfur has been demonstrated. In practice the conversion of lime-sulfur to hydrogen sulfide is rapid, and even more so at higher temperatures, so that the maximum effect would occur shortly after application.[36] It was suggested that heat liberated during oxidation plays an important role in causing injury to the plant. The effect of low pH because of formation of sulfuric acid may be another important factor in phytotoxicity. Thus, the sulfur application to lemon trees results in pH often lower than the isoelectric points of the proteins involved. These two factors may help to explain the observation that lime-sulfur and lead arsenate bring about internal structural changes in leaves. According to Berry, the greatly altered palisade tissue could account for decreases in the rate of photosynthesis.

C. Organic Compounds
The organic fungicides appear to have a more profound influence on the physiological processes of plants than do the inorganic fungicides. The dithiocarbamates increase the

chlorophyll content considerably. Chlorophyll content of McIntosh apple foliage increased from 25 to 48%; on spraying with various iron, manganese, and zinc dithiocarbamates and their mixtures, to which a small amount of thiram was added, the thicknesses of palisade parenchyma of treated apple leaves were thicker as compared to those of unsprayed leaves. The dry matter production of these leaves was ten times as high as that of the control.[35,36] Moreover, the leaves grew for a longer period of time and senescence was delayed. Delayed senescence in potatoes after dithiocarbamate spraying has been repeatedly observed.[37] The delay involved the risk of prolonged exposure to infection of some storage microflora. This stimulating side effect of dithiocarbamate spraying cannot be attributed to fungicidal properties. Side effects also occur in plants which remain apparently free from fungus infection. The mechanism of action to produce these side effects of dithiocarbamates on the metabolism of higher plants has already been demonstrated.[38] Nishnianidze and Gogiberdize[39] observed that zineb and polycarbacin increased photosynthesis for 7 days, and then the process slowed down to a level still higher than in untreated controls in citrus; euparen, benomyl, and copaz initially reduced photosynthesis, which then returned to normal. In all variants transpiration increased to maximum after 12 to 17 days. Photosynthesis increased with the zineb and zineb + copper treatment and decreased with Bordeaux mixture in grapevine.[32] Growth and yield were increased by zineb + copper in almost all the varieties, but in one variety Bordeaux mixture increased the yield. This indicates that the grapevine varieties respond to fungicides differently.

Lindahl[40,41] and McMillan and Riedhart[42] reported that nabam, thiram, and hydrocarbon fungicides inhibit photosynthesis in an alga *(Enteromorpha linza)* and citrus, but did not affect endogenous respiration in algae. Almost all the petroleum sprays applied to citrus leaves inhibited or stimulated respiration, depending on the type of hydrocarbon oil used. Inhibition of photosynthesis by hydrocarbon oil is thought to be mechanical interference of gaseous exchange[42] but Lindahl[40] did not observe inhibition of respiration in *E. linza.* Zineb and ferbam increased the rate of photosynthesis in peas and peaches, respectively.[37] Actidione and dyrene influence chlorophyll production on onions and rice.[43,44] Seed treatment with thiram and methoxyethyl mercury silicate (MEMS) at 200 and 250 g kg^{-1} increased the water uptake of hybrid maize; root growth was significantly slower and that of the coleoptile more vigorous.[45] In primary and secondary roots, the fungicides initially decreased dry matter — later they increased it; thiram increased while MEMS decreased energy of emergence. Sinzar[45] also observed that intake of O_2 and release of CO_2 were reduced. These parameters indirectly affect the photosynthesis process.

IV. PHOTOSYNTHESIS

Carlson[46] reported that detached barley leaves treated with systemic fungicides such as carboxin at low dosages remained green longer than untreated leaves, and the chlorophyll content of treated leaves was higher. However, at 15 ppm carboxin was highly inhibitory to photosynthesis: Mathre[47] confirmed the findings of Carlson and also showed that photosynthesis in other plants reacts similarly to carboxin while oxycarboxin and F 831 are less effective. However, at moderate levels ($10^{-4}\,M$) only carboxin would inhibit photosynthesis, to an extent of 22.33%. The leaves of treated seeds of barley and wheat contained a significantly higher content of chlorophylls, carotenoids, xanthophylls, and nucleic acids than did the control leaves.[48,49]

Side effects of benomyl foliar application at 25 ppm on stomatal frequency and rate of transpiration of sunflower *(Helianthus annus)*, cotton *(Gossypium barbadense)*, and cowpea *(Vigna sinensis)* were studied.[50] They observed that in benomyl-treated plants, whatever the change in the total number of stomata, the ratio of closed stomata to open stomata in either the upper or the lower epidermis was always influenced. The number of closed stomata always increased at the expense of area, irrespective of the concentration of benomyl applied; consequently, the rate of transpiration was significantly reduced with the rise of benomyl concentration. The effect was highly discernible in sunflower, low in cotton, and somewhat unpredictable in cowpea. Similar results were also recorded with some crop plants, especially when they were subjected to unfavorable conditions such as salinity[51,52] or drought.[53]

Nontarget effects of benomyl (25 ppm) on the photosynthetically active pigments, the relative photosynthesis activity ($^{14}CO_2$-light fixation), dry matter production, and the carbohydrates, as well as nitrogen contents, were studied.[50] In sunflower leaves, the contents of chlorophyll a, chlorophyll b, carotenoids, and consequent total pigments were considerably reduced; the ratio of chlorophyll a to b was also reduced, which indicates that biosynthesis of chlorophyll a was more suppressed than that of b. In addition, the recorded inhibition of carotenoid synthesis could result in chlorophyll photooxidation and degradation since the protecting character of carotenoids was disturbed.[54] This might also result in reduction of photosynthetic activity. Carbohydrate metabolism was increased with the increase in benomyl concentration, whereas nitrogen metabolism was decreased with the increase in benomyl level; this suggests that the C/N ratio increased with the increase in the benomyl level. Contrary to sunflower, the leaves of treated cotton and cowpea exhibited considerably higher contents of various pigment fractions of photosynthesis as compared to those of the control. The value of a/b ratios remained more or less unchanged, which means that the biosyntheses of both chlorophyll fractions were promoted concomitantly. Thus, the photosynthetic activity was enhanced with the rise in benomyl concentration.[50] A parallel increase was recorded in carbohydrate content, while nitrogen content remained more or less unchanged; consequently, the values of C/N ratios exhibited somewhat higher values than those of the control. Although plant growth is apparently not affected, some alterations in the physiological activity were observed. This can be attributed mostly to the enhancement of CO_2-photoassimilation. Similar growth-promoting effects of benomyl at the identical concentration were also recorded in sugar beet.[55,56]

Several biochemical effects of benzimidazole have been reported; these include an increase in the NADP and NAD ratio.[57] An increase in NADP and ATP levels in leaves[58] and the conservation of leaf protein and chlorophyll were observed.[59] Benzimidazole stimulates chlorophyll formation[60] and increases the number of intergrana and grana in the chloroplast lamella.[61]

Dimond and Rich[62] reported that benzimidazole fungicides possess cytokinin-like properties which are believed[63] to account for their antisenescent behavior. It is possible that these effects follow modification of ethylene production. Mukhopadhyaya and Bandopadhyaya[64] showed that carbendazim was approximately ten times less active in the preservation of green leaf than cytokinin. Staskawicz et al.[65] have confirmed this and demonstrated a substantial (80 to 96%) chlorophyll retention 3 days after treatment with 10 and 100 μg mℓ^{-1} carbendazim. This is well below the commercial application level. Cook[63] also reported that carbendazim and other fungicides have some cytokinin-like activity and it is possible that this is related to extended green leaf survival. Peat and Shipp[66] have observed unexpected yield increases in wheat after benomyl application and concluded that the effects of benomyl closely resemble those of kinetin on wheat[67] and have some similarities to those of benzyladenine on barley.[68] Therefore, Peat and Shipp concluded that benomyl, when applied to wheat, behaves as weak cytokinin.

Benzimidazoles (benomyl, carbendazim, thiophanates, and thiabendazole) have been successfully used in plant disease control. These systemic fungicides are absorbed by plants and come in direct contact with the symplast.[69] Given such a contact with the living protoplasm of the plants, these systemic fungicides can be expected to exert some physiological effects. Thus, benomyl and carbendazim, and their parent compound benzimidazole, show antisenescent effects similar to cytokinins.[59,70-74] They also show other cytokinin-like effects, such as activity in callus bioassay,[75] breaking of seed dormancy,[76] and activity in the *Amaranthus betacyanin* test.[76] Carbendazim is also active as an antisenescent compound in oat protoplasts.[65] Carbendazim delays senescence both in dicotyledons[72,78] and monocotyledons.[65]

Tripathi et al.[74] studied the physiological and biochemical effects of carbendazim on wheat leaves and whether the antisenescent action on wheat leaves was due to carbendazim molecules per se or its parent compound, benzimidazole, which has been identified as one of the metabolic products of carbendazim in plants.[79] They observed that carbendazim retarded the breakdown of chlorophyll, protein, and RNA in detached wheat leaves during senescence, and carbendazim as well as kinetin partially prevented the rise of acid protease, ribonuclease, and esterase activities. They further observed that at the antisenescent concentration kinetin stimulated, but carbendazim and benzimidazole interfered with, chlorophyll synthesis in etiolated leaves, a process that was completely inhibited by cycloheximide. Both carbendazim and kinetin stimulated protein synthesis but inhibited DNA and RNA synthesis in leaf segments. Martin and Thimann[80] also observed the antisenescent action of cycloheximide with oat leaves and found that some protein, probably a protease, is synthesized *de novo* and plays a role in senescence. The preservation of protein and RNA in carbendazim-treated leaves might be due to a retardation of the rise in hydrolytic enzymes usually accompanying senescence of leaves, including wheat leaves.[81] This may explain the effect of the fungicide on the preservation of protein and RNA and high levels of ribonuclease and deoxyribonuclease.[82] Several proteases in tobacco[83] and in oats[80,84] have been observed in senescing leaves. The reduction of these levels by cytokinins and other treatments which retard senescence suggests a fundamental role for such enzymes during plant development. The similarity of action of carbendazim, kinetin, and benzimidazole may be due to their related structures.[65,74] The antisenescent effect has been reported for a number of compounds including benomyl,[27] carbendazim + zineb,[85] zineb,[87,88] captafol, and tridemorph.[89] Staskawicz et al.[65] have shown that carbendazim retards the breakdown of chlorophyll in oat leaves, so that fungicides of the benzimidazole group may have a direct biochemical effect on senescence. A further possibility is that fungicides control leaf-surface fungi, which affect the rate of senescence. Experiments have shown that benomyl, thiophanate-methyl, captafol, and zineb reduce the populations of these fungi[90,91] and that some species affect the rate of senescence when inoculated into barley leaves.[92] Late application of fungicides to cereals is rapidly becoming standard practice in modern agriculture. Apart from their obvious importance in the control of pathogenic fungi, it has been established that these chemicals may inhibit the growth of saprophytic fungi on aerial parts of the plant.[93] Fungal saprophytes are important in several respects during the life of the plant. They may grow on the green tissues and affect the development of pathogens; as tissues reach maturity they may grow even more actively and accelerate senescence and growth on the senesced tissues.

Triadimefon, a member of the triazole group, exhibits considerable activity against plant diseases such as rusts, smuts, and powdery mildew.[97] Fisher et al.[98] and Buchenauer and Grossman[99] noted that clover, barley, tomato, and wheat, after seed treatment with triadimefon (0.15%), remained greener than did untreated plants. Buchenauer and Grossman reported a delay of senescence of detached barley leaves floated on triadimefon suspension. This increase in chlorophyll content may be due either to the favorable stabilizing effect of the fungicide on the integrity of the chloroplast or to an increased synthesis of chlorophyll. Gautam et al.[100] reported that soil treatment of soybean seedlings at 0.5 to 10 ppm triadimefon increased, at almost all the stages of growth, chlorophyll and b and total chlorophyll (a + b), with the increase in the concentration of the fungicide. They further observed that at each time interval, the ratio of chlorophyll a to chlorophyll b increased with the increase in the concentration of the fungicide. Carbendazim, which also delays senescence, maintains the integrity of the chloroplast and other cell organelles of wheat leaves, as shown by electron microscopy.[74] The higher chlorophyll content induced by the triadimefon treatment was reversed by simultaneous application of gibberellic acid (GA).[101]

Fletcher and Nath[101] observed that triadimefon has several physiological effects on crop plants, some of which are yet to be harnessed for production. The fungicide at 20 μg mℓ^{-1} reduced transpiration in wheat (Frederick), peas (Little Sweete), and soybeans (Hodgson) by 49, 29, and 19%, respectively, after soil application. The reduction prevented the leaves from wilting and becoming senescent. In another experiment, triadimefon was applied as a soil drench after flower initiation in soybean and peas and after swelling in Cherry Belle radish roots; after treatment, the water level was maintained at half field capacity. The treated plants were shorter, more compact, and had thicker green leaves than the controls. In radishes, the treatment also prevented flowering. These are desired attributes for producing higher yields of crops in the semiarid tropics, particularly in regions such as India and the Middle East. Triadimefon reduced transpiration; protected the plant from drought conditions; increased yield in soybean, wheat, and peas; and increased the root/shoot ratio in radishes.[101]

Barley seedlings raised from seed treated with 0.25 to 0.5 g kg^{-1} propiconazole, CGA 64 251, or diclobutrazole showed retarded development of coleoptile, primary leaves, and roots. The descending order of the activity of these compounds is as follows: diclobutrazole > CGA 64 251 > propiconazole. There was marked chlorophyll catabolism inhibition.[102] It was also observed that seed treatment with propiconazole and diclobutrazole improved frost resistance and that treated plants were more resistant to drought and salt deficiency.

Seed treatment with triazole fungicides, triadimenol, propiconazole, etaconazole, diclobutrazole, and paclobutrazole at 2.5 g kg^{-1} seed retarded elongation of the growth of wheat, barley, oats, and rye seedlings; the various compounds decreased in growth-regulating activity in the following descending order: paclobutrazole > diclobutrazole > etaconazole > propiconazole > triadimenol. The primary leaves in treated plants showed disturbed geotropism.[103] Elongation in the roots was less severely retarded by the triazole fungicides than that of the primary leaves. Because of the thickened and broadened primary leaves of the treated plants, fresh weight and dry weight of the shoots were less significantly reduced than elongation. It was observed that the roots of seed-treated seedlings were partly thickened and contained more water than the control plants. Similar observations were also reported for triadimefon on wheat, soybean, peas, and radishes.[101] On a basis of leaf area, higher amounts of chlorophyll were found in primary leaves of barley seedlings treated with paclobutrazole, diclobutrazole, etaconazole, and propiconazole.[103] A single application of propiconazole and etaconazole at 0.15 kg a.i. (active ingredient) ha^{-1} had no effect on the net photosynthesis of pecan, but benomyl (0.15 kg a.i. ha^{-1}) reduced photosynthesis 1 day after treatment but had no effect after 9 days.[104]

Following root treatment, the triazole fungicides retarded the elongation of tomato shoots. The elongation of shoot growth and the morphological changes of the younger leaves caused by the fungicides were retarded to different degrees by simultaneous GA application; extracts of shoots of treated barley seedlings contained lower amounts of gibberellin-like compounds than untreated seedlings.[102,103]

Application of anilazine markedly increased the chlorophyll content in onions in newly developed second leaves. This shows that accelerated synthesis of chlorophyll in newly developed leaves may be due to anilazine application at 0.2%.[43] On the other hand, the content of chlorophyll, especially chlorophyll a, was reduced in the second leaves treated with cycloheximide, in comparison with those treated with zineb or the control. However, despite spraying, the chlorophyll content in the fourth leaves, i.e., old leaves, did not change, showing results similar to those in the control. This shows that fungicide spraying on comparatively old leaves does not influence chlorophyll synthesis markedly. Cycloheximide, however, seems to inhibit chlorophyll synthesis in younger leaves but does not accelerate its decomposition.[43] In the field, Shishiyama et al.[43] found that onion leaves turned dark

Table 1

EFFECT OF FOLIAR APPLICATION OF FUNGICIDES ON CHLOROPHYLL A AND B AND TOTAL CHLOROPHYLL CONTENT (mg g^{-1}) OF GROUNDNUT FRESH LEAVES AFTER 45 DAYS GERMINATION[107,108]

Fungicides	Concentration (%)	Chlorophyll a	Chlorophyll b	Total chlorophyll
Carbendazim	0.15	4.47	4.56	8.95
Thiophanate-methyl	0.15	5.24	4.85	10.15
Zineb	0.25	3.69	3.00	6.30
Thiram	0.25	3.70	2.50	6.40
Nil (control)		3.79	2.43	6.22
Least sum of differences (LSD) at 5%		0.139	0.015	0.12

Table 2

EFFECT OF FOLIAR APPLICATION OF FUNGICIDES ON VEGETATIVE GROWTH, PLANT HEIGHT (cm), AND DRY WEIGHT OF PLANT (g) OF GROUNDNUT PLANT AFTER 45 DAYS GERMINATION[107,108]

Fungicides	Concentration (%)	Plant height (cm)	No. leaves/plant	Dry weight/plant (g)
Carbendazim	0.15	27.64	140.00	3.52
Thiophanate-methyl	0.15	26.38	139.53	3.50
Zineb	0.25	20.13	102.85	2.28
Thiram	0.25	20.25	102.56	2.30
Nil (control)	—	18.25	101.25	2.23
LSD at 5%	—	1.094	0.670	0.067

green after they were sprayed with anilazine. This might be caused by the increase of chlorophyll synthesis in leaves. They further observed that at 25 ppm concentration, chlorophyll and DNA synthesis were inhibited in onions by cycloheximide, but RNA synthesis was increased twofold.

Triarimol and pyracarbolid inhibited respiration at concentrations which did not produce visible symptoms. Pyracarbolid caused 40% inhibition of photosynthesis 10 days after treatment in wheat at a concentration considerably below the field rate, and maneb induced a slight and temporary reduction in photosynthesis.[105] An increase in the reduction of sugar in cereals following benomyl treatment has been recorded.[106] If the inhibition of senescence is related to ethylene effects,[62] it is possible that the effects of fungicides on respiration and senescence are related.

The foliar application of the fungicides carbendazim, thiophanate-methyl, zineb, and thiram at the recommended rates of application for plant disease control results in a significant increase in the chlorophyll content of groundnut plants (Table 1). Since the amount of chlorophyll in the plant is the precursor of food material, its increase is bound to increase the yield and the final product.[107,108]

Plants treated with systemic fungicides, both carbendazim and thiophanate-methyl, showed greater height, number of leaves per plant, and dry weight than those treated with nonsystemic fungicides (Table 2) or the control plants. Nonsystemic fungicides, however, increased plant height over that of control plants with no change in the number of leaves per plant or dry weight. Hence it is suggested that fungicides should be used judiciously where soil is not highly fertile.[107,108]

V. BIOCHEMICAL PROCESSES

A. Sterol

Ancymidol and triarimol are synthetic pyrimidine-5-yl-methanol analogues with retardant activity. Ancymidol is used as a growth retardant but is also weakly fungitoxic; triarimol is a fungicide but also retards the growth of higher plants.[109] Sherald et al.[110] found that two mutants of the fungus *Cladosporium cucumerinum* selected for their resistance to triarimol were resistant to ancymidol also. Several effects have been observed in bean plants when triarimol or ancymidol was applied at high dosages or early after germination. These include inhibition of stem and root growth, root swelling, bulbous stem swelling, inhibition of the hypocotyl hook opening, leaf epinasty, and abscission of leaves and flowers. All these processes have been induced by the addition of ethylene to plant tissues.[111]

Shive and Sisler[112] reported that when ancymidol and triarimol were incorporated into agar upon which the *Tan-ginbozu* (a bioassay cultivar of *Oryzae sativa*) seedlings were grown, they inhibited root and leaf sheath elongation. The root length of seedlings treated with 10 μg mℓ^{-1} ancymidol were 43% of the control length and the second leaf sheath length was 25%. *Tan-ginbozu* is reported to contain no detectable gibberellin-like activity.[113] The growth retardation in *Tan-ginbozu* by these retardants was possibly due to the fact that (1) the seedlings require an undetectable amount of gibberellins and these retardants inhibit the biosynthesis or action of these gibberellins or (2) these retardants facilitate growth retardation through some means other than inhibition of gibberellin biosynthesis or action, possibly by inducing ethylene synthesis. Douglas and Paleg[114] have identified sterol-synthesis inhibition as a possible mode of action for other growth retardants; however, Shive and Sisler[112] demonstrated growth retardation due to these retardants in tissues where there is no net sterol synthesis. Therefore, they exluded this possibility. Shive and Sisler demonstrated that both retardant treatments drastically reduce the gibberellin-like activity in bean extracts. They have also shown that application of gibberellin can completely relieve the dwarfing effects of these retardants. Thus they have concluded that the primary mode of action of the retardants in relation to growth retardation is the inhibition of gibberellin-induced response(s), and the point at which this interference mostly probably occurs is during the biosynthesis of gibberellin. Shive and Sisler further observed that sterols and gibberellins are apparently involved in membrane activity, and the growth-retardant and fungicide modes of action may be linked somehow to the maintenance of membrane structure and function. Ancymidol (and triarimol) is now known to have effect(s) on higher plant growth hormone(s). Hormones affecting cell expansion in fungi are being discovered;[115] therefore, the ancymidol and triarimol modes of action in fungi and higher plants may be linked through the control of growth hormones.[112]

The growth-retardant effect of ancymidol has sometimes been encountered in the use of fungicidal analogues, such as triarimol, fenarimol, and nuarimol, as plant protectants.[116] The growth-retardant action at low concentrations of these chemicals was annulled, fully or in part, by application of GA. This suggests that the growth retardation is due primarily to interference with GA-dependent processes in the plant. However, at higher inhibitory concentrations, GA may be only slightly active or ineffective in reversing growth retardation. In pea plants, growth was reduced by root application of 10 ppm ancymidol or nuarimol, but not by the same concentrations of triarimol or fenarimol.[117] This inhibition can be reversed by application of GA to the shoot tips. Application to 100-nℓ concentrations of these inhibitors results in growth retardation, which can be partially annulled by GA in the case of triarimol or fenarimol but not in the case of ancymidol or nuarimol.[117] At this higher inhibitory concentration, it appears that growth retardation involves processes in the plant, in addition to those dependent on GA.

Buchenauer[118] studied the effects of nuarimol, fenarimol, triadimefon (each 2.5 mg/250 mℓ soil), and imazalil (5 mg/250 mℓ soil) on the growth, gibberellin activity, and lipid metabolism in wheat, tomatoes, and barley. The fungicides strongly retarded elongation of tomato shoots, but treatment with GA (2 mg/250 mℓ soil) markedly stimulated shoot elongation, and the fungicide effect was relieved by simultaneous application of GA. When lower quantities of fungicidal agents had been applied, GA not only annulled the retarding effect, but also promoted shoot elongation which exceeded that of the treated control plants. In wheat the root system was more severely affected than the coleoptiles and primary leaves. At comparatively low concentrations, triadimefon and imazalil caused weaker inhibitory effects than nuarimol. GA (10 μg mℓ^{-1}) stimulated elongation of coleoptiles and primary leaves but was ineffective on root growth. Growth inhibition induced by the concentration of the chemicals used was only partially relieved by the presence of GA. However, at lower concentration of the fungicides GA either annulled retardation or promoted elongation as compared to nontreated control plants. The fungicides did not affect seed germination, but similar growth-retarding effects were obtained when the fungicides were applied as seed treatments. Similar results were obtained with barley.[102,103] Although growth of primary leaves of wheat and barley plants was promoted by indole acetic acid (IAA) (10 mg mℓ^{-1}) as compared to wheat treated with GA (10 μg mℓ^{-1}), IAA did not alleviate growth retardation. Kinetin (10 μg mℓ^{-1}) showed a slight antagonistic activity. The fungicides interfere with gibberellin biosynthesis, and gibberellin-like activity was more severely reduced in wheat shoots by nuarimol than by triadimefon. Imazalil exhibited the least effect.[102,103,118]

Experiments on the effect of fungicides on lipid metabolism of barley indicated that radioactivity of the fraction of ^{4}C-des-methyl sterols was reduced by triadimefon (49%), imazalil (40%), and nuarimol (76%). The migration properties of this fraction corresponded to authentic sterols; on the other hand, radioactivity in sterols migrating to ^{4}C-methyl and 4,4-dimethyl sterols increased in extracts of treated plants. The fraction of 4,4-dimethyl sterols was less affected by the fungicides than was the ^{4}C-methyl sterol fraction. Radioactivity of triglycerides was greatly increased in leaves treated with nuarimol, whereas the effect of triadimefon on the synthesis of triglycerides was weaker than that of nuarimol: Imazalil was ineffective. The chemicals also induced a higher synthesis of GA (about 20%) and of sterol esters (15 to 25%).[118] The triazoles inhibited incorporation of ^{14}C-acetate in the ^{4}C-des-methyl sterol fraction of detached primary leaves of barley, and the activity of ^{4}C, 4-dimethyl, ^{4}C-methyl, and ^{14}C-methyl sterol fractions was increased in treated shoot sections. Gas chromatography analysis revealed reduced contents of campesterol, stigmasterol, and β-sitosterol in extracts of shoots of treated barley seedlings.[103] Barley plants were affected by fungicides similar to the way fungi are affected. Among the compounds used, nuarimol most severely interfered in phytosterol biosynthesis. However, in comparison to sensitive fungi *(Ustilago avenae)*, much higher concentrations of the chemicals were necessary to affect the GA and sterol biosynthetic pathway in wheat plants, which finally resulted in growth retardation.[118] Since the compounds are able to affect GA and sterol biosynthesis in plants, it is likely that the fungicides act in plants at the membrane level. The chemicals increased formation of membranes and affected membrane permeability.[119,120]

Most evidence identifies gibberellin biosynthesis as the primary target of pyrimidine-5-yl-methanols in growth regulation. Shive and Sisler[112] showed that triarimol and ancymidol both markedly reduce the GA content of bean *(Phaseolus vulgaris)* seedlings. This result is consistent with the findings of Coolbaugh et al.[121,122] who demonstrated that ancymidol blocks the GA biosynthetic pathway between kaurene and kaurenoic acid in *Marah macrocarpus* and *M. oreganus.* The three cytochrome P-450-dependent reactions involved in this conversion (the oxidations of kaurene, kaurenol, and kaurenal) displayed a similar degree of sensitivity to ancymidol.[121,122] These reactions are very sensitive; the kaurene inhibition (K_i) of ancymidol for kaurene oxidation by microsomes of *M. macrocarpus* was estimated to be 2 nℓ.[122] Ancymidol interacts with cytochrome P-450 but it apparently does not inhibit the microsomal electron-transport system reducing the cytochrome.[122] Comparing the activity of ancymidol and a variety of analogues (including triarimol, fenarimol, and naurimol) as inhibitors of the elongation of pea plants and as inhibitors of kaurene oxidation by microsomes of *M. macrocarpus,* it was observed that the relative effectiveness of an analogue for the two processes was similar; ancymidol and pyrimidol were highly effective as inhibitors of both processes, whereas the fungicides were poor inhibitors of both. Nuarimol was the only fungicide with detectable growth-regulatory activity to roots of plants at a concentration of 10 nℓ. The apparent LD_{50} of ancymidol and several isopropyl analogues for microsomal kaurene oxidation ranged between 1 and 100 nℓ, whereas the LD_{50} for triarimol, fenarimol, and nuarimol ranged between 1 and 100 nℓ. Thus, the analogues regarded as growth regulators and those regarded as fungicides fall into two activity groups, insofar as inhibition of higher plant elongation and kaurene oxidation are concerned.[122,123] When the three fungicides were compared with growth regulators as inhibitors of growth, GA biosynthesis, and microsomal oxidation of kaurene in the fungus *Gibberella fujikuroi,* the activities of the fungicides and growth regulators were in reverse order to that observed in higher plants.[123] In accordance with the findings of Shive and Sisler[112] for triarimol and ancymidol, Coolbaugh et al.[123] did not observe a clear delineation between the sensitivity of fungal growth and GA biosynthesis to pyrimidine-5-yl-methanol fungicides and growth regulators.

It seems likely that GA biosynthesis in higher plants is more sensitive than demethyl sterol biosynthesis to ancymidol. On the other hand, the sensitivity of these two processes to triarimol, fenarimol, and nuarimol may not differ appreciably. While the growth-retardant action of triarimol could be alleviated by GA in *Phaseolus vulgaris* seedlings,[112] this was achieved at an early stage of seedling growth when large reserves of demethyl sterols were present. Long-term growth retardation might possibly involve inhibition of demethyl sterol biosynthesis. The poor reversal of growth-retardant action by GA in soybean plants treated with 100 nℓ nuarimol, triarimol, and ancymidol suggests that processes such as sterol synthesis, in addition to GA synthesis, may be affected. Fenarimol at 7 μg mℓ^{-1} has, in fact, been shown to inhibit demethyl sterol synthesis in bramble cell cultures and nuarimol has been shown to have a similar effect on barley shoots. An interesting problem which has yet to be resolved is the relative sensitivity of GA and sterol biosynthesis in higher plants to compounds such as fenarimol, nuarimol, and ancymidol.[112]

Triazoles such as triadimefon and triadimenol are highly specific, and their mode of action revealed that both toxicants markedly inhibited ergosterol biosynthesis.[124] Both the compounds are systemically active, and several physiological effects have been reported. After root application, triadimefon strongly retarded elongation of tomato shoots.[99] The growth-retarding effects were partly or completely overcome by an application of GA. It was further reported that the effect of triadimefon on the induction of higher chlorophyll content of leaves of cotton plants was also annulled after simultaneous application of GA. On the other hand, triadimefon exerted an effect similar to GA in delaying the senescence of detached leaves in the dark. Triadimefon exhibited cytokinin-like activity in *Amaranthus caudatus* seedlings. When sections of primary barley leaves were floated in triadimefon and triademenol in the dark, senescence was retarded. Compared to that in untreated plants, phenylalanine ammonia lyase (PAL) activity in shoots of 12-day-old wheat plants was reduced by triadimefon and triadimenol.[124]

Buchenauer and Grossmann[99] reported that triadimefon interferes with GA activity in plant growth and does not affect GA actions that are involved in nongrowth processes such as retardation of senescence and stimulation of α-amylase synthesis in barley endosperms. The findings of Buchenauer and Grossmann[99] were confirmed in experiments on biochemical effects on cereal aleurone following GA treatment.[120] Before synthesis of hydrolytic enzymes was stimulated by GA, the hormone induced an increased formation of membranes (particularly in endoplasmic reticulum) and this affected membrane permeability. The accumulation of free fatty acids may also contribute to the fungal toxicity of triadimefon.[99] Since triadimefon and triadimenol[124] exhibit a primary mode of action in sporidia of *U. avenae* similar to that of triarimol in sporidia of *U. maydis*,[125,126] (i.e., inhibition of sterol ^{14}C oxidative demethylation during ergosterol biosynthesis), it may be assumed that triarimol as well as triadimefon and triadimenol interfere with GA biosynthesis. Buchenauer and Rohner[124] reported that shoots of barley seedlings grown in the presence of triadimefon and triadimenol contained both lower amounts and altered relative proportions of [4]C and 4-des-methyl sterols than did shoots of control seedlings. Incorporation studies indicated that the sterol accumulates nuarimol, fenarimol, and imazalil interfered with sterol biosynthesis in primary leaves of barley seedlings similar to triadimefon and triadimenol.[127] Schmitt and Beneveniste[128] reported profound changes in the sterol biosynthesis of cell cultures of bramble cells *(Rubus fructicosus)* in the presence of fenarimol. In the treated cells, campesterol and sitosterol markedly decreased, but other sterols accumulated a major one of which was obtusifoliol. Schmitt and Beneveniste suggested that fenarimol may act as a specific inhibitor of enzymes involved in 14-demethylation reactions, and it appears likely that triadimefon and triadimenol might interfere with sterol biosynthesis in barley seedlings at the same target site which corresponds to the primary mode of action of both chemicals in *U. avenae*.

Sterols are major compounds in numerous fungi and are closely involved in membrane structure and function; the primary action of GA in plants is also at the membrane level.[119] The altered ratios of sterols in shoots of triadimefon- and triadimenol-treated plants may cause physiological effects and affect growth and development processes, such as stunting, thickening, bulbous formation, etc. Shive and Sisler[112] are also of the view that the mode of action in fungi and higher plants may be linked through the control of growth hormones.

B. Protein

Several workers have reported the effects of fungicides on cell metabolism. Cook[63] and Sijpesteijn[130] concluded that nucleic acid and protein synthesis seem to be related to the mode of action of fungicides. The members of the benzimidazole group (carbendazim and benomyl) are believed to act particularly on tubules and microtubules in cell division. Other possible related effects have also been observed. Adele et al.[106] observed that soil application of 100 mℓ benomyl (125 and 250 μg mℓ^{-1}) to pot-grown wheat in the glasshouse reduced the protein level in treated plants after 18 days to about half the level in the untreated control. In benomyl-treated wheat plants the ribonuclease activity was 2 to 3 times higher than untreated plants.[131] The DNA synthesis and transcription process was inhibited by seed treatment, thus decreasing RNA synthesis and leading to a decrease of protein synthesis and cell division. However, Staskawicz et al.[65] found that 10 μg mℓ^{-1} carbendazim slightly suppressed ribonuclease and deoxyribonuclease activity initially, although pretreatment led to apparent decreases in RNA and DNA synthesis. One of the primary toxic effects of benomyl is an inhibition of protein and RNA synthesis by 50% at 15 and 10 μg mℓ^{-1}, respectively.[132] Direct effects of carbendazim on RNA synthesis have also been reported in the tobacco mosaic virus.[133]

Benzimidazole has long been recognized as a purine analogue because its toxicity to *Saccharomyces cerevisiae* is largely reversed by adenine and guanine.[134] Therefore, conversion of benzimidazole to a nucleotide in biological systems would not be unexpected in view of the results obtained with other purine analogues.[135] In wheat leaves, the benzimidazole nucleotide has been identified as a product of benzimidazole metabolism.[136] According to Kapoor and Waygood[136] the nucleotide may be incorporated into nucleic acids because labeled benzimidazole is released as [14]C. In addition to this, the benzimidazole nucleotide may also be incorporated into a coenzyme in wheat plants. Further, it was observed[137] that enzyme preparations made from wheat embryos catalyze the substitution of the benzimidazole-adenine dinucleotide. This group of fungicides delays senescence in detached wheat leaves and thus appears to have beneficial effects.[138]

Gautam et al.[100] reported that there was a progressive decrease in DNA content in leaves with an increase in the concentration of triadimefon. Maximum decreases of 35.65% in 10 days, 26.96% in 20 days, 19.48% in 30 days, 17.68% in 40 days, and 21.29% in 50 days were noticed at 10 ppm triadimefon. The minimum decrease of only 1.80% was seen 50 days after treatment with 0.5 ppm triadimefon. The RNA content decreased with the increase in the concentration of the fungicide, with maximum decreases of 9.66% in 10 days, 8.12% in 20 days, 6.70% in 30 days, and 5.64% in 40 days noticed at 10 ppm. However, at 50 days, the minimum RNA content was found in plants treated with 5.0 ppm triadimefon.[100] It was further observed that protein content at 10 days after treatment decreased with the increase in the concentration of triadimefon. However, in 20-, 40- and 50-day-old plants treated with 0.5 to 2.5, 0.5 to 5.0, and 0.5 to 10 ppm triadimefon, respectively, the protein content was higher than that of untreated plants of corresponding ages. At 30 and 40 days, the protein content of plants treated with 5 to 10 and 10 ppm, respectively, was lower than that of corresponding control plants. In general, it was observed that the increase in protein content percentage was inversely related to the concentration of the fungicide. These findings are partly in agreement with those of Buchenauer[139] who found that 8-hr treatment of the fungus *U. avenae* with 1.5 and 2.0 ppm triadimefon reduced RNA by 18 and 34%, DNA by 12 and 18%, and protein by 7 and 14%, respectively. An inhibitory effect on RNA and DNA content and a stimulation of protein content also occur in the petiolar base of cabbage treated with carbendazim.[140]

The exudation of organic compounds from plant roots is the main nutrient source for microorganisms inhabiting the rhizosphere. Various environmental factors influence both the quality and the quantity of root exudates. Foliar application of various chemicals has been shown to modify the spectrum of root exudation as well as the microbial population of the rhizosphere.[141,142] Balasubramanian and Rangaswamy[143] reported that the amino acid exudates of sunhemp *(Crotolaria juncea)* after leaf application of Dithane Z-78 at 200 ppm markedly reduced glycine, histidine, and alanine. In sorghum, however, the quantities of glutamic acid, lycine, proline, threonine, tyrosine, and others were not significantly affected. Jalali and Domosch[144] studied the effect of dichlofluanid, triforine, and tridemorph at 0.075, 0.05, and 0.2% after foliar application in wheat seedlings (3 weeks old) on the amino acid spectrum of wheat root exudates. They observed that triforine and tridemorph brought about an appreciable shift in amino acid exudation, although liberation of some individual acids was noticeably enhanced. The exudation of aromatic amino acids (viz., phenylalanine and tyrosine) was totally or partially suppressed; other than suppression of glycine and aspartic acid exudation, dichlofluanid had few effects. Triforine and tridemorph also affected the exudation of the closely interrelated amino acids valine and leucine as well as threonine, isoleucine, and aspartic acid. The reduced exudation is partially compensated for by the increased amounts of other amino acids such as proline, serine, and lysine. The available data, however, exclude the possibility of pathway interpretations; in the case of the dichlofluanid, the overall metabolic disturbances are far less pronounced.[144]

Zsoldos[23] studied the effect of Kitazin on the free amino acid content of the roots; at 10^{-3} *M* Kitazin, there is a very marked decrease of the free amino acid content, presumably due to damage which increases cell membrane permeability. This results in leakage of the free amino acids or a change in metabolism leading to flower formation. It is remarkable that at 10^{-5} *M* Kitazin there are more of some of the free amino acids (e.g., glutamic acid, aspartic acid, and amides) than in the control, which might be due to fungicidal treatment resulting in a favorable effect on nitrogen incorporation. Jaiswal et al.[145] found that the beneficial effects of fungicides on the growth and development of sugar cane are due to more efficient utilization of soil and fertilizer nitrogen. Also, some pesticides are known to increase the respiratory rate of plant tissues, which is associated with increased ion uptake.[146] Stimulation of nitrogen metabolism (nitrate absorption, nitrate reductase induction, amino acid and

Table 3
EFFECT OF FOLIAR
APPLICATION OF
CARBOXIMIDES ON
PROTEIN CONTENT OF
WHEAT (cv. BISON)[147]

Fungicide	Quantity of fungicides applied in (lb/acre)	
	2.67	5.33
Carboxin	14.00	15.50
Oxycarboxin	13.80	15.10
Control	12.40	

Note: Results measure percentage of protein content of wheat.

protein synthesis) was reported for wheat plants[147] and for tomato plants[148] which had been treated with oxathiin and benzimidazole fungicides. Treatments with benzimidazole (200 ppm) also caused a significant increase in incorporation of labeled glycine and adenine into the protein and the nucleic acid bases of bean seedlings.[149]

Reyes et al.[147] reported that foliar application of carboxin and oxycarboxin in the boot stage of wheat increased protein content (Table 3). The data present in Table 3 indicate that there is a considerable increase in grain protein in treated wheat plants. It was also reported that seed treatment at 1 to 5 oz/cwt had no effect on the protein content of harvested grain. In growth chambers, nitrogen metabolism was stimulated in wheat, corn, sorghum, and soybean plants by carboximides applied in nutrient solutions. These carboximides increased nitrate absorption and nitrate reductase induction when nitrate supply was low and/or temperatures were suboptimum.[147] Such benefits of carboximides may be due to their effect on growth-regulating hormones and other physiological processes which resulted in the growth and the higher yield.

Application of benzimidazole fungicides causes biochemical changes in plant species. Singh and Kang[150] noted increased production of chlorophyll, total sugars, phenols, proteins, and lipid contents in groundnut plants treated with carbendazim (0.15%), which are all yield-contributing factors.

Singh and Kang[151] studied the effect of foliar application of carbendazim (at 0.05, 0.10, and 0.15%) on the amino acid composition of groundnut cultivar M-13 leaves at various growth stages. They observed that application at shorter intervals tended to increase the contents of γ-methylene glutamic acid (γ-MGA) and γ-methylene glutamine (γ-MG), which are the most important acids and amides in the groundnut plant. Glycine, serine, leucine, and isoleucine were not detected in plants sprayed at 14-day intervals as compared to plants sprayed at 21-day intervals; β-alanine was more evident after 75 days in leaves of plants sprayed at 14-day intervals as compared to the control, but it disappeared totally in the 105- and the 135-day-old crop. They further observed that at 105 days, the quantity of different amino acids, such as aspartic acid, glutamic acid, γ-MGA, proline, γ-amino butyric acid and the amide γ-MG, was greater in the unsprayed leaves. However, the plants treated with higher dosages of carbendazim showed a lesser content of the amino acids and amide. This might be due to the enhanced metabolism of plants treated with carbendazim, and the different amino acids might have been diverted towards protein synthesis.[151]

The quantitative and qualitative changes in the amino acids exuded from roots are among the symptomless effects that fungicides can have when applied below the phytotoxic level.

Table 4

EFFECT OF FOLIAR SPRAY OF FUNGICIDES ON TOTAL PHENOL, NITROGEN, AMINO ACID, AND TOTAL SUGAR CONTENT (mg g^{-1}) OF GROUNDNUT LEAVES AFTER 45 DAYS GERMINATION[107,108]

Fungicides	Concentration (%)	Total phenol (fresh leaves)	Nitrogen (dry leaves)	Amino acid (fresh leaves)	Total sugar (fresh leaves)
Carbendazim	0.15	3.95	3.80	18.99	35.92
Thiophanate-methyl	0.15	3.81	3.86	18.96	35.85
Zineb	0.25	2.80	2.49	21.31	26.16
Thiram	0.25	2.58	2.55	21.38	26.00
Nil (control)		2.49	2.49	21.46	24.00
LSD at 5%		0.080	0.301	0.620	0.761

It is also apparent that many chemicals used in plant protection could have a measurable influence on host metabolism without producing visible effects. Since these influences could include changes in root metabolism and in the sensitive symbiotic relationship, such as those involved in formation of an endogone mycorrhiza, these might deviate from their typical pattern of development.

Swinburne[152] reported that benomyl, carbendazim, thiabendazole, and tridemorph applied as seed dressing resulted in reduction of up to 30% in PAL activity in 10-day-old barley *(Hordeum vulgare)* seedlings; thiophanate-methyl reduced PAL activity by 10% but carboxin had no significant effect. In glasshouse and field experiments, PAL activity in plants treated with benomyl, through seed treatments or soil drenches, decreased with increasing concentrations; however, total phenols are not affected significantly with any of the fungicides. PAL activity in bean hypocotyl was reduced by applying benomyl to the soil. PAL was also increased in wounded hypocotyls excised from plants treated with benomyl, but the increases were much smaller than those observed in untreated plants and did not result in significant increases in phenol concentration. The significance of these findings in relation to the effect of these fungicides on plant disease resistance has been discussed.[152] The benzimidazoles (carbendazim and thiophanate-methyl) and dithiocarbamates (zineb and thiram) are the most common compounds used in Indian agriculture for the control of plant diseases. Hence, a study was undertaken on their effects on the quantitative changes in groundnut plants (cv. Jyoti) at the recommended rates of their application (Table 4). It was observed that such application results in significant increase in total phenol, nitrogen, amino acids, and total sugars over those in the control.[107,108]

C. Nuclear Function

Fungicides inhibit cytokinesis or mitosis in plants, according to Richmond and Phillips.[153] They also observed that benomyl and carbendazim induced abnormalities in cell division in onion *(Allium cepa)* root tip cells at 8 μg mℓ$^{-1}$; after 6-hr treatment, formation of the cell plate was inhibited and cells were seen to have two nuclei. Another abnormality produced was the logging of chromosomes at anaphase bridges. After 3 days, chromosomes sometimes occurred as amorphous clumps and more than two nuclei were found in the treated cells. Griseofulvin produced similar results in cells. Effects such as these were also observed in *Botrytis cinerea* by the authors. Seeds and young seedlings of maize were treated with 0.031 to 0.5% benomyl at 25°C and pH 7.[154] All the concentrations had a more or less lethal or very toxic effect on seeds, which caused blockage of mitosis. Treatment at 0.5% for 5 hr inactivated the root spindle, blocked mitosis, disoriented chromosomes, and almost completely stopped the mitotic process. Spasojevic[154] reported that onion is more sensitive to benomyl than is maize.

The carboximides carboxin and oxycarboxin at 5000 ppm (a very high concentration not used in plant disease control) for 1 hr treatment of onion meristematic somatic cells at 26°C produced cytological effects.[155] The effects resemble the aberrations caused by radiation and mutagenic treatment. Both chemicals caused a lowering of the mitotic index and differential concentration of chromosomes: oxycarboxin treatment resulted in a greater frequency of abnormal anaphase and telophase cells compared to that of carboxin, and in the induction of polyploidy and binucleate cells.[155] However, such effects by these fungicides have not yet been observed in fungi. Pandit[156] reported that carbendazim application on the root cells of *Allium cepa* var. *cepa* produced cytological effects in the form of mitotic chromosome aberrations at low concentrations (500 ppm) and produced the least clastogenic or turbagenic effects, whereas higher concentrations (2000 ppm) applied at 24 hr or more increased both the number of cell divisions and the chromosomal aberration. Kozer and Klein[157] observed a depressive effect on mitosis in onion root meristematic cells induced by zineb and concluded that the effect resembles colchicine inhibition of the karyokinetic spindle. This fungicide may become a very important tool in breeding because of its mutagenic effects.

The effect of carbendazim (50% wettable powder) at 0.5, 1.0, and 2.0 kg ha^{-1} on the cell cycle in the root meristem of wheat and faba bean was studied[158] using photometric determination of the relative DNA content in cell nuclei. It was observed that in wheat the effect of carbendazim prolonged the G_1-phase, delayed the S- and G_2-phases, and lowered the mitotic rate; in faba bean the G_2-phase was delayed. Mitotic activity was reduced in both the crops.[158]

D. Carbohydrates

Studies were conducted on the effects of NH_4^+ and NO_3^- on the growth and chemical composition of creeping bent grass and the influence of nitrification-suppressor fungicides such as benomyl, dyrene, and maneb at 90, 90, and 135 kg ha^{-1} a.i., respectively, in soil.[159] It was found that the plant receiving NO_3^- as its sole source of nitrogen generally contained a lesser amount of soluble carbohydrates and a greater amount of total nitrogen than the plant receiving nitrogen as NH_4^+; the soluble carbohydrate content of leaf and stolan tissue was less in the plant treated with benomyl than in that treated with dyrene. Similar results were observed in comparisons of tissue from benomyl and maneb treatments. Maneb-treated tissues contained more carbohydrates during early growth, but at later periods the differences in carbohydrate content were not significant. This suggests that fungicides influence carbohydrate metabolism. Additional evidence for fungicidal effects on plant metabolism was obtained in turf-quality ratings.[159]

Singh and Kang[160] reported that application of carbendazim at 0.05 to 0.15% at shorter intervals resulted in significant increase in total sugar content of groundnut leaves at all stages of crop growth. They also observed a higher level of chlorophyll in this treatment. This might have helped to increase synthesis of carbohydrates due to the enhanced photosynthesis process, which may have resulted in the increased sugar content. Soil application of triadimefon at 0.5 to 10 ppm caused progressive increases in carbohydrate content of leaves from plants at all time intervals, with the highest increase of 94.45% at 30 days after treatment with 10 ppm triadimefon. This increase might reflect the stimulated photosynthetic rate in plants treated with triadimefon.[100]

Acidity modifies the sweetness of an apple; the balance of the acid and sugar content is one of the main factors which determine the taste, and thus the quality, of the apple. Moreover, acidity enhances the flavor. The effects of acids on the flavor vary, each acid having a slightly different tartness. Fumaric, malic, and citric acids have been described as metallic, smooth, and clean, respectively. The acids also vary in their effects on aftertaste. The acidic taste of some acids is retained and stimulates the taste buds for a longer period of time, as does malic acid relative to citric acid.[161] Rouchaud et al.[162] reported that selected fungicides such as benomyl and captan at the operational rates of treatment increased the total free sugar concentrations in Golden Delicious, but had little or no effect in Jonagold. In continuation an unexpected effect of these fungicides was observed on the acid contents of some apple cultivars.[161] It was observed that fungicidal application decreased the content of malic, syringic, and the total free acids plus their salts in Golden Delicious and did not change, or slightly increased in Jonagold the content of citric acid plus its salts, although it had no effect on fumaric acid plus its salts.

The results indicate that fungicides have nontarget effects on the total acid and free sugar content of apples. They can alter these chemicals qualitatively as well as quantitatively. Since each acid and sugar has its own taste, the nontarget effects of fungicides used for the control of plant disease can modify the taste and the quality of an apple. This unexpected effect varies according to cultivars.

E. Phenols

Phenols are involved in the resistance and susceptibility of plants. Application of systemic fungicides on foliage results in qualitative and quantitative changes in the phenol content of plants.

Foliar application of carbendazim on groundnut at the rate of 0.05 to 0.15% resulted in more of an increase in total phenol content in treated leaves than in untreated ones.[160] Similarly, Gautam et al.[100] also reported an increase in total phenol content after treatment by soil application of triadimefon with 0.5 to 10 ppm at 10 and 20 days; they observed that the increase in phenol content corresponds to the increase in the concentration of the fungicide. However, the increase was observed 10 and 20 days after treatment, not in 40- and 50-day time intervals. In 30 days after the treatment, the phenolics in plants treated with 0.5 and 1.0 ppm triadimefon increased by 4.46 and 8.11%, respectively, but decreased in plants treated with 2.5, 5.0, and 10.0 ppm by 4.34, 7.44, and 12.21%, respectively.[100] The increase in phenol content may correspond to the action of carbendazim both as a fungicide and as a growth hormone.[163] Besides being directly fungitoxic to plant pathogens, the presence of excess amounts of total phenol might be responsible for providing resistance against early and late leaf spot infection to the groundnut plant via the greater accumulation of phenols in treated plant tissues as compared to controls.

F. Fats and Oil

The systemic fungicide benomyl and the nonsystemic fungicide chlorothalonil were highly effective against *Cercospora arachidicola*, which causes leaf spot in groundnut. Rapid adoption of these materials in cultural practices has resulted in a marked change in the physiological condition of plants during the latter period of fruit set and fruit development.[164] Under these conditions the leaf canopy is frequently retained intact until harvest, the period of fruit development is considerably lengthened, and yield is greatly increased. These fungicides produced the most pronounced effects on fatty acid composition of peanut oil by foliar sprays; the fungicides decreased levels of stearic acid (18:0) and increased levels of linoleic acid (18:2) as compared to those in the control. These treatment effects produced a maximum increase over the control in the calculated iodine number of approximately 1.7 units. Worthington and Smith[164] observed significant differences on oil stability (autooxidation induction period), but effects on the interaction between oil stability and fatty acid composition were not significant. They also observed significantly different effects of fungicides on fatty acid composition of varieties of plants. Increase in oil content of groundnut by sulfur and chlorothalonil application has been reported by Laurence et al.[19]

Beuchat et al.[165] observed an increased oil content of groundnut kernels in benomyl-, chlorothalonil- and tin-hydroxide-sprayed plants. Similarly, Rajgopal and Vidhyasekaran[166] also reported that chlorothalonil, carbendazim, and mancozeb sprays increased the oil content and further observed an increase in the saponification value of the oil. They also reported reduction in the free fatty acid content of the kernel. Saponification value indicates the amount of low molecular weight fatty acids in the soil, which are more odorous.[167] Captafol and carbendazim sprays increased the iodine value of the oil.[167] The higher the iodine value, the greater the degree of unsaturation in the fatty acid content of the oil.[168] Vegetable oils, which are high in unsaturated fatty acids and iodine, decrease blood cholesterol and hence decrease the risk of heart disease.[167]

The source of these differences is unknown, but it may be associated with an extended period of plant vigor. Thus the period of fruit development is lengthened considerably due to delayed maturity.[19,164]

VI. FUTURE OUTLOOK

Each plant species has a particular nutrient level for normal growth and any change from the norm results in changes in the uptake pattern. The application of fungicides on plants results in a variety of interactions which ultimately affect the uptake and distribution of nutrients. However, the effects of the fungicides may vary under different circumstances, which include differences in the responses of different cultivars. Very few studies were conducted on the interaction of fungicides and nutrients in the plant. Some of the results are difficult to interpret because of inadequate information on the relationship between the mineral content of a crop plant and its harvested yield. It is noteworthy that there have been no attempts to determine the changes in the nutrient status of vegetable crops on which fungicides have been used extensively for the control of plant diseases and for economic reasons. All too often, experimenters are concerned only with assessing the yield response to disease-control treatment and do not evaluate the results on the basis of changed nutrient response. Perhaps this is due to the fact that in developed countries crops are often grown with such high levels of fertilization that the effect of fungicides is not apparent. This is not true in developing countries or even in other parts of the world where there are increased costs and water pollution problems caused by unused and leached fertilizers. Certainly, although available evidence is scarce, it is enough to suggest that future research into fungicide/plant interactions will prove profitable both economically and scientifically.

The application of systemic fungicides also affects the physiological and biochemical processes of plants. These processes cause changes in the developmental pattern of plants, particularly young plants, leading to earlier initiation of tiller buds, and therefore to an increase in the number of emerged tillers. In older plants, fungicides effected an increase in assimilate movement to unemerged, and normally dormant, tiller initials and profuse vegetative growth. Some fungicides change the normal patterns or apical dominance and hence allow more rapid lateral growth. Therefore, fungicides should be included in cropping schedules to augment productivity.

REFERENCES

1. **Hance, R. J.**, Effects of pesticides on plant nutrition, *Residue Rev.*, 78, 13, 1981.
2. **Martin, J. P.**, Side effects of organic chemicals on soil properties and plant growth, in *Organic Chemicals in the Soil Environment*, Vol. 2, Goring, C. A. I. and Hamakar, J. W., Eds., Marcel Dekker, New York, 1972, 733.
3. **Cox, R. S. and Winfall, J. P.**, Observation on the effect of fungicide on gray mould and leaf spot and on chemical composition of strawberry plant tissue, *Plant Dis. Rep.*, 43, 755, 1959.
4. **Emge, R. G. and Linn, M. B.**, Effect of spraying with zineb on the growth and zinc content of tomato plant, *Phytopathology*, 42, 133, 1952.

5. **Beyers, E.,** Dithane M-45 as a nutrient spray, *Deciduous Fruit Grower,* 14, 91, 1964.

6. **German, P., Kierstead, L. G., and Mathes, W. T.,** Quality of apples as affected by sprays, *Conn. Agric. Exp. Stn. New Haven Bull.,* 576, 46, 1953.

7. **Hillborn, M. T., Boulanger, L. W., and Cooper, G. R.,** The effect of some pesticides on chemical composition of McIntosh apple leaves, *Plant Dis. Rep.,* 42, 776, 1958.

8. **Ross, R. G. and Longley, R. P.,** Effect of fungicides on McIntosh apple trees, *Can. J. Plant Sci.,* 43, 497, 1963.

9. **Ross, R. G., Crow, A. D., and Longley, R. P.,** Performance of McIntosh apple trees sprayed with captan, dodine or pre-cover and cover sprays of dichlone, *Can. J. Plant Sci.,* 49, 655, 1969.

10. **Ross, R. G. and Webster, D. H.,** Effect of fungicides on the nutrient content of apple leaves, *Can. J. Plant Sci.,* 56, 1001, 1976.

11. **Heeny, H. B., Ward, G. M., and Rutherford, W. M.,** Zinc deficiency in eastern Ontario orchards, *Can. J. Plant Sci.,* 44, 195, 1964.

12. **Mason, J. L., Gardener, R. P., and Sanderson, P. C.,** Dimethyldithiocarbamate sprayes of Zn, Fe and Mn as a nutrient source for apple trees, *Can. J. Plant Sci.,* 52, 110, 1972.

13. **Chamel, A. R., Marcelle, R. D., and Eloy, J. F.,** Cuticular retention *in vitro* and localization of zinc after foliar application of zinc containing fungicides, *J. Am. Soc. Hortic. Sci.,* 107, 804, 1982.

14. **Cole, H., Mackenzie, D., Smith, C. B., and Bergman, E. L.,** Soil incorporated fungicides and nematicides on macro and micro element constituents of *Zea mays* and *Phaseolus vulgaris, Bull. Environ. Contam. Toxicol.,* 3, 116, 1968.

15. **Tewfik, M. S. and Hamdi, Y. A.,** Efficiency of *Rhizobium leguminosarum* as affected by certain herbicides and nematicides, *Zentralbl. Bakteriol. Parasitenkd. Infektionskr. Hyg. Abt. 2,* p. 130 and 725, 1975.

16. **Cummins, D. G. and Smith, D. H.,** Effect of cercospora leaf spot of peanut on forage yield and quality and on seed yield, *Agron. J.,* 65, 919, 1973.

17. **Elik, E. and Kecskes, M.,** The effect of some seed treatments using micro-elements and fungicides on vetch plants grown from seeds inoculated with rhizobia, Proc. Symp. Soil Microbiology, Budapest, June 1970, 431.

18. **Taha, S. M., Mahmoud, S. A. Z., and Salem, S. H.,** Effect of pesticides on Rhizobium inoculation and symbiotic nitrogen fixation of some leguminous plants, Proc. Symp. Soil Microbiology, Budapest, June 1970, 423.

19. **Laurence, R. C. N., Gibbons, R. W., and Young, C. T.,** Changes in yield, protein, oil and maturity of groundnut cultivars with the application of sulphur fertilizers and fungicides, *J. Agric. Sci.,* 86, 245, 1976.

20. **Klingensmith, M. L.,** The effects of benzimidazole on cation uptake by plant roots, *Am. J. Bot.,* 48, 711, 1961.

21. **Dyar, J. J.,** Certain effects of benzimidazole on young tobacco plants, *Plant Physiol.,* 43, 477, 1968.

22. **McClurg, C. A. and Bergman, E. L.,** Influence of selected pesticides on leaf elemental content and yield of garden beans *(Phaseolus vulgaris* L.), *J. Environ. Qual.,* 1, 200, 1972.

23. **Zsoldos, F.,** Uptake and efflux of ions in fungicide-treated rice plants, *Plant Soil,* 41, 41, 1974.

24. **Epstein, E.,** *Mineral Nutrition of Plants: Principles and Perspective,* John Wiley & Sons, New York, 1972, 468.

25. **Dilz, K. and Schepers, J. H.,** Nitrogen fertilizing of cereals. XXIV. Effects of fungicides on the nitrogen response of winter wheat and spring wheat with and without application of chlormequat, *Stikstof,* 6, 452, 1972.

26. **Dilz, K. and Schepers, J. H.,** Nitrogen fertilizing of cereals. XXV. Effects of control of ripening disease with benomyl on nitrogen response of winter wheat and spring wheat with and without application of chlormequat, *Stikstof,* 7, 110, 1974.

27. **Ellen, J. and Spiertz, J. H. J.,** The influence of nitrogen and benlate on leaf area, duration, grain growth and pattern of N, P, and K uptake of winter wheat *(Triticum aestivum), Z. Acker Pflanzenbau,* 141, 231, 1975.

28. **Moyer, J. L. and Paulsen, G. M.,** Nitrogen use in winter wheat in response to systemic pesticides, *Agron. J.,* 69, 58, 1977.

29. **Penny, A. and Jenkyn, J. F.,** Results from two experiments with winter wheat comparing top dressing of a liquid nitrogen fertilizer either alone or with added herbicide or mildew fungicide or both and of nitro-chalk with or without the herbicide or fungicide or both, *J. Agric. Sci.,* 85, 533, 1975.

30. **Mulder, D.,** Report of the plant pathologists for 1960, *Tea Res. Inst. Ceylon Annu. Rep., 1960,* p. 75, 1961.

31. **Southwick, F. W. and Childers, N. F.,** Influence of Bordeaux mixture and its component parts on transpiration and apparent photosynthesis of apple leaves, *Plant Physiol.,* 26, 721, 1941.

32. **D'Armini, M. and Raggi, V.,** Productivity, vegetative development, photosynthesis and respiration in grapevine treated with Bordeaux, zineb and zineb + Bordeaux, *Riv. Patol. Veg. IV,* 10(1), 37, 1974.

33. **Llewelyn, F. W. M.,** The scorching of apple leaves by copper sulphate, *Ann. Appl. Biol.*, 45, 376, 1957.
34. **Agnew, E. E. and Childers, N. F.,** The effect of two sulphur sprays on photosynthetic activity of apple leaves, *Proc. Am. Soc. Hortic. Sci.*, 37, 379, 1940.
35. **Hoffman, M. B.,** The effect of lime sulphur spray on the respiration rate of apple leaves, *Proc. Am. Soc. Hortic. Sci.*, 33, 173, 1936.
36. **Berry, W. E.,** Spray injury studies. Progress Report II. The effect of time and temperature on the production of hydrogen sulphide during atmospheric decomposition of lime-sulphur, Rep. Agric. Hortic. Res. Stn. Bristol, 1939, 52.
37. **Guntz, M., Hascoet, M., Ventura, C., and Arnoux, J.,** The influence of some fungicidal products on yield of the potato: fungicidal and physiological action, *Ann. Epiphyt.*, 9, 517, 1959.
38. **Thorn, G. D. and Ludwig, R. A.,** *The Dithiocarbamate and Related Compounds,* Elsevier, Amsterdam, 1962, 298.
39. **Nishnianidze, N. O. and Gogiberdize, G. S.,** Some data on the physiological activity of fungicides in relation to citrus plants, *Tr. Nauchno-Issled. Inst. Zashch. Rast. Uzb. SSR,* 25, 50, 1973.
40. **Lindahl, P. E. B.,** Inhibition of growth and photosynthesis in submerged plants with tetramethylthiurandisulphide and sodium dimethyldithiocarbamate and reversal of this inhibition of photosynthesis, *Nature (London),* 191, 51, 1961.
41. **Lindahl, P. E. B.,** The effect of sodium dimethyldithiocarbamate and tetramethylthiurandisulphide on photosynthesis and endogenous respiration in *Enteromorpha linza* (L.), *J. Agric. Physiol. Plant,* 16, 644, 1963.
42. **McMillan, R. T. and Riedhart, J. M.,** The influence of hydrocarbons on photosynthesis of citrus leaves, *Proc. Fla. State Hortic. Soc.,* 77, 15, 1964.
43. **Shishiyama, J. J., Fukutomi, M., and Akai, S.,** Effect of some fungicides on synthesis of chlorophylls, deoxyribonucleic acid and ribonucleic acid in onion leaves, *Phytopathology,* 55, 844, 1965.
44. **Suzuki, T.,** Effect of dihydrostreptomycin spray upon the photosynthesis, respiration, and chlorophyll content of rice, *Ann. Phytopathol. Soc. Jpn.,* 22, 60, 1966.
45. **Sinzar, B.,** Contribution to the study on effects of fungicides (TMTD and MEMS) for seed dressing in the intensity of some physiological processes at the germination and emergence phase of maize, *Arh. Poljopr. Nauke,* 26, 57, 1974.
46. **Carlson, L. W.,** Effects of Vitavax on chlorophyll content, photosynthesis and respiration of barley leaves, *Can. J. Plant Sci.,* 50, 627, 1970.
47. **Mathre, D. E.,** Effects of oxathiins systemic fungicides on various biological systems, *Bull. Environ. Contam. Toxicol.,* 8, 311, 1972.
48. **Forster, H.,** Einfluß von Triadimefon auf den Pigment-, Nuklein-saure- und Proteingehalt junger Gerstennopflanzen, *Mitt. Biol. Bundesanst. Land. Forstwirtsch. Berlin-Dahlem,* 178, 159, 1977.
49. **Forster, H.,** Mechanism of action and side effects of triadimefon and triadimenol in barley plants, 3rd Int. Congr. Plant Patholology, Munich, August 1978, 365.
50. **Ahmed, A. M., Heikal, M. M., and Hindawy, O. S.,** Side effects of benomyl fungicide treatments on sunflower, cotton and cowpea plants, *Phyton (Horn, Austria),* 23(2), 185, 1983.
51. **Ahmed, A. M., Heikal, M. M., and Shaddad, M. A.,** Changes in some plant-water relation parameters of some oil producing plants over a range of salinity stresses, *Biol. Plant.* 21, 259, 1979.
52. **Heikal, M. M., Ahmed, A. M., and Shaddad, M. A.,** Salt tolerance of some oil producing plants, *Agriculture (Heverlee),* 28, 437, 1980.
53. **Kramer, P. J.,** *Plant and Soil Water Relationship: A Modern Synthesis,* Tata McGraw-Hill, New Delhi, 1969, 230.
54. **Melin, C., Aselsson, L., Ryberg, H., and Virgin, H. I.,** Chlorophyll accumulation in normal and SAN 9789 treated wheat leaves, 5th Int. Congr. Photosynthesis, Greece (Abstr.), 1980, 375.
55. **Hanker, I. and Kudelova, A.,** Effect of primicarb, dimethoate, and benomyl on photosynthesis and ^{14}C distribution in sugar beet, *Sb. UVTI (Ustav Vedeckotech. Inf.) Ochr. Rostl.,* 12, 109, 1976.
56. **Hanker, I., Kudelova, A., and Edger, J.,** The effect of the Folcidin (cypendazole) fungicide on the control of bound and free amino acids in crucifer plants, *Sb. UVTI (Ustav Vedeckotech. Inf.) Ochr. Rostl.,* 14, 285, 1978.
57. **Godawari, H. R. and Waygood, E. R.,** Nicotinamide adenine dinucleotide metabolism in plants. I. Intermediates of biosynthesis in wheat leaves and the effect of benzimidazoles, *Can. J. Bot.,* 48, 2267, 1970.
58. **Mishra, D. and Waygood, E. R.,** The effect of benzimidazole and kinetin on the nicotinamide nucleotide content of senescing wheat leaves, *Can. J. Biochem.,* 46, 167, 1968.
59. **Person, C., Samborski, D. J., and Forsyth, F. R.,** Effect of benzimidazole on detached wheat leaves, *Nature (London),* 180, 1294, 1957.
60. **Wang, D. and Waygood, E. R.,** Effect of benzimidazole and nickel on the chlorophyll metabolism of detached leaves of Khapli wheat, *Can. J. Bot.,* 37, 743, 1959.

61. **Waygood, E. R.,** Benzimidazole effect in chloroplast of wheat leaves, *Plant Physiol.,* 40, 1242, 1965.
62. **Dimond, A. E. and Rich, S.,** Effect of physiology on the host and on host/pathogen interactions, in *Systemic Fungicides,* 2nd ed., Marsh, R. W., Ed., Longman, London, 1977, 115.
63. **Cook, R. J.,** Unexpected effects of fungicides on cereal yields, *EPPO Bull.,* 11, 277, 1981.
64. **Mukhopadhyaya, A. N. and Bandopadhyaya, R.,** Cytokinin like activity of carbendazim, *Pesticides,* 11(7), 24, 1977.
65. **Staskawicz, B., Khar-Sawhney, R., Slaybaugh, R., Adams, W., and Galston, A. W.,** The cytokinin-like action of methyl-benzimidazole-2-yl-carbamate on oat leaves and protoplasms, *Pestic. Biochem. Physiol.,* 8, 106, 1978.
66. **Peat, W. E. and Shipp, D. M.,** The effects of benomyl on the growth and development of wheat, *EPPO Bull.,* 11, 287, 1981.
67. **Langer, R. H. M., Prasad, P. C., and Lauds, H. M.,** Effects of kinetin on tiller bud elongation in wheat *(Triticum aestivum* L.), *Ann. Bot. (London),* 37, 565, 1973.
68. **Williams, R. H. and Cartwright, P. M.,** The effect of application of synthetic cytokinin on shoot dominance and grain yield in spring barley, *Ann. Bot. (London),* 46, 445, 1980.
69. **Peterson, C. A. and Edgington, L. V.,** Entry of pesticides into the plant symplast as measured by their loss from an ambient solution, *Pestic. Sci.,* 7, 483, 1976.
70. **Samborski, D. J., Forsyth, F. R., and Peterson, C. A.,** Metabolic changes in detached wheat leaves floated on benzimidazole and the effect of these changes on rust reaction, *Can. J. Bot.,* 36, 591, 1958.
71. **Mishra, D. and Mishra, B.,** Retardation of induced senescence of leaves from crop plants by benzimidazole and cytokinins, *Exp. Gerontol.,* 8, 235, 1973.
72. **Pressman, E. and Palevitch, D.,** Cytokinin like activity of benomyl as a senescence inhibitor in broccoli heads, *Hortic. Sci.,* 8, 496, 1973.
73. **Williams, P. F.,** Cytokinin activity of benomyl, *Aust. Plant Pathol. Soc. Newsl.,* 4, 12, 1975.
74. **Tripathi, R. K., Vohra, K., and Schlosser, E.,** Effect of fungicides on the physiology of plants. III. Mechanism of cytokinin-like antisenescent action of carbendazim on wheat leaves, *J. Plant Dis. Prot.,* 87, 633, 1980.
75. **Skene, K. G. M.,** Cytokinin-like properties of the systemic fungicide benomyl, *J. Hortic. Sci.,* 47, 179, 1972.
76. **Thomas, T. H.,** Growth regulatory effects of three benzimidazole fungicides on the germination of celery *(Apium graveolens)* seeds, *Ann. Appl. Biol.,* 74, 233, 1973.
77. **Byrde, R. J. W. and Richmond, D. V.,** A review of the selectivity of fungicides in agriculture and horticulture, *Pestic. Sci.,* 7, 372, 1976.
78. **Tripathi, R. K. and Schlosser, E.,** Effect of fungicides on physiology of plants. II. Inhibition of adventitious root formation by carbendazim and kinetin, *J. Plant Dis. Prot.,* 86, 12, 1980.
79. **Rouchaud, R. P., Decallone, J. R., and Meyer, J. A.,** Metabolic rate of methyl-benzimidazole-2yl-carbamate in melon plants, *Phytopathology,* 64, 123, 1974.
80. **Martin, C. K. and Thimann, K. V.,** The role of protein synthesis in the senescence of leaves. I. The formation of protease, *Plant Physiol.,* 49, 64, 1972.
81. **Sodak, L. and Wright, S. T. C.,** The effect of kinetin on ribonuclease, acid phosphatase, lipase and esterase levels in detached wheat leaves, *Phytochemistry,* 6, 1629, 1969.
82. **Srivastava, B. I. S.,** Increase in chromatin-associated nuclease activity of excised barley leaves during senescence and its suppression by kinetin, *Biochem. Biophys. Res. Commun.,* 32, 533, 1968.
83. **Anderson, J. W. and Rowan, K. S.,** Activity of peptidase in tobacco leaf tissue in relation to senescence, *Biochem. J.,* 97, 741, 1965.
84. **Drivdahl, R. H. and Thimann, K. V.,** Protease of senescing oat leaves. I. Purification and general properties, *Plant Physiol.,* 59, 1059, 1972.
85. **Dickinson, C. H. and Walpole, P. R.,** The effect of late application of fungicides on the yield of winter wheat, *Exp. Husb.,* 29, 23, 1975.
86. **Rowley, N. K., Wainwright, A., and Chipper, M. E.,** The development of systemic fungicide triadimefon for the control of foliar disease in wheat and oats in the UK, *Proc. 1977 Br. Crop Prot. Conf. Pests and Diseases,* 1, 17, 1977.
87. **Jenkins, J. E. E., Melville, S. C., and Jemmett, J. L.,** The effect of fungicides on leaf diseases and yield in spring barley in south-west England, *Plant Pathol.,* 21, 49, 1972.
88. **Melville, S. C., Griffin, G. W., and Jemmett, J. L.,** Effects of fungicide spraying on brown rust and yield in spring barley, *Plant Pathol.,* 25, 99, 1976.
89. **Jenkyn, J. F.,** Effects of chemical treatments for mildew control at different times on the growth and yield of spring barley, *Ann. Appl. Biol.,* 88, 369, 1978.
90. **Jenkyn, J. F. and Prew, R. D.,** The effect of fungicides on the incidence of *Sporobolomyces* spp. and *Cladosporium* spp. on flag leaves of winter wheat, *Ann. Appl. Biol.,* 75, 253, 1973.
91. **Dickinson, C. H. and Wallace, B.,** Effects of late applications of foliar fungicides on activity of micro-organisms on winter wheat flag leaves, *Trans. Br. Mycol. Soc.,* 67, 103, 1976.

92. **Skidmore, A. M. and Dickinson, C. H.**, Effect of phylloplane fungi on senescence of excised barley leaves, *Trans. Br. Mycol. Soc.*, 61, 49, 1973.
93. **Dickinson, C. H.**, Interactions of fungicides and leaf saprophytes, *Pestic. Sci.*, 4, 563, 1973.
94. **Thorn, G. N.**, Physiological aspects of grain yield in cereals, in *The Growth of Cereals and Grasses*, Milthorpe, F. L. and Ivins, J. D., Eds., Butterworths, London, 1966, 560.
95. **Walpole, P. R. and Morgan, D. G.**, A qunatitative study of grain filling in *Triticum aestivum* L., cultivar Maris Widgeon, *Ann. Bot. (London)*, 34, 309, 1970.
96. **Sinclair, J. B.**, Fungicide spray for the control of seed-borne pathogens of rice, soybeans, and wheat, *Seed Sci. Technol.*, 9, 245, 1981.
97. **Buchenauer, H.**, Systemisch-Fungizide Wirkung und Wirkungs-Mechanisms von Triadimefon (MEB-6447), *Mitt. Biol. Bundesanst. Land Forstwirtsch. Berlin-Dahlem*, 165, 154, 1975.
98. **Fisher, D. J., Packard, J. A., and McKenzie, C. M.**, Uptake of systemic fungicide triadimefon by clover and its effect on symbiotic nitrogen fixation, *Pestic. Sci.*, 10, 75, 1979.
99. **Buchenauer, H. and Grossmann, E.**, Triadimefon: mode of action in plant and fungi, *Neth. J. Plant Pathol.*, 83(S), 93, 1977.
100. **Gautam, S. R., Thapliyal, P. N., and Tripathi, R. K.**, Effect of fungicide on physiology of plants. VIII. Effects of triadimefon on physiology of soybean plants, *Indian Phytopathol.*, 37, 600, 1984.
101. **Fletcher, R. A. and Nath, V.**, Triadimefon reduces transpiration and increases yield in water stressed plants, *Physiol. Plant.*, 62, 422, 1984.
102. **Buchenauer, H., Kohts, T., and Roos, H.**, Zur Wirkungsweise von Propiconazol (Desmel), CGA 64 251 und Diclobutrazol (Vigil) in Pilzen und Gerstenkeimlingen, *Mitt. Biol. Bundesanst. Land Forstwirtsch. Berlin-Dahlem*, 203, 310, 1981.
103. **Buchenauer, H., Kutzer, B., and Koths, T.**, Effect of various triazole fungicides on growth of cereal seedlings and tomato plants as well as on gibberellin content and lipid metabolisms in barley seedlings, *Z. Pflanzenkr. Pflanzenschutz.*, 91, 506, 1984.
104. **Wood, B., Gottwald, T., and Payne, J.**, Influence of single application of fungicides on net photosynthesis of pecan, *Plant Dis.*, 69, 427, 1984.
105. **Prusky, D., Dinoor, A., and Eshel, Y.**, Symptomless effects of experimental fungicides on wheat, *Phytopathology*, 64, 812, 1974.
106. **Adele, S., Baku, T., and Ceasusescu, S.**, Effects of some new fungicides on the amino acid, protein and sugar levels in wheat seedlings, *An. Univ. Bucuresti Biol. Veg.*, 22, 189, 1973.
107. **Anon.**, Studies on the Effects of Benzimidazoles and Dithiocarbamates Fungicides on the Nutrient Status of Groundnut besides Control of Tikka and Rust Diseases, Indian Council of Agricultural Research *(Ad hoc* scheme), J. N. Agricultural University, Jabalpur, India, 1986, 206.
108. **Thomas, M.**, Studies on the Effects of Benzimidazoles and Dithiocarbamates Fungicides On Nutritional Status of Groundnut, Ph.D. thesis, J. N. Agricultural University, Jabalpur, India, 1986, 156.
109. **Covey, R. P.**, Orchard evaluation of two new fungicides for the control of apple powdery mildew, *Plant Dis. Rep.*, 55, 514, 1971.
110. **Sherald, J. L., Ragsdale, N. N., and Sisler, H. D.**, Similarities between the systemic fungicides triforine and triarimol, *Pestic. Sci.*, 4, 719, 1973.
111. **Burg, S. P.**, Ethylene, plant senescence and abscission, *Plant Physiol.*, 43, 1503, 1968.
112. **Shive, J. B. and Sisler, H. D.**, Effects of ancymidol (a growth retardant) and triarimol (a fungicide) on the growth, sterols and gibberellins of *Phaseolus vulgaris*, *Plant Physiol.*, 57, 640, 1976.
113. **Suge, H. and Murakami, Y.**, Occurrence of a rice mutant deficient in gibberellin-like substances, *Plant Cell Physiol.*, 9, 411, 1968.
114. **Douglas, T. J. and Paleg, L. G.**, Plant growth retardant as inhibitors of sterol biosynthesis in tobacco seedlings, *Plant Physiol.*, 54, 238, 1974.
115. **Torrey, J. G.**, Hormonal regulation of plant growth and morphogenesis, *Science*, 181, 1075, 1973.
116. **Sisler, H. D., Ragsdale, N., and Waterfield, W. F.**, Biochemical aspects of the fungitoxic and growth regulation action of fenarimol and other pyrimidine, 5 yl-methanol, *Pestic. Sci.*, 15, 167, 1984.
117. **Coolbaugh, R. C., Swanson, D. K., and West, C. A.**, Comparitive effects of ancymidol and its analog on growth of peas and ent-kaurene oxidation in cell free extracts of immature *Marah macrocarpus* endosperm, *Plant Physiol.*, 69, 707, 1982.
118. **Buchenauer, H.**, Mode of action of triadimefon in *Ustilago avenae*, *Pestic. Biochem. Physiol.*, 7, 309, 1977.
119. **Luttage, U. K., Bauer, K., and Kohler, D.**, Fruhwirkungen von Gibberellin-Saure auf Membranetransporte in jungen Erbsenpflanzen, *Biochim. Biophys. Acta*, 150, 452, 1968.
120. **Jones, R. L.**, Gibberellic acid and the fine structure of barley aleurone cells. II. Changes during the synthesis and secretion of L-amylase, *Planta*, 88, 73, 1969.
121. **Coolbaugh, R. C. and Hamilton, R. H.**, Inhibition of ent-kaurene oxidation and growth by α-cyclopropyl-α-(p-methoxyphenyl)-5-pyrimidine methyl alcohol, *Plant Physiol.*, 57, 245, 1976.

122. **Coolbaugh, R. C., Hirono, S. S., and West, C. A.**, Studies on the specificity and site of action of α-cyclopropyl-α(p-methoxyphenyl)-5-pyrimidine methyl alcohol (Ancymidol), a plant growth regulator, *Plant Physiol.*, 56, 571, 1978.
123. **Coolbaugh, R. C., Heil, D. R., and West, C. A.**, Comparative effects of substituted pyrimidines on growth and gibberellin biosynthesis in *Gibberella fujikuroi, Plant Physiol.*, 69, 712, 1982.
124. **Buchenauer, H. and Rohner, E.**, Effect of triadimefon and triadimenol on growth of various plant species as well as on gibberellin content and sterol metabolism in shoots of barley seedlings, *Pestic. Biochem. Physiol.*, 15, 58, 1981.
125. **Ragsdale, N. N. and Sisler, H. D.**, Mode of action of triarimol in *Ustilago maydis, Pestic. Biochem. Physiol.*, 3, 20, 1973.
126. **Ragsdale, N. N. and Sisler, H. D.**, Specific effects of triarimol on sterol biosynthesis in *Ustilago maydis, Biochim. Biophys. Acta*, 380, 81, 1975.
127. **Buchenauer, H.**, Untersuchungen zur Wirkungsweise und zum Verhalten verschiedener Fungizide in Pilzen und Kulturpflanzen, *Habilitationsschrift*, p. 354, 1979.
128. **Schmitt, P. and Beneveniste, P.**, Effect of fenarimol on sterol biosynthesis in suspension cultures of bramble cells, *Phytochemistry*, 18, 1659, 1979.
129. **Barnes, M. F., Light, E. N., and Long, A.**, The action of plant growth retardants on terpenoid biosynthesis inhibition of gibberellic acid production in *Fusarium moniliforme* by CCA and Amo-1618. Action of the retardants on sterol biosynthesis, *Planta*, 88, 172, 1969.
130. **Sijpesteijn, A. K.**, Effects on fungal pathogens, in *Systemic Fungicides*, 2nd ed., Marsh, R. W., Ed., Longman, New York, 1977, 131.
131. **Tyuterev, S. L. and Khalesova, L. A.**, Effects of systemic fungicides on content and functional activity of nucleic acid in plants, *Fiziol. Rast.*, 26, 187, 1979.
132. **Rankin, P. W., Surak, J. G., and Thompson, N. P.**, Effect of benomyl and benomyl hydrolysis products on *Tetrahymena pyriformis, Food Cosmet. Toxicol.*, 15, 187, 1977.
133. **Fraser, R. S. S. and Whenham, R. J.**, Chemotherapy of plant virus disease with methyl-benzimidazole-2yl-carbamate: effects on plant and multiplication of tobacco mosaic virus, *Physiol. Plant Pathol.*, 13, 51, 1978.
134. **Woolley, D. W.**, Some biological effects produced by benzimidazole and their reversal by purines, *J. Biol. Chem.*, 15, 225, 1944.
135. **Hitching, G. H. and Elion, G.**, Purine analogues, in *Metabolic Inhibitors*, Vol. 1., Hochster, R. M. and Quastel, J. H., Eds., Academic Press, London, 1963, 215.
136. **Kapoor, M. and Waygood, E. R.**, Metabolism of benzimidazole in wheat. I. Formation of benzimidazole nucleotide, *Can. J. Biochem.*, 43, 153, 1965.
137. **Kapoor, M. and Waygood, E. R.**, Metabolism of benzimidazole in wheat. II. Formation of benzimidazole adenine dinucleotide and its products, *Can. J. Biochem.*, 43, 165, 1965.
138. **Sisler, H. D.**, Effect of fungicides on protein and nucleic acid synthesis, *Annu. Rev. Phytopathol.*, 7, 311, 1969.
139. **Buchenauer, H.**, Mechanism of action of Bayleton (Triadimefon) in *Ustilago avenae, Pflanzenschutz Nachr.*, 29, 281, 1976.
140. **Tripathi, R. K. and Schlosser, E.**, Effect of fungicides on physiology of plants. III. Mechanism of cytokinin-like antisenescent action of carbendazim on wheat leaves, *J. Plant Dis. Prot.*, 87, 631, 1979.
141. **Helleck, F. E. and Cochrane, V. W.**, The effect of fungistatic agents on bacterial flora of rhizosphere, *Phytopathology*, 40, 715, 1950.
142. **Agnihotri, V. P.**, Effect on Aspergilli. XIV. Effect of foliar spray of urea on the Aspergilli of the rhizosphere of *Triticum vulgare* L., *Plant Soil.*, 20, 364, 1964.
143. **Balasubramanian, A. and Rangaswamy, G.**, Influence of foliar application of chemicals on the root exudations and rhizosphere microflora of *Sorghum vulgare* and *Crotalaria juncea, Folia Microbiol. (Prague)*, 18, 492, 1973.
144. **Jalali, B. L. and Domosch, K. H.**, Effect of some fungitoxicants on the amino acid spectrum of wheat root exudates, *Phytopathol. Z.*, 90, 22, 1977.
145. **Jaiswal, S. P., Saini, N. K., and Sharma, S. K.**, The effect of some persticides in conjuction with nitrogen sources on growth and carbohydrate and nitrogen constituents of sugarcane during formative phase, *Plant Soil*, 38, 33, 1973.
146. **Audus, L. J.**, *The Physiology and Biochemistry of Herbicides*, Academic Press, London, 1964, 680.
147. **Reyes, J. C., Moyer, J. L., Hansing, E. D., and Paulsen, G. M.**, Increased protein in winter wheat grain following use of oxathiin fungicides, *Phytopathology*, 59, 1046, 1969.
148. **Thorn, G. D. and Richardson, L. T.**, Effect of various fungicides on the synthesis of amino acids in tomato roots, *Phytopathology*, 59 (Abstr.), 1053, 1969.
149. **Pozsar, B. L., Kiraly, Z., and Hammady, M. Ei.**, Cytokinin activity of benzimidazole, *Acta Bot. Acad. Sci. Hung.*, 13, 169, 1967.

150. **Singh, P. P. and Kang, M. S.,** Physiological response of groundnut to Bavistin application for the control of Tikka disease, Symp. Plant Disease Problems Indian Society of Plant Pathology, October 1 to 3, 1978, 13.
151. **Singh, P. P. and Kang, M. S.,** Effect of carbendazim on amino acid composition of groundnut, *Indian Phytopathol.,* 37, 725, 1984.
152. **Swinburne, T. R.,** The effect of benomyl and other fungicides on phenylalanine-ammonia lyase in *Hordeum vulgare* and *Phaseolus vulgaris, Physiol. Plant Pathol.,* 5(1), 81, 1975.
153. **Richmond, D. V. and Phillips, A.,** The effect of benomyl and carbendazim on mitosis in hyphae of *Botrytis cinerea* and *Allium cepa, Pestic. Biochem. Physiol.,* 5, 367, 1975.
154. **Spasojevic, V.,** Effect of fungicide Benlate on the mitosis of maize, *Arh. Poljopr. Nauke,* 27, 13, 1974.
155. **Mohan, S. T.,** Cytological effects of fungicides, Plantvax and Vitavax on somatic cells of *Allium cepa, Curr. Sci.,* 44, 813, 1975.
156. **Pandit, T. K.,** Cytological effects of Bavistin, *Nucleus (Calcutta),* 24, 106, 1981.
157. **Kozer, W. and Klein, M.,** Effect of Cynkoton (zineb) on mitotic activity and mitosis in the root meristem of seedlings of *Allium cepa* L., *Hodowla Rosl. Aklim. Nasienn.,* 19, 321, 1975.
158. **Lazur-Keul, G. and Keul, M.,** Effect of carbendazim on the cell cycle in the root meristem of *Triticum aestivum* and *Vicia faba, Rev. Roum. Biol. Ser. Bot.,* 28, 131, 1983.
159. **Mazur, A. R. and Hoghes, T. D.,** Chemical compositions and quality of pennocross creeping bent grass as affected by ammonium nitrate and several fungicides, *Agron. J.,* 68, 721, 1976.
160. **Singh, P. P. and Kang, M. S.,** Effect of Bavistin on biochemical composition of groundnut, *Res. Punjab Agric. Univ.,* 201, 479, 1983.
161. **Rouchaud, J. P., Marechal, G., and Meyer, J.,** Unexpected effects of selected fungicide treatments on the acid content on "Golden Delicious" and "Jonagold" apples, *Sci. Hortic.,* 22, 67, 1984.
162. **Rouchaud, J., Moons, C., and Meyer, J. A.,** Effects of selected fungicide treatments on the sugar content of "Golden Delicious" and "Jonagold" apples, *Sci. Hortic.,* 20, 161, 1983.
163. **Mercer, P. C.,** The control of Cercospora leaf spot on groundnut in Malawi and its effect on the parameters of yield, *PANS (Pest Artic. News Summ.),* 19, 201, 1973.
164. **Worthington, R. E. and Smith, D. H.,** Effects of several foliar fungicides on the fatty composition and stability of peanut oil, *J. Agric. Food Chem.,* 21, 619, 1973.
165. **Beuchat, L. R., Smith, D. H., and Young, C. T.,** Effects of foliar fungicides on aflatoxin, oil and protein content and maturing rate peanut kernels, *J. Sci. Ed. Agric.,* 25, 477, 1974.
166. **Rajgopal, R. and Vidhyasekaran, P.,** Effect of fungicidal sprays on the quality of groundnut kernel and its oil content, *Indian Phytopathol.,* 36, 52, 1983.
167. **Potter, N. N.,** *Food Science,* AVI Publishing, Westport, Conn., 1973, 706.
168. **Woodman, A. G.,** *Food Analysis,* McGraw-Hill, New York, 1941, 607.

Chapter 4

GROWTH, YIELD, AND OTHER CHARACTERISTICS OF PLANTS

I. INTRODUCTION

The efficiency of fungicides for controlling specific diseases is well known throughout the crop industry. Scientists, however, have very little information regarding the beneficial and deleterious nontarget effects of fungicides on plant growth, yield, and other growth-promoting factors. These side effects, or nontarget effects, continue to be the least understood aspects of fungicide use. It is apparent that some side effects do result from use of fungicides and that their beneficial effects are likely to help offset the negative ones. Protagonists who recommend fungicide usage rightly emphasize yield responses rather than disease control. Nevertheless, increased yield following fungicidal treatment is normally assumed to result from disease control and this assumption is rarely questioned. Sometimes, however, fungicides treatments, in ostensibly disease-free situations, produce significant increases in yield. In these circumstances, unless some unrecognized diseases are identified it is difficult to believe that fungicides do not have some stimulatory phytotonic effect independent of normal disease control. The difficulties of detecting and measuring such effects, however, are great.

It is clear that if the beneficial effects could be identified and exploited and the deleterious effects minimized, the art of crop disease management would certainly be a profitable venture in the foreseeable future. This chapter highlights some nontarget effects of fungicides on growth, yield, and other factors responsible for improving the quality and quantity of plant products.

II. GROWTH AND YIELD

A series of reports is available that discusses the application of a particular fungicide that resulted in a yield increase in excess of that expected on the basis of the amount of disease and the degree of control achieved.[1] This is normally accounted for by the control of well-defined diseases but there are occasions when this explanation seems inadequate, for example, when the amount of disease may seem to be trivial or its control insufficient to have produced the observed yield responses. These responses have been achieved by the application of fungicides at the foliage, grain-ripening, late, or preharvest phase. Since growth and yield are not mutually exclusive factors, they will be discussed simultaneously.

A. Nonsystemic Fungicides
1. Inorganic (Sulfur)
Generally soluble forms of sulfur have a more pronounced effect on vegetative growth than do sulfur dust or suspension. Turrell[2] observed that sulfur reduced the growth of leaves of some citrus varieties. The new vegetative growth of apple trees treated with lime-sulfur was less than those treated with colloidal sulfur.[3] Corroborating observations have been made by other investigators who used different criteria for evaluation.[4,5]

Groundnut cultivars reacted differently to sulfur-based fungicides with respect to yield, oil, and protein content. Laurence et al.[6] applied sulfur fungicide at 120 kg ha^{-1} (six applications at 2-week intervals) in dust form or a fertilizer banded at identical rates and times and chlorothalonil at 1.25 kg ha^{-1} (six applications at 2-week intervals) on groundnut and showed improvement in groundnut kernel yield. Improved fertility of sulfur-deficient soils contributes to such gains when sulfur dust is used as the fungicides. Application of sulfur alone to deficient soils lowered the oil/protein ratio of kernels, whereas a single fungicidal application raised the oil content, probably through delayed plant maturity. Chlorothalonil also delayed maturity of groundnut and raised kernel yield and oil content in soils where sulfur is in sufficient supply. Foliar application of sulfur dust may lead to improved yields but kernel protein production is little affected.[6]

2. Copper

More information is available about the effect of copper fungicides on plant growth than about the effect of other fungicides, possibly because of their wider use and greater phytotoxicity. A number of detailed spray evaluation programs revealed that Bordeaux mixture (BM) has more pronounced effects on vegetative growth than other organic fungicides. An enhancement of growth and development of peas, French beans, and broad beans due to the use of copper fungicides has been reported.[7] Malformation and atrophy of coffee tree roots have been attributed to BM. In these studies, the extent of injury caused by BM and several other copper fungicides used for the control of *Rhizoctonia* spp. in seed beds was proportional to the concentration of copper in the fungicides.[8] This suggests that the incorporation of various micro- and macronutrients, besides reducing the toxic effect of copper fungicides, would also promote better growth and development.

In the absence of any significant observable disease, a single spray of BM applied at the beginning of the long rains (March through May) was found to increase substantially the yield of berries of coffee in the following years. The origin of the discovery is obscure, but it attracted much attention and a detailed study was made in the years 1934 to 1940. This showed that the mean annual increase with only one spray in March was over 100%.[8] Yield increases of this magnitude could not be ignored and tonic spraying became a standard practice; explanations of the mechanism involved, however, are still incomplete. One visual effect of tonic sprays was due to the retardation of natural leaf fall in the dry seasons (June through October and December through January) so that the trees tended to be well foliated throughout the year. Yield increase is directly linked to leaf retention because the number of leaves present on the tree during the periods of most intense flower initiation.[9] Although copper fungicides were used initially for the tonic sprays, the same effects are also produced by dithiocarbamates.[10,11]

Copper is toxic to the development of tomatoes.[12,13] The decrease in yield of tomatoes treated with BM is generally attributed to its adverse effect on plant growth. A greater degree of phytotoxicity to tomato plants with 4:4:40 BM than with 4:2:40 BM indicated that the higher concentration of lime was probably operative in causing phytotoxicity. This is further supported by the observations of Shutak and Christopher[14] who also found that lime content in BM is an important factor in causing toxicity. An interesting observation has been made by Horsfall and Turner[15] in the case of potatoes, in which the application of BM of blight-free plants caused a reduction in yield of tubers. This reduction in yield is attributed to the injurious effect of lime. However, such reduction was not observed in organic fungicides not containing lime. These workers recognized this relationship and attempted to separate the foliage effects of BM on treated and untreated blight-affected plants.

Contradictory reports have been written regarding the effects of BM spray and resistance of various plants to drought and frost conditions. Copper oxychloride, when sprayed on coffee, delayed fruit ripening but had the beneficial effect of increasing resistance to drought. On the other hand, application of BM to tomatoes growing under drought conditions resulted in a marked reduction in plant growth, yield, and fruit size.[16] Tomatoes and potatoes sprayed with copper fungicides were damaged more by frost than those sprayed with organic fungicides containing zinc.[17] These observations suggest that resistance, especially in dry seasons, and susceptibility to drought and frost after copper sprays, varied from plant to plant. However, copper sprays in general increased the susceptibility of the plants to frost. These conclusions will require further confirmation.

3. Carbamates

Unexpected benefits, i.e., yield improvements not related to disease levels, have been noted in cereals by many workers. Jenkins et al.[18] conducted a series of trials which provided the basis for yield-loss assessment methods for barley. They applied zineb (0.25%) at 10- to 14-day intervals from tillering until senescence for the control of *Puccinia hordei* and *Selenophoma denacis* and observed that although the disease was almost absent yield increase was obtained. Observations that single sprays of zineb (1.85 kg active ingredient (a.i.) ha^{-1}) have affected yield increases have been made in several experiments.[19] The mean responses were largest in experiments sown where brown rust *(P. hordei)* was present (yield increase 9%, 0.39 t ha^{-1}) after the first of April compared with 0.14 t ha^{-1} (3%) where no diseases were detected. On crops sown before March 30 responses were only 2.6% (0.12 t ha^{-1}) and 0.5% (0.02 t ha^{-1}). Comparison of zinc sulfate with zineb indicated that prevention of zinc deficiency was not the cause of yield response. There was a consistent increase in grain nitrogen following the zineb treatment.

An increase in yield of apple treated with ferbam[20] has been reported. Zineb showed growth-inhibiting action on lupin seedlings at a concentration of 10^{-3} to 10^{-4} M and delayed emergence of lateral roots at 10^{-5} M concentration. The purified chemical, however, had a stimulatory effect on root development at 10^{-6} to 10^{-8} M concentration. None of these effects could be demonstrated with corresponding concentrations of ziram, while nabam was toxic at 10^{-5} M concentration.[21] Zineb sprays in combination with thiram led to hardening and stunting of tomato plants.[21] McKeen[22] reported that maneb was detrimental to tomato and pepper seedlings in glasshouse conditions but not under field conditions. Maneb may retard the growth of cruciferous crops.[23] The response of potatoes to zineb and other zinc-containing fungicides (increased tuber yield) is well established.[15,24] Higher yields have been attributed to the absence of lime phytotoxicity in BM[15] and the presence of carbon and sulfur in dithiocarbamates.[25] The organic sulfur fungicides (carbamates) showed varied effects on the vegetative growth phase in different crops because of their organic nature and the absence of injurious lime and copper ions associated with BM which make them comparatively safer.[7]

4. Captan

Captan and its analogue folpet are representative of this group of organic fungicides. Captan is one of the most commonly used fungicides because of its broad-spectrum nature. Captan mainly acts as a protectant but in some cases it is claimed to have acted systemically.[7] Application of captan produced better quality fruit and increased the yield of apple.[26] It has been suggested that the increased yield of trees sprayed with captan might be brought about by its stimulatory effect on photosynthesis.

5. Miscellaneous Fungicides

Pentachloronitrobenzene (PCNB) does not fall into any of the groups described above. PCNB is basically a soil fungicide. It is marketed with the trade names Quintozene and Brassicol. Reports have explored the undesirable effects on PCNB-treated potatoes[27] and on starch and total solid content.[28] Mondy[27] postulated that the flavor changes reported result from the accumulation of compounds produced by altered metabolism. Sweeney et al.[29] studied palatability and composition of potatoes grown in soil treated with active PCNB at 30 and 50 lb acre^{-1}. There was evidence in some cases of unusual flavors, odors, and colors resulting from PCNB treatment at 50 lb acre^{-1}. However, differences were not detected consistently. It was observed that adverse effects of PCNB usually diminished with storage. PCNB treatments did not lead to a significant effect on tyrosinase, total phenols, starch, total solids, specific gravity, pH, titrable acids, citric acid, malic acid, or ascorbic acid.[29]

B. Systemic Fungicides

Systemic fungicides applied simultaneously with other pesticides may have synergistic or antagonistic effects on the yield of crops.[22] Although not much work has been conducted on this pattern, some references are available which merit special mention here. Carboximides in some circumstances appear to be directly beneficial to plants. Oxathiins or carboximides (carboxin and oxycarboxin) showed unique growth-stimulatory properties when applied to either seed, soil, or foliage of wheat, barley, peanuts, sorghum, and broad beans.[30,31] The effects of growth stimulation include increased plant height of beans from foliar sprays, greener foliage of wheat and barley, increased germination vigor of peanut, and increased yield of grain from seed treatment. Especial growth stimulation was achieved with carboxin. They further observed that both carboxin and oxycarboxin increase protein in wheat seed through foliar application. Carboxin-treated wheat and barley seeds, when planted in low fertility soil, showed virtually no symptoms of nutrient deficiency, whereas the untreated control plants were markedly affected from lack of nutrients. The detached leaves when treated with carboxin remained green longer than untreated leaves. The chlorophyll content of treated leaves was also higher.[30]

In 135-day-old peanut plants, the content of different amino acids was generally lower in the plants treated with the higher dose of carbendazim; this may be indicative of the fact that the metabolic processes were still actively taking place in these plants, probably due to the delayed maturity of the crop induced by carbendazim.[32] Preharvest application of benomyl or thiabendazole reduces the incidence or severity of chilling injury in grapefruit.[33-35]

Leaves of plants treated with benzimidazole compounds appear greener, possibly due to an increased number of intergrana and grana lamellae of the chloroplast,[36] and they are more closely spaced. Benzimidazole-compound treatment to tobacco plants increased fresh weight and dry weight of leaves; the chemicals also enhanced the uptake of potassium considerably. The leaves of young tobacco plants showed overall increase in width, length, and thickness and were darker green than untreated controls.[37] Carbendazim has been also found to increase the number of female flowers and consequently fruits in vegetables, especially on cucurbitaceous crops like bottle gourd and musk melon; these effects could be due to cytokinin-like activity or some other growth-promoting factors of carbendazim.[38]

While conducting experiments on the effects of fungicidal seed treatment on the control of Rhizoctonia root rot *(Rhizoctonia bataticola)* of vegetable seedlings, a unique growth promotion/stimulation phenomenon was observed in brinjal, cowpea, tomato, and French bean seeds after treatment with carbendazim and thiram and their combinations at 0.25% in different ratios.[39] The seed treatment increased root and shoot length and fresh and dry weight. The side effects resulted from the increased uptake of nutrition from the soil after seed treatment. These are yield-contributing factors in the plants. Any favorable effect on plant growth with seed treatment is very likely to affect the yield of the plant favorably.[39]

Application of fungicides to cereal crops generally increases grain yield. Griffiths and Scott[40] illustrated this through experimental data from pot tests. Barley plants were sprayed with benomyl, ethirimol, tridemorph, and drazoxolon for the control of powdery mildew in their early development. The only evidence of the disease on unsprayed plants was a brief appearance at the end of tillering which rapidly declined to trace amounts due to high temperatures in the greenhouse. The greatest amount of mildew recorded did not exceed 5% over plants treated with fungicides, all of which effected complete control of mildew. The increase in the yield over control by benomyl, ethirimol, drazoxolon, and tridemorph was 14.80, 31.12, 21.43, and 2.04%, respectively.[40]

In barley, Griffiths[41] demonstrated under controlled epidemics of powdery mildew that (1) there is a strong correlation between the severity and duration of mildew and reduction in green leaf area (GLA) integrated over time and (2) the grain yield of plants is highly correlated with values of GLA integrated for the period from seedling emergence to anthesis. The data show that retranslocation of stored carbohydrate produced before anthesis plays an important role in grain filling and indicate that postanthesis mildew may have little effect on yield. The main implication of these studies is that fungicide treatments which increase GLA prior to anthesis in barley are likely to enhance yield. Evidently this may be achieved by different means, such as by control of weak pathogens or by directly extending the functional life of leaves. The end result in terms of grain yield, however, will be the same.[40,41]

Application of fungicides to tolerant crops generally increases grain yield. Wheat *(Triticum aestivum)* crop treated with carbendazim and carbendazim-producing fungicides has often shown increases in yield in the absence of any detectable disease.[19,42,43] Peat and Shipp[42] reported that the major effect of benomyl at concentrations in the range of 200 to 5000 ppm was to increase the development of tillers. Spraying at any stage of development caused an increase in the number of tillers about a week later. They further observed that this was due to more rapid initiation of tillers and not to more rapid growth following initiation. Associated with this earlier initiation was an increase in the incorporation of ^{14}C-labeled assimilates into newly initiated tiller buds, but not into older ones. An increase in the number of tillers could lead to an increase in yield. These responses can be explained by suggesting that benomyl changes the normal pattern of apical dominance and hence allows more rapid lateral growth.[19,42,43]

Fehrmann et al.[44] reported that spring application of carbendazim and carbendazim-generating fungicides or captafol to control eyespot in winter wheat increased yield by 2 to 3% in fields with only slight or no incidence of *Cercosporella herpotrichoides* in control plots. They observed no indication that the phenomenon was due to the control of any other pathogen at the haulm base.[44]

The yield of several varieties of soybeans was increased significantly when sprayed with the fungicides benomyl, triphenyltin hydroxide, and thiabendazole at 0.25% a.i. kg ha^{-1}. Yields of Davis and Lee-68 were increased significantly only when sprayed with benomyl.[45] The yield increases were obtained when the fungicides were applied at late bloom to early pod set with a second application 2 weeks later or at full green pod stage. The grower, therefore, has a long span of time during which to apply the fungicide.

Yield differences between fungicide-treated plants and controls were attributed to increased seed weight rather than to an increase in seed number,[45] although seed number appeared to be more important than seed weight in which case both could be responsible for increased yield. Significant additive effect on yield, seed weight, and seed and pods per plant was recorded when benomyl and carbendazim (0.15%) were sprayed on soybeans at the flowering and pod formation stages.[46] It was concluded that the increase is due to the control of *Aspergillus, Rhizopus,* and *Cladosporium* sp. In all of the varieties tested, the weight of the seeds from the benomyl-treated plants was significantly higher than that of controls, except in two of the seven cases in which the seed number significantly increased.[45] Thapliyal and Singh[47] did not obtain significant increases in seed weight of soybean cultivar Bragg. It is certain that the experimental conditions in both places were the same. Horn et al.[48] found that fungicide-treated soybean plants always remained green longer than the controls or the inoculated plants. This delay in harvest induced by the fungicide could contribute to yield increases simply because it provides a few more days for seeds to fill. The sprayed plants in the field were easily distinguishable from the unsprayed plants by the time the latter were ready for harvest. This was irrespective of systemic or nonsystemic fungicides.[48] However, they could not obtain yield increase in this study in the absence of disease control.

Priestley[49] demonstrated that winter wheat and spring barley differ widely in their yield response to standard fungicide treatment. Regression analyses reveal that yield response is partly the result of mildew reduction, so that cultivars which respond well tended to be those which were relatively susceptible to mildew. Cook[50] drew similar conclusions from a different set of winter trials in which carbendazim and captafol had been applied. However, the positive constants in the multiple regression equations indicate yield increases of 1.6 to 6.8% in spring barley and 2.9 to 15.8% in winter wheat even when there was no disease reduction.

Similar unexplained yield effects have been reported by Fehrmann et al.[44] and Cook[50] following the application of carbendazim and other compounds to winter wheat. Furthermore, the measurements made on winter wheat cultivars by Priestley and Bayles[51] showed that the increase in the percentage of leaf area that remained green was substantially larger than the corresponding reduction in disease. Taken together, the evidence suggests that there are two ways by which fungicide treatment increases the proportion of GLA: disease reduction and delayed senescence. Since the yield is closely correlated to GLA duration,[52] it follows that a yield response could result from either process. Yield response was more closely related to disease reduction than to change in GLA, suggesting that most of the response was due to disease reduction. However, at other sites, yield response was more closely related to delayed senescence. At still other sites, yield response was not significantly correlated with either factor.

Delayed senescence of wheat crop has been reported in response to the application of a number of fungicides.[51,53-56] Studies have also been made, however, of fungicide effects on photosynthetic activity and senescence in relation to grain growth.[57] Kettlewell et al.[58] reported that in a season characterized by high levels of foliar disease and moisture stress during the grain-filling period, application of the fungicide propiconazole (125 and 250 g a.i. ha^{-1}) produced an increase in yield of wheat which could not be directly related to disease development of the flag leaf. There was no significant effect on the GLA, but fungicide treatment prolonged the photosynthetic activity of the flag leaf. This was further explored by frequent field measurements of GLA, photosynthesis, and disease development during grain growth in wheat cultivars.[57] It was observed that the maintenance of GLA and photosynthetic activity by propiconazole (125 and 250 g a.i. ha^{-1}) was not noticeable until grain growth was almost complete; thus the delay of senescence did not contribute substantially to final grain weight. As a result, the duration of green flag area showed no relationship to grain growth or grain yield in this study. Other work[52] also does not support the correlation between yield and postanthesis leaf area duration above the flag leaf node.[59] One of the reasons for the poor correlation may be the independence of the senescence of the flag leaf and the cessation of grain fill. The flag leaf may remain green after grain growth has ceased, especially at low temperatures[60,61] or after high nitrogen application.[62] Fungicide treatment in the presence of low levels of observable diseases can also prolong flag leaf life beyond the end of grain growth.[57] It has been suggested that there will be no yield benefit in the presence of low levels of disease from extension of green flag leaf area through application of late-season fungicides unless these are effective in earlier stages of grain fill. However, at earlier stages of grain fill, there are likely to be ample supplies of photosynthate,[63,64] unless early leaf senescence has reduced the photosynthate supply sufficiently to limit grain growth.[65] Thus, prolongation of GLA with fungicide may only be beneficial if it can offset early senescence, as has been achieved by applying propiconazole.[57]

The lack of significant differences in disease control between treatments indicates that the fungicides act indirectly.[57] Physiological effects of fungicides on the senescence of cereal leaves have been demonstrated in other works, either through the maintenance of the integrity of cell membranes[66] or through the retardation of chlorophyll degradation.[67] Some leaf surface microorganisms can accelerate cereal leaf senescence[68,69] and some fungicides may influence senescence by affecting such microflora populations.[70]

It is never certain whether or not unexplained yield responses are the result of fungicides controlling diseases other than the ones identified. Although it is quite possible that eyespot control may have contributed to yield response in some winter wheat trials[49] it is most unlikely that this is the entire explanation for the observed increase in GLA, since Fehrmann et al.[44] have reported unexplained yield responses even when eyespot control is taken into account.

Studies of the use of late-season fungicide sprays between ear emergence and harvest revealed that they decreased most components of superficial microflora and may control seed-borne pathogens and storage microflora.[53,70-72] Phylloplane mycoflora are important in several respects during the life of the plant. They grow on the green tissues and affect the development of pathogens as tissues reach maturity. They may grow even more actively, accelerate senescence and growth of the senescent tissues, and result in deterioration of grain quality. They can, in extreme instances, lead to discoloration and the head-blackening syndrome. The field and storage fungi also develop during the preharvest period. The field fungi may disappear during drying but the storage fungi continue to grow. These effects can, therefore, result in both direct and indirect reduction in yield and quality of the sample.[53] In a series of experiments, Dickinson and Wallace[72] demonstrated an increase in yield by dithiocarbamates and benzimidazoles. In these trials, the activity of the phylloplane microflora was also monitored and the largest yield increases were shown to result from treatments which had the most severe effects on germination and development of a number of filamentous fungi. These fungi, particularly *Alternaria alternata* and *Cladosporium cladosporoides*, become especially active on senescing plant tissue. The largest increase in yield was recorded in fungicide-treated plots where the greatest reduction in fungal colonization of the glumes was observed. This suggests that these filamentous fungi are causal agents of crop loss.

Cook[73] showed that single sprays with benomyl or benomyl + maneb applied between flag leaf and ear emergence gave leaf and ear protection which resulted in increased yield. It was implied by Cook but not mentioned that there would be fewer seed-borne *Septoria nodorum* and *S. tritici* resulting from such treatments since the ears were protected. Dickinson and Walpole[53] reported that a single application of a mixture of carbendazim (1.12 kg ha^{-1}) and zineb (4.48 kg ha^{-1}) shortly before anthesis gave yield increase of 18 to 21% in wheat. Lescar et al.[74] demonstrated that treatment with a single spray of carbendazim-generating fungicides with or without maneb at full ear emergence could give significant increases in wheat yield. However, the benefits were not always economical and varied according to site, year, and cultivar. Allison et al.[75] also showed that fungicide applied between the start of flag leaf emergence and full ear emergence increased wheat yield.

Fehrmann et al.[44] reported an average yield increase of 2 to 3% following application of benzimidazole fungicides or captafol at stem erect stage (GS 30-31) for the control of eyespot (*Pseudocercosporella herpotrichoides*), regardless of disease level. On spraying of fungicides between the start of flag leaf emergence and milky ripe stage of grain development, although disease was generally slight, an average yield response of 0.185 t ha^{-1} (3.31%) was recorded. Spray applied before the start of flowering gave the largest response in yield, up to 0.218 t ha^{-1} in wheat.[50] However, negative effects have been observed in comparatively dry weather.

Similarly, spraying a boot stage in barley increases the yield significantly, but no mention has been made of the effects of soil-borne *Rhynchosporium secalis*, cause of leaf blotch.[76] Jacobson[77] reported that wheat fields sprayed with benomyl, carbendazim, or benomyl + mancozeb reduced the wheat scab *(Fusarium graminearum)* from 24% in unsprayed fields to 12% in plants treated with the fungicides.

Field experiments with barley have generally shown a good relationship between disease and yield; however, in some experiments the yield response is greater than would be expected from the level of disease after the late application of fungicides. Fraselle[78] reported improvement in yields of winter barley with carbendazim-generating fungicides at the beginning and end of ear emergence, and this was accounted for by the control of powdery mildew and rust at the heading time. On spring and winter barley, however, late applications of thiophanate-methyl alone and in combination with maneb repeatedly resulted in yield increases which could not satisfactorily be linked to visible fungal attack.[79] Late application of carbendazim-generating fungicides alone or in mixture can produce yield increases of up to 12% in winter barley.[80] The resulting yield increases could not always be correlated with the control of leaf diseases. These observations are based on the use of fungicides BAS 32500 F and BAS 39700 F which were applied separately at growth stages 5 to 7 and 10.1 to 10.5 on winter barley. The effect of sprays on the micoflora of flag leaves, glumes, and awns of barley was studied by Mappes and Hampel,[80] who found a reduction in the main components of the microflora on all plant organs in the treated plants as compared to untreated controls. Similar results have also been reported.[81-83] Hill and Lacey[84] reported that late application of fungicides benomyl, benodonil, tridemorph, and captafol on the ripening barley at or after anthesis decreased most components of the superficial microflora; captafol was most effective overall in decreasing numbers of fungi. However, benomyl decreased *Cladosporium* inoculum more than captafol but was ineffective against *Alternari* which became more numerous on benomyl-treated grain than on untreated. However, as was found previously,[85] *Alternaria* was tolerant of benomyl treatment and was even increased by it, perhaps filling niches left by the control of *Cladosporium*. Apart from this, effects were much greater with the protectant and systemic fungicides than with those used to control specific diseases. Yield increases of up to 4% were obtained by late application of fungicide, with the least increase in a hot and dry year. Germination was also increased by the same treatment up to 8%.[84]

The late application of fungicides such as benomyl, carbendazim + zineb, and tridemorph at the field rates shortly before or after anthesis gave yield increases of up to 21%.[53] Cook[50] also observed similar results in wheat. Hill and Lacey[84] observed an increase in grain weight, yield, and germination of barley but yield responses were much smaller than those from similar experiments.[53] This could result from climatic differences, since it was found that benefits of captafol or benomyl application were larger in years with adequate rain than in dry years.

Many taxa of storage fungi and actinomycetes were found on barley grain before harvest, although in small numbers. Contamination of *Aspergillus* and *Penicillium* spp. was similar to that previously reported.[86,87] Other taxa have not been reported previously on grain before storage although they have sometimes been observed from pasture. Application of fungicides before harvest results in the control of these fungi; thus quality of stored seed is increased. Direct suppression of the buffering capacity of the leaves by broad-spectrum fungicides may allow fungi normally regarded as saprophytes, such as *Alternaria* spp., to become weak pathogens at the end of the season. This possibility has been discussed.[1]

Several fungi frequently develop on the necrotic spots which appear beneath pollen on wheat leaves. During wet weather the dead pollen provides a nutrient base which may allow limited colonization of leaves and occasionally ear tissue. Development of fungi such as *Alternaria* spp., *Cladosporium* spp., *Didymella exitialis, Botrytis cinerea,* and *Fusarium* spp. is likely to be encouraged by a wet and prolonged ripening period. Late application of fungicides on wheat results in the control of these pathogens and hence an increase in the yield.[50]

Chinn[88] reported that seed treatment with imazalil at 0.2 and 0.3 g a.i. kg^{-1} not only reduced the severity of common rot of wheat, but the treatment also increased the diameter of subcrown internodes. Other morphological changes were noted such as a decrease in the length of subcrown internodes that inversely affects the depth of the crown and a stimulation of coleoptile node tillers (CNT). CNT arise from the axil of the first node at the base of the coleoptile adjacent to the seed and are a second source of tillering for the plant. Chinn et al.[89] studied the imazalil seed treatment in four wheat cultivars (Cypress, Glenlea, Neepawa, and Wascana) at 0.2 and 0.3 g a.i. kg^{-1} and found that Cypress had the longest internodes, followed in descending order by Glenlea, Neepawa, and Wascana. Generally, plants from imazalil-treated seed had significantly shorter subcrown internodes. Only 2 few plants in nontreated seed produced CNT and many of these were soft or aborted, whereas the treated seed produced a number of CNT and many of these were firm tillers. It was further observed that imazalil significantly reduced the length of subterranean plant parts. It appears that cultivars of wheat and barley with shorter internodes, and hence deeper crowns, are better able to escape damage by root rot than those with opposite attributes. Apart from the yield benefits obtained, the application of the effective fungicides produced noticeably clean grain samples due to control of several seed-borne fungi.[53]

Yield increases are not normally obtained by fungicidal treatments unless very high levels of disease are present. Stoker and Dewey[90,91] reported that seed treatment with carboxin (0.15%) gave 100% control of loose smut *(Ustilago nuda)* on barley. However, the beneficial effect of carboxin in controlling loose smut was partially offset by an apparent direct adverse effect on yield. At very low levels of infection, the yield reduction due to carboxin treatment exceeded the benefit derived from the smut control. As infection level rose above 5%, carboxin treatment increased yields; however, this increase did not reach statistical significance until loose smut levels approached 20%. Carboxin-treated plots had significantly more tillers than untreated plots.[90,91]

Metalaxyl application (1.12 kg a.i. ha^{-1}) on pepper *(Capsicum annum)* and cabbage *(Brassica oleracea)* immediately after seeding increased growth.[92] Root length of pepper at 39 days after seeding and shoot length at both 25 and 39 days after seeding were increased by metalaxyl. The fresh weight of marketable pepper transplants was increased by metalaxyl in all fall soil treatments, whereas the fresh weight of total pepper transplants was not significantly increased with metalaxyl in plastic-covered treatments. The average fresh weight of marketable and total cabbage transplants was increased with the fungicides by 1.4 and 1.5 g per transplant, respectively. Since the treated parts appeared to be healthy in both treated plants and in controls, the growth response very likely was not related to pathogen control.[92]

Studies were conducted on the effects of fungicides on apple yield. Application of fungicides captan, thiram, dithianon, bionocapryl, chinomethionate, and thiocarbamate increased fruit set and yield.[93] Spraying of captan at 0.2% three times during flowering had no unfavorable effect on fruit set and yield.[94] Other reports also found no effects on fruit set and yield.[95,96] Dancs and Kiss[97] reported that spraying of several fungicides at flowering time reduces fruit set due to inhibition of pollen germination; this inhibition is partly or entirely compensated for by meteorological factors and retarded dehiscence of anther. The amount of yield will not be reduced; it may even increase. They observed that increase in yield relates to the fungicidal effect of the plant protectives applied. The inhibition of pollen germination may be influenced not only by the active ingredient, but also by the accompanying aggregates, emulgators, and other additives.[95-97]

Not all responses are positive. In experiments on the control of spring barley powdery mildew *(Erysiphe graminis),* yield reduction was recorded in about 20% of the experiments in which a single spray of tridemorph (0.7 ℓ ha^{-1}) was applied at a range of growth stages. The data suggested that the reduction was not a result of experimental error or poor mildew control; the toxic spray in general was associated with better than average mildew control. Therefore, precaution should be taken that tridemorph + maneb + carbendazim formulations should not be used before night frost.[44]

Robinson and Hodges[98] reported that an increase in the rate of benomyl caused a decrease in dry weight of *Agrostis palustris* when the fungicide was used for the control of *Ustilago striiformis.*

Application of Kitazin at 10^{-3} and 10^{-5} M concentration results in general growth inhibition. Although root length was the same from the beginning to the end of the experiment, the formation of new roots is prevented, thereby demonstrating the influence of Kitazin on morphogenesis.[99] The dry matter content of rice plants was the same.

III. POLLEN GERMINATION AND POLLEN TUBE ELONGATION

Fungicide application before, during, and after bloom is a common practice. This may impair fruit setting by reducing pollen viability or the receptivity of the stigma and may interfere with the fertilization process. Reports relating to the effects of fungicidal sprays on pollen germination and seed setting applied to plants at the time of anthesis are incomplete, particularly with regard to the systemic fungicides which translocate upward and downward in plants. In addition, information on the effect of germination of pollen anthers, undehisced or recently dehisced, is conflicting.[97,100-102]

McDaniels and Furr[103] claimed that the presence of sulfur on the stigma before pollen germination resulted in failure of fertilization in the apple. Germinated pollen on the stigmatic surface was unaffected by sulfur application. It was further demonstrated that sulfur reduced or inhibited pollen germination in vitro. Rich[100] studied the effects of various fungicides applied during bloom on apple pollen germination and fruit set. He observed a reduction in pollen germination by application of sulfur (99%), dichlone (50%), captan (50%), and ferbam (76%) at the rate of 2.7, 0.9, 0.225, and 0.94 kg/450 ℓ, respectively. He concluded that in vitro germination methods were too severe in their effects on pollen germination and that none of the fungicides used reduced pollen germination on fruit set when applied to apple trees in bloom. With the same treatments Eaton[104] observed significant reduction (53, 47.1, and 40%) in pollen germination after applications of sulfur, dichlone, and ferbam, respectively, in the field. Captan almost entirely prevented pollen germination and arrested the elongation of pollen tubes. Rich[100] and Eaton[104] found that ferbam reduced apple pollen germination in a sucrose medium but found sulfur to be without effect. Differences between fungicides were not observed by Rich but were by Eaton. The reason was probably that different methods were used for testing the materials. Rich attempted to germinate apple pollen in media containing the fungicides in the field dose, rather than as constituents of media, in an attempt to simulate natural conditions more closely. In view of the higher percentage of pollen germination in ferbam and sulfur fungicides, these fungicides do not seem to be hazardous to fruit setting.

Schmidt[105] evaluated numerous fungicides (lime-sulfur, ziram, captan, copper oxychloride, etc.) on apples and pears but none appeared to reduce stigma receptivity. However, Braun and Schonbeck[106] observed an inhibition of pollen germination by nabam (0.25%) on apples and pears. In subsequent studies, they reported that pollen germination in vitro was reduced to the highest extent by the fungicides captan, thiram, and dithianon, while binocapryl, chinome thionate, and thiocarbamate had a slight toxic effect on pollen of apple trees.[106]

The extent of pollination of apple trees when treated with a combination of zineb and karathane and a range of organic fungicides did not decrease.[107] Liebster[95] in his 2-year experiment did not observe early abscission as a reaction to treatments with fungicides and stated that commercial organic fungicides had no unfavorable effect on fruit set. Minoiu et al.[108] investigated fruit set in apples, pears, and sour cherries after treatments with BM (4:4:50) and found that copper had an unfavorable effect on pollen germination. It was further pointed out that all plant protectants had an unfavorable effect on pollen germination in every fruit species.

Captan at 40 ppm in vitro arrested development of sweet cherry pollen at various stages of germination.[104] Captan sprays at 1 kg/50 ℓ water to several cultivars of apple anthers until runoff significantly reduced pollen germination, although tnot equally for all cultivars. In addition, captan inhibited pollen tube elongation of cultivars. Rich[100] found sufficient viable pollen in counts following captan sprays in bloom and this finding is true for most varieties he studied. However, the reduction in pollen germination and tube growth might sometimes be important, particularly when other conditions are unfavorable to pollination.[109]

There are different effects of captan on pollen of apple and cherry cultivars, which suggests that this is a potentially useful tool for the study of the biochemistry of pollen germination, as well as the mode of action.[109] Captan also inhibited strawberry pollen germination in vitro; however, pollen germination was not affected when it was sprayed on dehisced anther except at higher rates.[110] Apple inflorescences that had been hard-pollinated were either bagged or sprayed with captan.[111] Pollen-seeded agar slides were suspended in the trees; pollen germination and tube elongation were suppressed on the slides and fruit set was significantly reduced on the inflorescences that were sprayed.[111] These results suggest that captan could interfere with pollination under field conditions.

Legge and Williams[111] and Church and Williams[112] also reported a reduction in apple pollen germination on the stigma. Spraying anthers before dehiscence caused less damage to pollen than spraying dehisced anthers.[113] Cultivars[106,109] or species[105] may differ in their reaction to a fungicide. Church and Williams[114] conducted several tests of various fungicides applied to flowers of several apple cultivars before or after dehiscence; in vitro studies indicated that captan and dinocap applied to undehisced anthers reduced germination of pollen from all the cultivars used, while dinocap liquid, dithianon, sulfur, and triforine caused a reduction only in some cultivars. Other fungicides did not significantly decrease germination. When dehisced anthers of Crawley Beauty, a cultivar not used in the predehiscence treatments, were sprayed with captan, dinocap, dithianon, sulfur, or triforine, pollen germination was considerably reduced. Although precise measurements have not been made, it has been reported that all the sprays containing captan or dinocap powder reduced pollen tube growth. The dinocap liquid formulation shortened tubes in cultivars where germination was significantly reduced. Triforine caused no obvious reduction in tube length and sulfur appeared to reduce the length of only some tubes of cultivar Worcester Pearmain.[114] The application of fungicides also changes the color and amount of pollen. Church and Williams[114] and Burth and Ramson[113] reported that benomyl was not toxic but captan and dinocap were. The latter workers reported harmful effects of dodine which Church and Williams[114] could not confirm.

Treatment of cranberry flowers with phaltan and maneb at the rate of 0.66 and 1.5 kg/250 ℓ, respectively, by spray reduced pollen germination while treatment with captan and ferbam at 1 and 0.75 kg/250 ℓ did not;[115] the results are similar to those of Rich.[100] Apple and cranberry pollen seeded on agar treated with fungicides failed to germinate.[115] Maneb and phaltan may exert an inhibiting influence by absorption through the anther walls, since it is unlikely that the fungicide came in direct contact with the pollen prior to dehiscence; captan and ferbam do not appear to be absorbed through anther tissue to the same degree as maneb and phaltan since they did not reduce pollen germination appreciably when applied to undehisced flowers.[115] The effect of zineb, ziram, and ferbam on germination of lowbush blueberry pollen and number of seeds per berry was studied.[102] The results indicated that zineb had no effect on the germination of pollen whereas ferbam and ziram inhibited it; the concentration of fungicide applied was 2.9 kg ha^{-1}; it was further observed that the toxicity of the fungicides was higher under wet conditions than dry. The fungicides had no adverse effect on the number of seeds per berry.

The toxic effects of fungicides on pollen germination could result partly from formulating substances other than the fungicide, as surfactants may damage pollen.[116] Church and Williams[114] observed that although both powder and liquid forms of dinocap were toxic, the powder formulation was more so.

Church and Williams[114] indicated that cultivars differed in their response to dinocap liquid, dithionon, sulfur, and triforine; the germination of pollen from Aldenhamensis and Winter Gold was unaffected by these fungicides. They also indicated that pollen and viability can be affected by some fungicides, most of which are approved for use in apple orchards. Consideration should therefore be given to the relative importance of disease control and fruit set before open flowers of the pollen donor cultivar are sprayed with these fungicides.[114]

IV. FRUIT SET

Butt et al.[117] studied the fungitoxic and phytotoxic effects of fungicides (benomyl, pyrazofos, triarimol, triforine, and Dowco 199) controlling powdery mildew on apple and compared their performance to the traditional fungicides such as binapicryl, dinocap, and elemental sulfur on eight apple cultivars: Cox's Orange Pippin, Golden Delicious, Fortune, Jonathan, Worcester, Cox, Tydeman's, and Sunset. The fungicidal sprays were applied at the rate of 250 ℓ ha^{-1} at 14-day intervals from pink bud (late April/May) and it was observed that these compounds were more phytotoxic without substantially or consistently improving the control of powdery mildew *(Podosphaera leucotricha)*. Fruit russet was increased and the crop and fruit size of some cultivars was reduced. In comparison with captan or dinocap/captan, benomyl russeted Cox, Fortune, Golden Delicious, Sunset, and Worcester. When benomyl was used at half strength in a mixture with dinocap, phytotoxicity was reduced; triarimol russeted Cox, Jonathan, and Fortune; Dowco 199 russeted Fortune, Jonathan, and Sunset; and pyrazofos and triforine badly damaged Cox.[117]

Some inconsistencies in the causes of spray injury are evident. Benomyl did not russet Fortune or Jonathan in 1968,[118] but it russeted one in 1969 and both cultivars in 1970. Triarimol russeted Cox in 1969 but not in 1970 and Dowco 199 russeted Cox, Tydeman's, and Worcester in 1968[118] but not in 1969. These differences between year are probably associated with rainfall[119] and minimum temperatures[120] at the time the crop is most sensitive, which is in spring and early summer.[121,122] May and June were cool, dull, and wet in 1968, warm in 1969, and very warm and dry in 1970.[117]

Increased russeting of fruits was not the only undesirable side effect of the alternative fungicides. Although these did not alter the total crop or fruit size of Golden Delicious, benomyl and Dowco 199 reduced the weight of the 1969 Fortune crop by 22 and 30%, respectively, in comparison with the standard treatment.[117] Benomyl, triarimol and pyrazofos reduced the Cox crop by 39 and 34%, respectively; pyrazofos also reduced the Cox crop by 39%. In contrast, benomyl increased the 1969 Cox crop by 56%, when seven preblossom frosts and only fair pollinating weather resulted in a generally poor set. This role of benomyl has been investigated.[123] When blossoming was profuse and pollinating conditions almost perfect, benomyl had no effect on the Cox crop. The poor control of fruit tree red spider mite, *Panonychus ulmi,* by these fungicides must be considered in the interpretation of the yield response.[124]

Effects on fruit size were not just compensatory adjustments to changes in total crop. Butt et al.[117] reported that the 72% increase of Cox crop attributable to Dowco 199 + Urbacid in 1969 was accompanied not by smaller apples but by a 5% increase in fruit size. In contrast, although the weight of the Worcester crop was reduced 38% by triarimol in 1970, the fruits were 13% smaller; in some instances the fungicide altered fruit size without detectably changing the total crop weight or the number of fruits, as with the increased size of Worcester fruits after treatment with Dowco 199 + Urbacid and the smaller Jonathan fruits after treatment with benomyl or triarimol.[114] Incorporation of captan or Urbacid in different ratios has been suggested as an aid to alleviate the russeting effects of the fungicides. Although noticeable alleviation of russet has been achieved in some instances by adding the adjuvant fungicides to the sprays, these benefits have been too sporadic to recommend their use in the immediate future. There is, therefore, a continuing need for new products that are safer as well as more fungitoxic.

Various fungicides have been reported to increase fruit set of apples on some occasions.[125-127] Church et al.[128] conducted studies on the changes in fruit set, leaf size, and shoot growth of apple (cv. Cox) caused by benomyl, pyrazofos, tradimefon, and sulfur sprayed at the rate of 0.5, 0.15, 0.125, and 3.8 g a.i. ℓ^{-1}; they observed that out of four fungicides sprayed on apple flowers before flowering or at petal fall stage, triadimefon increased fruit set, but reduced leaf size. Strydom and Honeyborne[126] also reported that cultivar Starking Delicious of apple can respond to triadimefon in the same way as cultivar Cox and demonstrated that this effect can be produced by two prebloom and three postbloom sprays applied at the rate of 25 and 100 ppm. The higher concentration significantly increased (5% level) fruit set per cluster and number of seeds per fruit. They did not observe any detrimental effects. The higher rate of triadimefon increased the average yield per tree (kilogram per centimeter trunk circumference) by about 30% with no detrimental effect on fruit size. No difference in fruit firmness, total soluble solids, or total acidity, was found. Effects on vegetative growth were observed with the chemicals which affected set, though no causative link was established.[128]

Church and Williams[127] studied the effect of preblossom fungicidal sprays on the ability of Cox's Orange Pippin apple flower to produce fruit. The fungicides used were benomyl, binapicryl, bupirimate, captan, dinocap, dithianon, dodine, mancozeb, sulfur, thiophanate-methyl, and triforine in liquid, colloidal, and wettable powder in one concentration. They observed that preblossom sprays on individual clusters in 1976 and 1977 caused no significant changes in the number of flowers per cluster which subsequently opened, and there was no obvious damage to flowers or leaves; in 1978, benomyl and sulfur gave a higher percentage and dodine a lower one. It should be noted that all these sprays were high volume, which probably deposited more fungicide on the cluster than a low volume spray would have.[128] However, most commercial fungicide programs, though at low volume, involve several applications and the total fungicide application might not be very different. These observations do support Chollet's evidence[129] that preblossom applications of dodine at high volume can reduce fruit set. This suggests that high-volume dodine sprays are not advisable when fruit set could be a crop-limiting factor. The mechanism involved is not yet clear but does not appear to directly affect either pollen germination of pollen tube growth. The increase in fruit set and seed number following the use of sulfur were unexpected, particularly as it can damage the leaves of some cultivars. However, Abbott[130] showed that the suppression of bourse shoot growth can increase the fruit set potential of blossom cluster, and the sulfur could be checking the growth of primary superior bourse leaves. (The bourse is the swollen stem at the base of the inflorescence.)

V. FRESH OR PROCESSED PRODUCTS

Fungicides may have a profound effect on the appearance of fresh and processed food products. The direct effects of injury can be quite serious in terms of visible surface defects on fresh market fruits and vegetables. The copper sprays increased the total solids, decreased the size, darkened the color, and hastened the maturity of raw fruit as compared to ferbam sprays on apple. The soluble solids of sour apple increased significantly as harvest season progressed, as a result of copper sprays, and the size of the fruit changed very little. It was also indicated that fruit became less firm during the first 2 weeks of harvest. Firmness increased in the third week.[131]

VI. FUTURE OUTLOOK

That fungicides have many nontarget effects on plants and their processes is readily apparent. The effects are sometimes too slight to be noticeable but great enough to increase the quality of the finished plant products. For example, grain samples are noticeably cleaner following application of fungicides, fruit set of treated apple and cherry is significantly higher than untreated, etc. The increases in growth, yield, and other factors were mainly attributed to prolonged photosynthetic activity rather than to control of specific diseases. The mechanisms of phytotonic effects are discussed in Chapter 5. These nontarget effects merit more attention in planning and decision-making processes. If, for instance, a single application of fungicide is known to increase the quality in the absence of disease, such application should be undertaken irrespective of the cost factor.

REFERENCES

1. **Dickinson, C. H.,** Interactions of fungicides with minor pathogens on cereals, *EPPO Bull.,* 11, 311, 1981.
2. **Turrell, F. M.,** A study of the physiological effects of elemental sulphur dust on citrus fruits, *Plant Physiol.,* 25, 13, 1950.
3. **Llewelyn, F. W. M.,** The hardening effect of lime sulphur sprays on maiden apple trees, *Ann. Appl. Biol.,* 55, 495, 1964.
4. **Agnew, E. E. and Childers, N. F.,** The effect of two sulphur sprays on photosynthetic activity of apple leaves, *Proc. Am. Soc. Hortic. Sci.,* 37, 379, 1939.
5. **Brody, H. W. and Childers, N. F.,** The effect of dilute liquid lime sulphur sprays on the photosynthesis of apple leaves, *Proc. Am. Soc. Hortic. Sci.,* 36, 205, 1939.
6. **Laurence, R. C. N., Gibbons, R. W., and Young, C. T.,** Changes in yield, protein, oil and maturity of groundnut cultivars with the application of sulphur fertilizers and fungicides, *J. Agric. Sci.,* 86, 245, 1976.
7. **Nene, Y. L. and Thapliyal, P. N.,** *Fungicides in Plant Disease Control,* Oxford & IBH Publishing, New Delhi, 1979, 507.
8. **Gillet, S.,** Result and observations of spraying trials of Bordeaux mixture on coffee at the Scott agriculture laboratories, *Mon. Bull. Coffee Board Kenya,* 1, 30, 1942
9. **van der Vossen, H. A. M. and Browning, G.,** Prospects of selecting genotypes of *Coffee arabica* L. which do not require tonic sprays of fungicide for increased leaf retention and yield, *Kenya Coffee,* 43, 361, 1978.
10. **Mulinge, S. K. and Griffiths, E.,** Effects of fungicides on leaf rust berry disease, foliation and yield of coffee, *Trans. Br. Mycol. Soc.,* 62, 495, 1974.
11. **Rayner, R. S.,** Tonic spraying of coffee, *Proc. Nairobi Sci. Phil.,* 9, 12, 1957.
12. **Aberden, J. E. E.,** Investigation on phytotoxicity of Bordeaux mixture to tomato, *J. Agric. Sci.,* 9(1), 1, 1952.
13. **Wilson, J. D.,** Comparative control of blue mold rot of tomatoes by various fungicides, *Phytopathology,* 46, 511, 1956.
14. **Shutak, V. G. and Christopher, E. P.,** The influence of Bordeaux spray on the growth and yield of tomato plants, *Proc. Am. Soc. Hortic. Sci.,* 36, 747, 1939.
15. **Horsfall, J. G. and Turner, N.,** Injuries of Bordeaux mixture, *Am. Potato J.,* 29(12), 308, 1943.
16. **Childers, N. F.,** Some side effects of sprays on the growth and transpiration of tomatoes, *Proc. Am. Soc. Hortic. Sci.,* 33, 532, 1936.
17. **Wilson, J. D.,** Relation between spray treatments and frost damage to potatoes, *Ohio Agric. Exp. Stn. Res. Bull.,* 32, 77, 1947.
18. **Jenkins, J. E. E., Melville, S. C., and Jemmett, J. L.,** The effect of fungicides on leaf diseases and on yield in spring barley in south-west England, *Plant Pathol.,* 21, 49, 1972.
19. **Cook, R. J.,** Unexpected effects of fungicides on cereal yields, *EPPO Bull.,* 11, 277, 1981.
20. **Palmiter, D. H.,** Rust disease of apple and their control in Hudson Valley, *Bull. N.Y. State Agric. Exp. Stn.,* 756, 1, 1952.
21. **Hance, R. J.,** Effect of pesticides on plant nutrition, *Residue Rev.,* 78, 13, 1981.
22. **McKeen, W. L.,** *Rhizoctonia* sp. Nova associated with dry-berry disease of Loganberry, *Can. J. Plant Sci.,* 39, 82, 1959.
23. **Rogers, W. E.,** A micro-environment chamber for critical control of relative humidity, *Phytopathology,* 56, 980, 1966.
24. **Hoyman, W. G.,** Potato-fungicide experiments in 1948, *N.D. Agric. Exp. Stn. Bull.,* 11, 32, 1949.
25. **Emge, R. G. and Linn, M. B.,** Effect of spraying with zineb on the growth and zinc content of tomato plant, *Phytopathology,* 42, 133, 1952.
26. **Ross, R. G.,** Control of *Gleosporium album* rot and storage scab of apple with orchard fungicide, *Can. J. Plant Dis. Surv.,* 44, 109, 1964.
27. **Mondy, N. I.,** Pentachloronitrobenzene effect on enzymatic activity and phenolic content of potatoes, *Am. Potato J.,* 43, 147, 1966.

28. **Gustafsson, N.,** The fight against potato scurf *(Streptomyces scabies)* through disinfection of the soil with PCNB, *K. Skogs. Lantbruksakad. Tidskr.*, 101, 301, 1962.
29. **Sweeney, J. P., Chapman, V. J., and Hepner, P. A.,** Effect of pentachloronitrobenzene on quality of pototoes, *Am. Potato J.*, 45, 383, 1968.
30. **Carlson, L. W.,** Effects of Vitavax on chlorophyll content, photosynthesis and respiration of barley leaves, *Can. J. Plant Sci.*, 50, 627, 1970.
31. **von Schmeling, B. and Clark, M.,** Oxathiin induced plant growth stimulation, 7th Int. Congr. Plant Protection (Abstr.), Paris, 1970, 227.
32. **Mercer, P. C.,** The control of Cercospora leaf spot on groundnut in Malawi and its effect on the parameters of yields, *PANS (Pest Artic. News Summ.)*, 19, 204, 1973.
33. **Wardowski, W. F., Albrigo, L. C., Grierson, W., Barmore, G. R., and Wheatson, T. A.,** Chilling injury and decay of grapefruit as affected by thiabendazole, benomyl and CO_2, *Hortic. Sci.*, 10, 38, 1975.
34. **Kokkalos, T. I.,** Thiabendazole reduces chilling injury (pitting) of cyprus grown grapefruit, *Hortic. Sci.*, 9, 456, 1974.
35. **Schiffman-Nadel, M., Chalutz, E., Wake, J., Latter, J. S., and Dagan, M.,** Reduction of pitting and rots of grapefruits by thiabendazole and benomyl during long term storage, *Res. Summ.*, p. 30, 1973.
36. **Waygood, E. R.,** Benzimidazole effect in chloroplast of wheat leaves, *Plant Physiol.*, 40, 1242, 1965.
37. **Dyar, J. J.,** Certain effects of benzimidazole on young tobacco plants, *Plant Physiol.*, 43, 477, 1968.
38. **Sohi, H. S. and Sokhi, S. S.,** Further studies on the efficacy of different fungicides for control of anthracnose disease of bottle gourds and musk melons, *Pesticides*, 10(4), 30, 1976.
39. **Vyas, S. C., Andotra, P. S., and Joshi, L. K.,** Effect of systemic fungicides on control of root rot of vegetables caused by *Rhizoctonia bataticola* and plant growth, *Pesticides*, 15(11), 22, 1981.
40. **Griffiths, E. and Scott, S. W.,** Possible phytotonic effects of fungicides on barley, in *Crop Protection Agents — Their Biological Evaluation*, McFarlane, N. R., Ed., Academic Press, London, 1977, 466.
41. **Griffiths, E.,** Role of fungicides in maximizing grain yield of barley, *EPPO Bull.*, 11, 347, 1981.
42. **Peat, W. E. and Shipp, D. M.,** The effects of benomyl on the growth and development of wheat, *EPPO Bull.*, 11, 287, 1981.
43. **Fehrmann, H.,** Modern developments in fungicides use on cereals, *EPPO Bull.*, 11, 259, 1981.
44. **Fehrmann, H., Reinecke, P., and Weihofen, U.,** Yield increase in winter wheat by unknown effects of MBC-fungicides and captafol, *Phytopathol. Z.*, 93, 359, 1978.
45. **Horn, N. L., Lee, F. N., and Carvar, R. B.,** Effect of fungicides and pathogens on yield of soybean, *Plant Dis. Rep.*, 59, 724, 1975.
46. **Julio, V. L., Lam-Samchez, A., Nakamura, K., Nakamura, A. M., and Don-Krona, S.,** Effect of fungicide sprays on the soybean [*Glycine max* (L) Merrill] crop, *Cientifica*, 10, 119, 1982.
47. **Thapliyal, P. N. and Singh, K. P.,** Soybean *(Glycine max)* rust: *Phakospora pachirhizi*, *Fungic. Nematic. Tests*, 29, 94, 1974.
48. **Horn, N. L., Whitney, G., and Fort, T.,** Yields and maturity of fungicide sprayed and unsprayed disease free soybean plants, *Plant Dis. Rep.*, 62, 247, 1978.
49. **Priestley, R. H.,** Fungicide treatment increases yield of cereal cultivar by reducing disease and delaying senescence, *EPPO Bull.*, 11, 357, 1981.
50. **Cook, R. J.,** Effects of late season fungicide sprays on yield of winter wheat, *Plant Pathol.*, 29, 21, 1980.
51. **Priestley, R. H. and Bayles, R. A.,** Effect of fungicide treatment on yield of winter wheat and spring barley cultivars, *Plant Pathol.*, 31, 31, 1982.
52. **Thorne, G. M.,** Physiology of grain yield of wheat and barley, *Rothamsted Exp. Stn. Rep. for 1973, II*, 5, 1974.
53. **Dickinson, C. H. and Walpole, P. R.,** The effect of late application of fungicides on the yield of winter wheat, *Exp. Husb.*, 29, 23, 1975.
54. **Rowley, N. K., Wainwright, A., and Chipper, M. E.,** The development of systemic fungicide triadimefon for the control of foliar diseases in wheat and oats in the U.K., *Proc. 1977 Br. Crop Prot. Conf. Pests and Diseases*, 1, 17, 1977.
55. **Harris, R. G. and Barnes, G.,** Prochloraz: the control of net blotch and *Septoria* in winter cereals, *Proc. 1981 Br. Crop Prot. Conf. Pests and Diseases*, 1, 267, 1981.
56. **Jarrett, G. A. and Allard, P. M.,** Control of *Septoria* spp. in winter wheat with chlorothalonil, *Proc. 1981 Br. Crop Prot. Conf. Pest and Diseases*, 1, 259, 1981.
57. **Davies, W. P., Kettlewell, P. S., Helen, J. G., Hocking, T. J., and Jarai, M. M.,** Senescence and net photosynthesis of the flag leaf and grain growth of wheat in response to fungicides, *Ann. Appl. Biol.*, 105, 303, 1984.
58. **Kettlewell, P. S., Davies, W. P., and Hocking, T. J.,** Disease development and senescence of the flag leaf of winter wheat in response to propiconazole, *J. Agric. Sci.*, 99, 661, 1982.
59. **Welbank, P. J., French, S. A. W., and Witts, K. J.,** Dependence of yields of wheat varieties on their leaf area durations, *Ann. Bot. New Ser.*, 30, 291, 1966.

60. **Sofield, I., Evans, L. T., Cook, M. G., and Wardlaw, I. F.,** Factors influencing the rate and duration of grain filling in wheat, *Aust. J. Plant Physiol.,* 4, 785, 1977.

61. **Spiertz, J. H. J.,** The influence of temperature and light intensity on grain growth in relation to the carbohydrate and nitrogen economy of the wheat plant, *Neth. J. Agric. Sci.,* 25, 182, 1977.

62. **Vos, J.,** Effects of Temperature and Nitrogen Supply on Post-floral Growth of Wheat: Measurement and Stimulations, Agricultural Research Report 911, Pudoc, Wageningen, 1981, 45.

63. **Evans, L. T. and Rawson, H. M.,** Photosynthesis and respiration by the flag leaf and components of the ear during grain development in wheat, *Aust. J. Biol. Sci.,* 23, 245, 1970.

64. **Watson, D. J.,** Size, structure and activity of the productive systems of crops, in *Potential Crop Production,* Wareing, P. F. and Cooper, J. P., Eds., Heinemann, London, 1971, 76.

65. **Gallagher, J. N. and Biscoe, G.,** A physiological analysis of cereal yield. II. Partitioning of dry matter, *Agric. Prog.,* 53, 51, 1975.

66. **Tripathi, R. K., Tandon, K., Schlosser, E., and Hess, W. M.,** Effect of fungicides on the physiology of plants. IV. Protection of cellular organelles of senescent wheat leaves by carbendazim, *Pestic. Sci.,* 13, 395, 1982.

67. **Buchenauer, H., Kohts, T., and Roos, H.,** Zur Wirkungsweise von Propiconazol (Desmel), CGA 64 251 und Diclobutrazol (Vigil) in Pilzen und Gerstenkeimlingen, *Mitt. Biol. Bundesanst. Land Forstwirtsch. Berlin-Dahlem,* 203, 310, 1981.

68. **Skidmore, A. M. and Dickinson, C. H.,** Effect of phylloplane fungi on senescence of excised barley leaves, *Trans. Br. Mycol. Soc.,* 60, 107, 1973.

69. **Ebrahim-Nesbat, F. and Fehrmann, H.,** Electron microscopical evidence of pathogenicity of *Cladosporium herbarum* in wheat leaves, Proc. 4th Int. Congr. Plant Pathology, Melbourne, Australia, 1983, 129.

70. **Dickinson, C. H.,** Interactions of fungicides and leaf saprophytes, *Pestic. Sci.,* 4, 563, 1973.

71. **Jenkyn, J. F. and Prew, R. D.,** The effect of fungicides on the incidence of *Sporobolomyces* spp. and *Cladosporium* spp. on flag leaves of winter wheat, *Ann. Appl. Biol.,* 75, 253, 1973.

72. **Dickinson, C. H. and Wallace, B.,** Effects of late application of foliar fungicides on activity of microorganisms on winter wheat flag leaves, *Trans. Br. Mycol. Soc.,* 67, 103, 1976.

73. **Cook, R. J.,** Effect of timed fungicide sprays on yield of winter wheat in relation to *Septoria* infection periods, *Plant Pathol.,* 26, 30, 1977.

74. **Lescar, L., Bouchet, F., and Faivre-Dupaigre, R.,** Fungicidal control of winter wheat foliage diseases in France, *Proc. 7th Br. Insectic. Fungic. Conf.,* 1, 97, 1973.

75. **Allison, D. A., Peake, R., and Tomlin, J.,** Trials in the UK to evaluate benomyl, carabendazim and DDT mixtures for yield response and control of *Septoria* eyespot and mildew in winter wheat, *Proc. 8th Br. Insectic. Fungic. Conf.,* 1, 305, 1975.

76. **James, W. C., Jenkins, J. E. E., and Jemmett, J. L.,** The relationship between leaf blotch caused by *Rynchosporium secalis* and losses in grain yield of spring barley, *Ann. Appl. Biol.,* 62, 273, 1968.

77. **Jacobson, B. J.,** Effect of fungicides on Septoria leaf and glume blotch, Fusarium scab grain yield and test weight of winter wheat, *Phytopathology,* 67, 1412, 1977.

78. **Fraselle, J.,** Etude de diverses modalités de traitement contre les maladies des parties aeriennes de l'escourgeon. Résultats des essais du réseau realises en 1974, *Parasitica,* 30, 167, 1974.

79. **Hampel, M.,** Zusammenhinge zwischen Getreidezuchtung und Agrarchemie, *BASF Agric. Bull.,* 8, 67, 1975.

80. **Mappes, C. J. and Hampel, M.,** Yield responses in winter barley to late fungicide treatments, *Proc. 1977 Br. Crop Prot. Conf. Pests and Diseases,* 1, 49, 1977.

81. **Skidmore, A. M. and Dickinson, C. H.,** Effect of phylloplane fungi on senescence of excised barley leaves, *Trans. Br. Mycol. Soc.,* 60, 107, 1973.

82. **Diem, H. C.,** Microorganisms of the leaf surface: estimation of the microflora of the barley phyllosphere, *J. Gen. Microbiol.,* 80, 77, 1974.

83. **Mishra, R. R. and Srivastava, V. B.,** Leaf surface microflora of *Hordeum vulgare, Acta Soc. Bot. Pol.,* 43, 203, 1974.

84. **Hill, R. A. and Lacey, J.,** The microflora of ripening barley grain and the effects of pre-harvest fungicides application, *Ann. Appl. Biol.,* 102, 455, 1983.

85. **Edgington, L. V., Khew, K. L., and Barron, G. L.,** Fungitoxic spectrum of benzimidazole compounds, *Phytopathology,* 61, 42, 1971.

86. **Flannigan, B. and Campbell, I.,** Pre-harvest mould and yeast flores on the flag leaves, bracts, and caryopsis, *Trans. Br. Mycol. Soc.,* 69, 485, 1977.

87. **Flannigan, B.,** Primary contamination of barley and wheat grain by storage fungi, *Trans. Br. Mycol. Soc.,* 71, 37, 1978.

88. **Chinn, S. H. F.,** Influence of seed treatment with imazalil on common root rot and the size of the subcrown internode of wheat, *Phytopathology,* 68, 1662, 1978.

89. **Chinn, S. H. F., Verma, P. R., and Spurr, D. T.,** Effects of imazalil seed treatment on subcrown internode lengths and coleoptile node tillering in wheat, *Can. J. Plant Sci.,* 60, 1467, 1980.

90. **Stoker, G. L. and Dewey, W. G.**, Effect of loose smut infection and Vitavax seed treatment on barley yields, *Crop. Sci.*, 9, 586, 1969.

91. **Stoker, G. L. and Dewey, W. G.**, Effect of loose smut infection and Vitavax treatment on barley seed, *Utah Sci.*, 31, 46, 1970.

92. **Jaworski, C. A., Phatak, S. C., and Csinos, A.**, Effects of metalaxyl on phycomycetous fungi and yield of pepper, tomato and cabbage, *J. Am. Soc. Hortic. Sci.*, 107, 911, 1982.

93. **Braun, H. and Schonbeck, F.**, Untersuchungen über den Einfluß verschiedener Pflanzenschutzpraparate auf die Befruchtung von Apfelbaumen, *Erwerbsobstbau*, 7, 26, 1965.

94. **Kaspers, H.**, Sind Blutenspritzungen mit organischen Fungiziden im Kernobstbau bedenklich, *Erwerbsobstbau*, 7, 28, 1965.

95. **Liebster, G.**, Ergebnis zwei jähriger Untersuchungen über den Einfluß während der Blute ausgebrachter Fungizide auf Junifruchtfall und Fruchtbehang beim Apfel, *Z. Pflanzenr. Pflanzenschutz.*, 72, 325, 1965.

96. **Donoho, C. W.**, Influence of pesticide chemicals on fruit set, return bloom, yield and fruit size of the apple, *J. Am. Soc. Hortic. Sci.*, 85, 53, 1964.

97. **Dancs, Zs. and Kiss, A.**, Effect of fungicides applied at the time of flowering on the pollen germination and fruit set of Jonathan apple trees, *Agron. Acad. Sci. Hung. Tomus*, 19, 313, 1970.

98. **Robinson, R. W. and Hodges, C. F.**, Effect of benomyl on eradication of *Ustilago striiformis* from *Agrostis palustris* and on plant growth, *Phytopathology*, 62, 533, 1972.

99. **Zsoldos, F.**, Uptake and efflux of ions in fungicide-treated rice plants, *Plant Soil*, 41, 41, 1974.

100. **Rich, A. E.**, Effects of various fungicides applied during bloom on apple pollination and fruit set, *Agric. Chem.*, 12(6), 64, 1957.

101. **Shawa, A. Y., Doughty, C. C., and Johnson, F.**, Effect of fungicides on "McFarlin" cranberry pollen germination and fruit set, *Proc. Am. Soc. Hortic. Sci.*, 89, 255, 1966.

102. **Lockhart, C. L.**, Effect of fungicides on germination of lowbush blueberry pollen and on number of seeds per berry, *Can. Plant. Dis. Surv.*, 47, 72, 1967.

103. **McDaniels, L. H. and Furr, J. R.**, The effect of dusting sulphur upon the germination of pollen and the set of fruit of apple, *Cornell Agric. Exp. Bull.*, 13, 499, 1930.

104. **Eaton, G. W.**, Germination of sweet cherry *(Prunus aviume)* pollen in vitro as influenced by fungicides, *Can. J. Plant Sci.*, 1, 740, 1961.

105. **Schmidt, T.**, A turfgrass leaf spot disease of Irises, *Pflanzenarzt*, 9(8), 78, 1965.

106. **Braun, H. and Schonbeck, F.**, Studies on the effect of plant protection preparation on the germination of apple and pear pollen, *Erwerbsobstbau*, 5(9), 170, 1965.

107. **Schafer, H.**, Blutenspritzungen im Obstbau, *Obst. Weinb.*, 101, 258, 1965.

108. **Minoiu, N., Alexander, A. V., Baicu, T., and Marin, A.**, Cereetasi cu privire la efectul produselor fitofarmacentice aplicate in timpul inflorituleu'la vita de vie si pomii fructiferi, *An. Sect. Prot. Plant Inst. Cent. Cercet. Agric. Bucharest*, 1, 255, 1965.

109. **Eaton, G. W.**, Germination of apple pollen as influenced by captan sprays, *Proc. Am. Soc. Hortic. Sci.*, 83, 101, 1963.

110. **Eaton, G. W. and Chen, L. I.**, The effect of captan on strawberry pollen germination, *J. Am. Soc. Hortic. Sci.*, 94, 558, 1969.

111. **Legge, A. P. and Williams, R. R.**, Adverse effects of fungicidal sprays on the pollination of apple flowers, *J. Hortic. Sci.*, 35, 190, 1975.

112. **Church, R. M. and Williams, R. R.**, The toxicity of apple pollen of several fungicides as demonstrated by in vitro and in vivo techniques, *J. Hortic. Sci.*, 52, 429, 1977.

113. **Burth, U. and Ramson, A.**, Über den Einfluß organischen Fungizide auf die generative, Leistung von Apfelbaumen, *Arch. Gartenbau*, 22, 239, 1974.

114. **Church, R. M. and Williams, R. R.**, Fungicide toxicity to apple pollen in the anther, *J. Hortic. Sci.*, 53, 91, 1978.

115. **Shawa, A. Y., Doughty, C. C., and Johnson, F.**, Effect of fungicides on McFarlin cranberry pollen germination and fruit set, *J. Am. Soc. Hortic. Sci.*, 89, 255, 1967.

116. **Shivanna, K. R.**, Effect of nonionic surfactants on pollen germination and pollen tube growth, *Curr. Sci.*, 41, 609, 1972.

117. **Butt, D. J., Kirby, A. H. M., and Willianum, C. J.**, Fungitoxic and phytotoxic effects of fungicides controlling powdery mildew on apple, *Ann. Appl. Biol.*, 75, 217, 1973.

118. **Kirby, A. H. M., Butt, D. J., and Williams, C. J.**, Trials in 1968 of new fungicides on apple and pear, *East Malling Res. Stn. Maidstone Engl. Rep.*, p. 171, 1970.

119. **Dalbro, K.**, Lagttagelser og forsog vedr. Skrubben frugt og bladpletter hos Cox's Orange, *Tidsskr. Planteavl*, 62, 112, 1958.

120. **Mitchell, A. E. and Kretchman, D. W.**, Relationship of spray chemicals and cell injury to russeting of Golden Delicious apples, *Trans. Ill. State Hortic. Soc.*, 91, 43, 1958.

121. **Walter, T. E.**, Russeting and cracking in apples; a review of world literature, *East Malling Res. Stn. Maidstone Engl. Rep.*, p. 83, 1967.

122. **Faust, M. and Shear, C. B.,** Russeting of apples, an interpretive review, *Hortic. Sci.,* 7, 233, 1972.
123. **Anon.,** Statutory Grading. Guides to the Graders, No. 2: Apples, Ministry of Agriculture, Fisheries and Food, London, 1972, 1.
124. **Butt, D. J. and Graham, J. E.,** Effects of apple mildew spray programme on the control of fruit free red spider mite, *Plant Pathol.,* p. 22, 1973.
125. **Byrde, R. J. W., Hutcheon, J., and Harper, C. W.,** Side effects of benomyl: effect on apple yield, *Long Ashton Res. Stn. Univ. Bristol Rep. for 1975,* p. 123, 1976.
126. **Strydom, D. K. and Honeyborne, G. E.,** Increase in fruit set of Starking Delicious apple with triadimefon, *Hortic. Sci.,* 16, 51, 1981.
127. **Church, R. M. and Williams, R. R.,** The effects of pre-blossom fungicide sprays on the ability of Cox's orange Pippin apple flowers to produce fruit, *J. Hortic. Sci.,* 58, 169, 1983.
128. **Church, R. M., Copas, L., and Williams, R. R.,** Changes in fruit set, leaf size and shoot growth of apple caused by some fungicides, insecticides and a plant growth regulator, *J. Hortic. Sci.,* 59, 161, 1984.
129. **Chollet, P.,** Repercussion das traitements fongicides appliques lors de les preflorances de la floration du poirier sur la fructification, *Sci. Agron. Rennes,* 22, 231, 1970.
130. **Abbott, D. L.,** The bourse shoot as a factor in the growth of apple fruits, *Ann. Appl. Biol.,* 48, 434, 1960.
131. **Struble, F. B. and Morrison, L. S.,** Control of apple blotch with fungicides, *Plant Dis. Rep.,* 45, 441, 1961.

Chapter 5

MECHANISM OF PHYTOTONIC EFFECTS OF FUNGICIDES ON PLANTS

I. INTRODUCTION

Many theories have been advanced to explain unexpected nontarget effects of fungicides on plants and plant organs. Any explanation of these apparently phytotonic effects of fungicides on plants must at present be largely speculative, but there are at least three ways in which they might be produced:[1-4]

1. The fungicide per se may directly affect physiology and growth.
2. Control of weak pathogens (even small amounts of recognized diseases) may at critical times have large effects on yield.
3. Leaf surface microflora may contribute to producing these effects.

II. DIRECT EFFECTS

Direct nontarget effects of fungicides on plant growth are difficult to assess in pot and field experiments which have previously been carried out. When a dithiocarbamate is involved, it has been suggested that the associated metal ion can alleviate a deficiency of this element in the soil.[4] In another case, Griffiths[5] reported that yield of coffee can be increased by copper sprays under disease-free conditions by 50 to 100%. Since yield increases of this magnitude could not be ignored, tonic spraying became a standard practice. However, the mechanism involved is still unclear.[6]

Carbendazim and carbendazim-generating compounds such as benomyl and thiophanate-methyl have as part of their chemical structure the benzimidazole molecule that resembles kinetin (Structures 1 and 2).

BENZIMIDAZOLE

KINETIN

STRUCTURE 1   STRUCTURE 2

Thus, they have cytokinin-like (6-furfuryl aminopurine) activities in plants.[7] Benomyl breaks down to carbendazim in aqueous solutions and plants,[8] but limited evidence is available that carbendazim may break down further in plants.[9] The fate of carbendazim in plants is of interest because the benzimidazole portion of the compound which resembles kinetin has been shown to increase protein synthesis[10] and to maintain chlorophyll biosynthesis in detached leaves.[11,12] The fungicide has certain growth-promoting effects in cereals such as wheat[13-15] and barley.[16] When detached leaves of wheat were floated in benzimidazole solution, not only was breakdown of proteins and chlorophyll prevented but the leaves were resistant to rust infection.[17-19] Mukhopadhyaya and Bandopadhyaya[20] reported cytokinin-like activity of benomyl and carbendazim in wheat leaves in cultivar RR-21 at 10, 25, 50, and 100 ppm; they observed that detached wheat leaves when floated in the carbendazim solution retained a greater amount of chlorophyll than the control. They also observed that carbendazim was approximately ten times less active in preserving green leaf than cytokinin, but that it delayed senescence of the chloroplast, thereby stimulating the growth of wheat. Carbendazim 1 μg mℓ^{-1} preserved chlorophyll and protein and stabilized cellular membranes, similar to the effect of kinetin.[21] Earlier, Richmond and Phillips[22] reported that carbendazim stabilized the cell membrane. Thus, the leakage of electrolytes, including nucleotides and amino acids, accompanying senescence was prevented. Staskawicz et al.[23] have confirmed the observation of Mukhopadhyaya and Bandopadhyaya[20] and demonstrated substantial (80 to 96%) chlorophyll retention in oat 3 days after treatment with 10 and 100 μg mℓ^{-1}. This is well below the level of commercial application. Carbendazim prevented the loss of chlorophyll and protein to a large extent and RNA slightly, but there was no effect on DNA content. Mukhopadhyaya and Bandopadhyaya[20] and Peat and Shipp[14] reported that carbendazim and benomyl, when applied to wheat, behave as weak cytokinins. Therefore, to achieve proper cytokinin-like activity, repeated applications should be made so that they may partly compensate for the weak activity under field conditions and thus considerably improve the quality and quantity of the crop yields. These responses in which chlorophyll retention is prolonged could be associated with the activity of cytokinin; the compounds with cytokinin-like activity are related to adenine which is believed to account for their antisenescent behavior.[21,24-28]

The mechanism by which fungicides delay senescence is unclear. The fungicides modify ethylene production; if the inhibition of senescence is related to ethylene production,[26] it is possible that the effects of fungicides on respiration and senescence are also related. Some of the reported biochemical effects of cytokinin on plant systems include stimulation of cell division and elongation, breaking of dormancy, delay of senescence accompanied by prevention of chlorophyll, protein and nucleic acid degradation, abolition of apical dominance, and induction of enzymes.[29] Peat and Shipp[14] also suggested that the effects may include changes in the normal pattern of apical dominance, which allows more rapid lateral growth and thus increases yield. Some of these properties, such as delay of senescence and breaking of seed dormancy, have been reported for benomyl and carbendazim.[25] These fungicides are active in typical cytokinin bioassay, e.g., soybean callus bioassay, radish cotyledons test,[24] and *Amaranthus* betacyanin test.[25] Cook[30] reported an increase in wheat yield after late application of fungicides such as carbendazim and benomyl. This might therefore be a response to delayed senescence of green leaves and consequently longer photosynthesis. However, yield increases were also noted following captafol treatment. The fungicide is nonsystemic and no effect on host metabolism is known.

Tripathi et al.[32] also studied the effect of carbendazim on the ultrastructure of cell organelles of leaves of wheat. They observed in induced senescent leaves that the organelles were completely disorganized and the contents aggregated; however, carbendazim treatment protected the cell organelles from disorganization. The intact cell wall, plasmalemma, chloroplasts, endoplasmic reticulum, ribosomes, and mitochondria were clearly seen in cells of treated leaves and the nucleus was also normal, without any evidence of disorganization. Benzimidazole, of which carbendazim is a derivative, has also been shown to protect the chloroplasts in wheat leaves.[33] The cellular organelles of leaves, floated on a solution of carbendazim containing 100 μg mℓ^{-1}, were partially or completely disorganized. Preservation of cellular organelles, including membrane structures, by carbendazim may explain why the fungicide checks the loss of chlorophyll and inhibits the efflux of ions and metabolites. It seems that the maintenance of membrane integrity is one of the major mechanisms underlying the antisenescent action of carbendazim.[31]

In a physiological study on the mechanism of the antisenescent activity of carbendazim on wheat leaves, it was observed that the fungicide inhibits the degradation of chlorophyll, RNA, and protein and checks the increasing activity of protease, ribonuclease, and esterase that accompanies senescence in leaves.[31] In a further study, Tripathi et al.[32] studied the mechanism of antisenescent properties of carbendazim: leaf segments floated on water and incubated in the dark turned yellow in 96 hr, whereas those on the carbendazim solution (20 μg mℓ^{-1}) remained green for up to 2 weeks. However, a higher concentration of 100 μg mℓ^{-1} was phytotoxic. The induced senescence of leaves leads to an efflux of ions and amino acids into the ambient solution; however, carbendazim (20 μg mℓ^{-1}) and kinetin (5 μg mℓ^{-1}) protected the leaves from this efflux and decreased the conductivity of solutions by 70 and 62%, respectively. Carbendazim and kinetin also decreased the efflux of amino acids by 64 and 71%, respectively. At 100 μg mℓ^{-1} (the phytotoxic concentration) carbendazim increased the efflux of ions and amino acids by more than 30% over water, suggesting disorganization of plasma membrane and possibly other membranes as well. Membrane-stabilizing effects of carbendazim and kinetin have also been observed in cabbage and cucumber.[21]

Seeds of most celery cultivars possess dormancy and require a light stimulus to germinate.[34] Several methods are being employed to overcome dormancy. One is the use of mixtures of gibberellins A_4 and A_7 ($GA_{4/7}$) with other plant growth regulators, particularly cytokinin, which are effective in breaking celery seed dormancy.[35] However, the use of cytokinins on celery seeds is not a practical proposition, for, although these growth regulators break dormancy, subsequent radicle growth of the seedlings is inhibited. Benomyl and BAS 3460 F stimulated the germination of celery seeds in the presence of $GA_{4/7}$, such seeds having been shown previously to respond to either the cytokinin or the growth retardant (0.2% of the formulated solution was taken and celery seeds were dipped for 24 hr at 22°C).[35] Thomas[34] treated celery seeds with benomyl and BAS 3460 F and observed that the germination was stimulated in the presence of $GA_{4/7}$ whereas thiabendazole was ineffective; the optimum concentration of benomyl was in the range of 50 to 500 mg ℓ^{-1} but some reduction of germination at concentrations above 100 mg ℓ^{-1} was apparent in some cultivars tested. The optimum concentration of carbendazim was difficult to establish since there appeared to be two germination maxima: in the absence of $GA_{4/7}$ the fungicides had no effect on germination of any cultivars. It was suggested that benomyl and carbendazim were active in these tests because they both possess cytokinin-like activity, but that the chemical structure of thiabendazole precludes any growth-regulatory activity even though all three compounds are benzimidazoles. Benomyl is hydrolyzed into carbendazim, whereas in thiabendazole the heterocyclic rings are joined by a carbon-carbon bond and carbendazim is not produced.[36] It has been suggested that the cytokinin-like activity of benomyl may be derived from carbendazim produced by breakdown,[34] but since different patterns of effects were obtained in the germination tests from benomyl and carbendazim, it seems more likely that the parent compound itself may possess true cytokinin-like activity. The practical implication of these results is that mixtures of $GA_{4/7}$ and some, but not all, benzimidazole-derived fungicides might be used as dual-purpose seed treatments to stimulate germination and to eradicate seed-borne diseases.

In later studies, a range of benzimidazole compounds were tested for their cytokinin-like activity[25] against celery seeds. The results showed that in addition to benomyl and carbendazim,[34] thiophanate was slightly effective over the range of concentrations. Thiophanate-methyl stimulated germination in the presence of $GA_{4/7}$ and in general the percentage of germination increased with increased fungicide concentrations. Fuberidazole did not stimulate germination, and at the highest concentration it was toxic; benomyl and NF 48 produced the highest germination response from seeds of both cvs. Pascal and Latham Blanching. However, these cultivars responded differently to the other cytokinin-active fungicides, thiophanate-methyl being more effective than carbendazim on cv. Pascal and carbendazim being more active (particularly at lower concentrations) on cv. Latham Blanching.[25] The relative cytokinin-like activity of the seven fungicides (benomyl, carbendazim, thiabendazole, thiophanate, thiophanate-methyl, NF 48, and fuberidazole) in the *Amaranthus* beta-cyanin bioassay was studied.[25] Thiabendazole, fuberidazole, and thiophanate showed little cytokinin-like activity whereas the others were active. Sijpesteijn[1] suggests that the different fungitoxic efficiencies of these three compounds may be due to their different rates of conversion into carbendazim, the rates of conversion of each compound being different and to some extent dependent on pH and their different rates of uptake into the plant. Hence, the difference in cytokinin-like activity of these fungicides could also be explained in the same way, since carbendazim itself was active in both the *Amaranthus* bioassay and in the seed germination tests. Neither thiabendazole nor fuberidazole was cytokinin-active and these compounds are not precursors of carbendazim; thiophanate stimulated seed germination only slightly and this low activity could be explained in terms of the lower cytokinin-like activity of its converted product, ethyl benzimidazole carbamate, which is less fungitoxic than carbendazim and also possesses lower cytokinin-like activity than carbendazim.[25] It was suggested that the hormonal properties of the cytokinin-active fungicides may not be due to carbendazim alone because sometimes they were more active than pure carbendazim.

Gross and Kenneth[37] proposed fungicide-induced hormonal imbalance as another possible mechanism for the observed effect.

III. CONTROL OF WEAK PATHOGENS

Control of plant disease at critical stages of plant development, if amount of disease is small, can influence yield potential. It is known that infection by aerial pathogens early in the development of crops can impair root development,[38] cause several diseases, reduce photosynthetic area, and change growth-regulatory activity. Such changes could be of great significance if they coincided with inflorescence differentiation, when they might affect yield potential. Such a possibility is of particular interest in relation to powdery mildew control experiments discussed by Griffiths and Scott.[39] They used benomyl, ethirimol, drazoxolon, and tridemorph and sprayed at growth stages 5 to 6 when powdery mildew was present in negligible amounts. These fungicides were very effective in controlling powdery mildew and also increased the numbers of grain per tiller. If mildew were also capable of restricting the development of inflorescence primordia, this would not have been true.[39] Evidently, effects of minor attacks of foliar diseases during early growth stages merit more detailed attention.

IV. ROLE OF LEAF SURFACE MICROFLORA

The possible effects of phyllosphere microorganisms other than recognized pathogens cannot be overlooked. It is now established that the surface of leaves is colonized by specialized microflora which increase in abundance as leaves mature and senesce.[39] The cultivars of the same plant also support a variety of microflora.[40] These microflora have been identified on many plant species[41] and their effect on plant growth has been recognized.[42] Studies[39,42,43] have, however, indicated that phylloplane organisms may accelerate senescence, that fungicide treatments both reduce the numbers of organisms and may accelerate senescence, and that fungicide treatments reduce the number of organisms present and prolong the functional life of leaves. These observations could well explain the results obtained by Jenkins et al.[43] in their experiments. Fungicidal treatments in the absence of disease increased the green leaf area (GLA); postanthesis and total grain yield were increased through the formation of larger grains.

Yield increases after treatment with fungicides could therefore be a response to a reduction in the mycoflora, which are weak pathogens. A further possibility is that fungicides control leaf surface mycoflora which affect the rate of senescence. Experiments have shown that benomyl and thiophanate-methyl reduce the population of these mycoflora[44-46] and some species affect the rate of senescence when inoculated into detached barley leaves.[42]

The influence of microorganisms on the photosynthetic activity of green tissues may be through their pathogenic activity, their saprophytic activity on the surface of the tissue, or their influence on senescence processes. The inhibition of both pathogen and saprophytes by fungicides may therefore effect a more substantial increase in photosynthetic activity beyond that which could be expected from the control of disease alone.[47] Microscopic examination of winter wheat treated late with the fungicides benomyl (1.12 kg ha^{-1}), zineb (4.48 kg ha^{-1}), and tridemorph (0.5 kg ha^{-1}) showed that fungicides had severely restricted the growth of a wide range of saprophytic fungi; the growth of such fungi has been shown[42] to result in an acceleration of senescence in detached barley leaf pieces in vitro. Previous work has shown that the flag leaf is the most important single organ in supplying the developing grain with assimilate.[48] The prolongation of the active life of the leaf, particularly during the latter stages of grain filling, as achieved by fungicide can be expected, therefore, to largely account for the yield increases recorded.[49] The photosynthetic area duration of the ear did not appear to respond to the same extent to fungicide treatments, but field assessments of this aspect are at present crude. These results suggest that fungicide spraying in field crops prior to harvest could produce better quality seed which would fetch higher prices, due to the control of organisms responsible for head blackening. The effects of fungicide spraying, then, should be figured into any yield increase in cost return analysis of the use of fungicides for plant disease control of field crops. Hill and Lacey[50] also reported that application of fungicides at or after anthesis on the ripening barley modified the grain microflora and yield increases of up to 4% were obtained by late application of fungicides. Among the fungicides used were benodanil, benomyl, captafol, and tridemorph at operational dosages. However, in the case of treatment with tridemorph, only minor effects on the microflora were seen.[50,51] Yield increases were observed by Mappes and Hampal[51] but not by Hill and Lacey.[50]

Foliar application of fungicides effected an increase in grain yield ranging from 18 to 43%,[2] depending on the fungicides used. The effect of spraying was an increase in the persistence of GLA. Significant increase in 1000 seed weight was the major source of the increased yield in wheat after foliar application of carbendazim.[30,52] This was also shown by Lescar et al.[53] and Allison et al.,[54] and Melville and Jemmette,[2] Spiertz,[55] and Cook[30] showed an increase with carbendazim of 1.9% in 1000 grain weight of wheat over unsprayed plants. The grain weight in cereals is largely determined by the extent and duration of green tissue after emergence and in wheat the flag leaf is especially important in this respect.[48] Melville and Jemmette[2] reported that effective spray treatment approximately doubled the area and duration of the green leaf tissue of the flag leaf after emergence as compared to controls. The prolongation of growth due to the sprayed treatments was very noticeable in the field, for the plant remained green for 1 to 2 weeks longer than the controls. At maturity, the sprayed plants showed brighter yellow and cleaner straw and the ears were much less heavily colonized by secondary molds such as *Cladosporium* and *Alternaria*.

Leaf surface or phyllosphere microflora accelerate leaf senescence. They have long been known to be important in the production of growth regulators[39] and it is likely, at least under some conditions, that they have effects on plant growth and differentiation. If this should prove true, fungicidal treatments, particularly systemic ones which penetrate plant tissues, would also be expected to have effects on growth and yield.[39]

The yield responses recorded here may therefore be the result of complex interactions between fungicides and hosts. It is possible that, besides controlling disease, fungicides influence the plant characteristic discussed above, such as the physiology of the host, nutritional requirements, and phyllosphere mycoflora which modify senescence and thus crop yields.

REFERENCES

1. **Sijpesteijn, A. K.**, Effects on fungal pathogens, in *Systemic Fungicides*, 2nd ed., Marsh, R. W., Ed., Longman, New York, 1977, 131.
2. **Melville, S. C. and Jemmett, J. L.**, The effect of glume blotch on the yield of winter wheat, *Plant Pathol.*, 20, 14, 1971.
3. **Griffiths, E.**, Role of fungicide application on cereals in France, *EPPO Bull.*, 11, 347, 1981.
4. **Dickinson, C. H.**, Interactions of fungicides with minor pathogens on cereals, *EPPO Bull.*, 11, 311, 1981.
5. **Griffiths, E.**, Negative effect of fungicides on coffee, *Proc. 6th Br. Insectic. Fungic. Conf.*, 3, 817, 1971.
6. **Griffiths, E.**, Iatrogenic plant diseases, *Annu. Rev. Phytopathol.*, 19, 69, 1981.
7. **Schruft, G.**, Die Wirkung des Botrytis Praparates Benomyl in Amaranthus Cytokinin-Test, *Wein-Wiss.*, 25, 329, 1970.
8. **Peterson, C. A. and Edgington, L. V.**, Quantitative estimation of the fungicide benomyl using a bioautograph technique, *J. Agric. Food Chem.*, 17, 898, 1969.
9. **Peterson, C. A. and Edgington, L. V.**, Transport of the systemic fungicide benomyl into bean plants, *Phytopathology*, 60, 475, 1970.
10. **Kapoor, M. and Waygood, E. R.**, Metabolism of benzimidazole in wheat. I. Formation of benzimidazole nucleotide, *Can. J. Biochem.*, 43, 153, 1965.
11. **Wang, D., Hao, M. S., and Waygood, E. R.**, Effect of benzimidazole analogues on stem rust and chlorophyll content, *Can. J. Bot.*, 39, 1029, 1961.
12. **Wang, D. and Waygood, E. R.**, Effect of benzimidazole on the chlorophyll metabolism of detached leaves of Khapli wheat, *Can. J. Bot.*, 37, 743, 1959.
13. **Langer, R. H. M., Prasad, P. C., and Laude, H. M.**, Effects of kinetin on tiller bud elongation in wheat (*Triticum aestivum* L.), *Ann. Bot. (London)*, 37, 565, 1973.
14. **Peat, W. E. and Shipp, D. M.**, The effects of benomyl on the growth and development of wheat, *EPPO Bull.*, 11, 287, 1981.
15. **Priestley, R. H.**, Fungicide treatment increases yield of cereal cultivars by reducing diseases and delaying senescence, *EPPO Bull.*, 11, 357, 1981.
16. **Williams, R. H. and Cartwright, P. M.**, The effect of application of synthetic cytokinin on shoot dominance and grain yield in spring barley, *Ann. Bot. (London)*, 46, 445, 1980.
17. **Person, E., Samborski, D. J., and Forsyth, F. R.**, Effect of benzimidazole on detached leaves, *Nature (London)*, 180, 1294, 1957.
18. **Samborski, D. J., Forsyth, F. R., and Person, E.**, Metabolic changes in detached wheat leaves floated on benzimidazole and the effect of these changes on rust reaction, *Can. J. Bot.*, 36, 591, 1958.
19. **Mishra, D. and Samal, B.**, Retardation of senescence in rice-leaves by benzimidazole and nickel chloride, *Proc. Indian Natl. Sci. Acad. Part B*, 40, 100, 1974.
20. **Mukhopadhyaya, A. N. and Bandopadhyaya, R.**, Cytokinin like activity of carbendazim, *Pesticides*, 11(7), 24, 1977.
21. **Tripathi, R. K. and Schlosser, E.**, Effect of fungicides on the physiology of plants. I. Effect of carbendazim on biochemical changes and membrane permeability of senescing cabbage and cucumber leaves, *Meded. Fac. Landbouwwet. Rijksuniv. Gent*, 42, 1073, 1977.
22. **Richmond, D. V. and Phillips, A.**, The effect of benomyl and carbendazim on mitosis in hyphae of *Botrytis cinera* and *Allium cepa*, *Pestic. Biochem. Physiol.*, 5, 367, 1975.
23. **Staskawicz, B., Khar-Sawhney, R., Slaybaugh, R., Adams, W., and Galston, A. W.**, The cytokinin-like action of methyl-benzimidazole-2yl-carbamate on oat leaves and protoplasms, *Pestic. Biochem. Physiol.*, 8, 106, 1978.
24. **Skene, K. G. M.**, Cytokinin-like properties of the systemic fungicide benomyl, *J. Hortic. Sci.*, 47, 179, 1972.
25. **Thomas, T. H.**, Investigations into cytokinin-like properties of benzimidazole derived fungicides, *Ann. Appl. Biol.*, 76, 237, 1974.
26. **Dimond, A. E. and Rich, S.**, Effect of physiology on the host and on host/pathogen interactions, in *Systemic Fungicides*, 2nd ed., Marsh, R. W., Ed., Longman, London, 1977, 115.
27. **Ellen, J. and Spiertz, J. H. J.**, The influence of nitrogen and Benlate on leaf area duration, grain growth and pattern of N, P and K-uptake on winter wheat (*Triticum aestivum*), *Z. Acker Pflanzenbau*, 141, 231, 1975.
28. **Dickinson, C. H. and Walpole, P. R.**, The effect of late application of fungicides on the yield of winter wheat, *Exp. Husb.*, 29, 23, 1975.
29. **Hess, D.**, *Plant Physiology*, Springer-Verlag, Berlin, 1975.
30. **Cook, R. J.**, Effects of late season fungicide sprays on yield of winter wheat, *Plant Pathol.*, 29, 21, 1980.
31. **Tripathi, R. K., Vohra, K., and Schlosser, E.**, Effect of fungicides on the physiology of plants. III. Mechanism of cytokinin-like antisenescent action of carbendazim, *J. Plant Dis. Prot.*, 87(10/11), 631, 1980.

32. **Tripathi, R. K., Tandon, K., Schlosser, E., and Hess, W. M.,** Effect of fungicides on the physiology of plants. IV. Protection of cellular organelles of senescent wheat leaves by carbendazim, *J. Plant Dis. Prot.,* 89, 1982.
33. **Waygood, E. R.,** Benzimidazole effect in chloroplast of wheat leaves, *Plant Physiol.,* 40, 1242, 1965.
34. **Thomas, T. H.,** Growth regulatory effect of three benzimidazole fungicides on the germination of celery *(Apium graveolens)* seeds, *Ann. Appl. Biol.,* 74, 233, 1973.
35. **Palevitch, D., Thomas, T. H., and Austin, R. B.,** Dormancy-release of celery seed-soak fungicides, *Natl. Veg. Res. Stn. Rep. for 1969,* 105, 1971.
36. **Kirby, A. H. M.,** Progress towards systemic fungicides, *PANS (Pest Artic. News Summ.),* 18, 1, 1972.
37. **Gross, Y. and Kenneth, R.,** The fate of carboxin, benomyl and thiabendazole in seed and soil treated plants as shown by *in vitro* and *in vivo* bioassay on some epiphytic yeast, *Ann. Appl. Biol.,* 73, 307, 1971.
38. **Little, R. and Doodson, J. K.,** The comparison of yields of some spring barley varieties in the presence and absence of mildew and when treated with fungicide, *Proc. 6th Br. Insectic. Fungic. Conf.,* 1, 91, 1971.
39. **Griffiths, E. and Scott, S. W.,** Possible phytotonic effects of fungicides on barley, in *Crop Protection Agents — Their Biological Evaluation,* McFarlane, N. R., Ed., Academic Press, London, 1977, 465.
40. **Gachichi, J. G.,** Studies on Effects of Fungicides on Phylloplane Mycoflora of Groundnut, M.Sc. (Ag) thesis, J. N. Agricultural University, Jabalpur, India, 1983, 1.
41. **Dickinson, C. H. and Preece, T. F.,** *Microbiology of Aerial Plant Surfaces,* Academic Press, London, 1976.
42. **Skidmore, A. M. and Dickinson, C. H.,** Effect of phylloplane fungi on the senescence of excised barley leaves, *Trans. Br. Mycol. Soc.,* 60, 107, 1973.
43. **Jenkins, J. E. E., Melville, S. C., and Jemmette, J. L.,** The effect of fungicides on leaf diseases and yield in spring barley in south-west England, *Plant Pathol.,* 21, 49, 1972.
44. **Jenkyn, J. F. and Prew, P. R.,** The effect of fungicides on the incidence of *Sporobolomyces* spp. and *Cladosporium* spp. on flag leaves of winter wheat, *Ann. Appl. Biol.,* 75, 253, 1973.
45. **Dickinson, C. H. and Wallace, B.,** Effects of late applications of foliar fungicides on activity of microorganisms on winter wheat flag leaves, *Trans. Br. Mycol. Soc.,* 67, 103, 1976.
46. **Fokkema, N. J., van der Laar, J. A. J., Nelis-Blombar, A. L., and Schippers, B.,** The buffering capacity of the natural mycoflora of rye leaves to infection by *Cochliobolus sativus* and its susceptibility to benomyl, *Neth. J. Plant Pathol.,* 81, 176, 1975.
47. **Dickinson, C. H. and Walpole, P. R.,** The effect of late applications of fungicides on yield of winter wheat, *Exp. Husb.,* 29, 23, 1975.
48. **Thorne, G. N.,** Physiological aspects of grain yield in cereals, in *The Growth of Cereals and Grasses,* Milthorpe, F. L. and Ivins, J. D., Eds., Butterworths, London, 1966, 88.
49. **Walpole, P. R. and Morgan, D. G.,** A quantitative study of grain filling in *Triticum aestivum* L., Cultivar Maris Widgeon, *Ann. Bot. (London),* 34, 309, 1970.
50. **Hill, R. A. and Lacey, J.,** The microflora of ripening barley grain and the effects of pre-harvest fungicide application, *Am. Appl. Biol.,* 102, 455, 1983.
51. **Mappes, C. J. and Hampal, M.,** Yield responses in winter barley to late fungicide treatments, *Proc. 1977 Br. Crop Prot. Conf. Pests and Diseases,* 1, 49, 1977.
52. **Cook, R. J.,** Effect of timed fungicide sprays on yield of winter wheat in relation to *Septoria* infection periods, *Plant Pathol.,* 26, 30, 1977.
53. **Lescar, L., Bouchet, F., and Faivre-Dupaigre, R.,** Fungicidal control of winter wheat foliage diseases in France, *Proc. 7th Br. Insectic. Fungic. Conf.,* 1, 97, 1973.
54. **Allison, D. A., Peake, R., and Tomlinson, J. A.,** Trials in the UK to evaluate benomyl, carbendazim and DDT mixtures for yield response and control of Septoria eyespot and mildew in winter wheat, *Proc. 8th Br. Insectic. Fungic. Conf.,* 1, 305, 1975.
55. **Spiertz, J. H. J.,** Effects of successive applications of maneb and benomyl on growth and yield of five wheat varieties of different heights, *Neth. J. Agric. Sci.,* 21, 282, 1973.

Chapter 6

IATROGENIC DISEASES

I. INTRODUCTION

Control of fungal plant disease (crop protection) is an important facet of modern crop production systems and one which frequently depends on the use of fungicides. The fungicides are, of necessity, compounds with high biological activity which, despite careful selection, tend to have some nontarget effects on organisms or physiological processes. A disturbing consequence has been the occurrence of diseases or abnormal conditions, the causes of which can be traced to the system of crop protection itself. Undesirable side effects, which include the induction of new diseases or, more commonly, the exacerbation of diseases already present, are common. It is observed on occasion that a fungicide allows a particular disease to become more severe or that a second disease occurs soon after a fungicide has been applied to control the initial or target disease. These occurrences are not always recognizable on uniformly treated plants. If recognized, they are not always brought to the attention of scientists. It has been a matter of concern that these results have not previously been reported on.

The incidence of increase of a pathogen or accentuation of disease after the application of a fungicide for the control of a target pathogen has been termed "disease trading" or "boomerang effect". This problem is part of the serious negative consequences of fungicide applications. Kreutzer[1] described the phenomenon in which a dominant pathogen is controlled by fungicidal application and a minor pathogen is stimulated, thus becoming the new dominant pathogen. Diseases which result from or are increased in severity by the use of specific fungicides may be referred to as "iatrogenic", a term derived from human medicine, but, as indicated by Horsfall,[2] equally applicable in plant pathology. Perhaps the classic case of iatrogenic disease is the well-known occurrence of powdery mildew on apple sprayed with captan in the U.S. Yarwood[3] discusses what he calls "manmade diseases". He indicates that diseases can be intensified by the use of fungicides. Examples of such exacerbation are common but information is frequently limited to the observation of diseases enhanced by a particular treatment. The subject has been reviewed by Griffiths and Berrie,[4] Horsfall,[5] and Griffiths.[6]

II. NONSYSTEMIC FUNGICIDES

It seems as if every pesticide since Bordeaux mixture was discovered has been found to stimulate one organism or another.[5] The rapid rise of organic fungicides in the late 1940s was accompanied by a rapid rise in iatrogenic diseases (Tables 1 and 3). Metallic and nonsystemic fungicides such as copper, organotin, and organomercury have relatively nonselective action against microorganisms. However, organic fungicides tend to have a narrower spectrum of activity and affect microorganisms less. They frequently affect one group of fungi more than others; such selectivity was even apparent with inorganic compounds.

Although they might be expected to have a greater impact on the soil microenvironment, they have not been seen to be involved in many undesirable effects on nontarget soil-borne disease organisms.

It has long been known that soil injection with carbon disulfide often effectively controls the rhizomorphs of the soil-borne pathogen *Armillaria mellea*. Bliss[7] showed that this was not always merely a direct effect of the toxicant on the pathogen, but often a reflection of its differential effects on *A. mellae* and the antagonistic fungus *Trichoderma viride,* which

Table 1
EXAMPLES OF IATROGENIC DISEASES DUE TO APPLICATION OF NONSYSTEMIC FUNGICIDES

Fungicide(s)	Target pathogens	Nontarget pathogens	Ref.
Carbon disulfide	*Armillaria mellea*	*Trichoderma viride*	7
Methyl bromide	*A. mellea*	*T. viride*	9, 10
Ethyl mercury phosphate	*Pythium ultimum, Rhizoctonia solani*	Antagonists	11
Copper fungicides	*Puccinia antirrhini*	*Fusarium roseum*	13
Zineb	Foliar diseases	*Botrytis cinerea*	17
Pentachloronitrobenzene (PCNB)	*R. bataticola, Sclerotium rolfsii*	*Pythium* sp., *Fusarium* sp., *Phytophthora* sp.	18
	R. solani	*Pythium* sp.	19, 20
Fenaminosulf, PCNB	*R. bataticola*	*Fusarium* sp., *Pythium* sp.	8
Chlorothalonil	*Cercospora arachidicola, Phaeoisariopsis personata, Puccinia antirrhini*	*Sclerotinia sclerotiorum*	25, 26
Captafol, chlorothalonil	*C. arachidicola*	*S. sclerotiorum*	27

was relatively unaffected. As a result of chemical treatment, the balance between the pathogen and the antagonist was shifted. The indication of this effect is observed when doses required to kill the pathogens must be higher in sterilized soil than in natural soil. In view of these observations, the approach to chemical control treatment of soil-borne pathogens should, rather than kill the pathogen, merely weaken it and render it susceptible to the attack of antagonists.[8] Ohr et al.[9] and Ohr and Munnecke[10] have observed this weakening effect for *A. mellea* after application of methyl bromide at low dosages: the antibiotic production of the pathogen was decreased so that *A. mellea* failed to keep away antagonists.

A little over 30 years ago, Gibson[11] observed that ethylenemercury phosphate, tested at low rates as a soil sterilant, actually increased damping-off in pine seedlings compared to untreated plots. He observed that the fungicide was not stimulatory to the growth of *Pythium ultimum* and *R. solani* and also had no phytotoxic effects on pine seedlings. He concluded that the nontarget increase in disease was brought about by proliferation of the pathogens in the soil environment under decreased competition.

Dieback or gummosis of apricot caused by *Euytpa armeniacae* emerged following the introduction of routine sprays with copper fungicides for the control of shot hole disease *(Clasterosporium carpophilum)*. Carter[12] argued that spraying caused a reduction of surface microflora containing species capable of rapidly colonizing pruning wounds, which are known to sustain *E. armeniacae*.[12] An iatrogenic disease requires integrated biological and chemical control. Copper was also shown to aggravate rust of antirrhinum because *Fusarium roseum* infection in uredial pustules was reduced while *Puccinia antirrhini* was unaffected.[13] The outbreak of coffee berry disease *(Colletotrichum coffeanum)* and rust *(Hemiliea vastatrix)* in Kenya in the 1960s was associated with copper sprays early in the season. Effective control of these diseases can be achieved but only by frequent fungicide application.[14,15] Parodoxically, in unsprayed crops berry disease was present in traces and unimportant, while rust was less damaging.[15,16] Both diseases can be classified as iatrogenic.[6] The application of copper sprays eliminated antagonistic microflora,[16] as in the report of Dimock and Baker[13] on antirrhinum rust. However, a switch to programs of captafol or copper sprays repeated throughout the rainy seasons brought the disease under control.[6]

A relatively minor disease of tomato, gray mold *(Botrytis cinerea)*, suddenly became a disease of great economic importance in the early 1950s. It was concluded that the gray mold increase may have been caused by the increased use of some zineb sprays used to control foliar disease.[17]

Pentachloronitrobenzene (PCNB) is used extensively for the control of soil-borne fungal pathogens such as *Rhizoctonia* spp., *Sclerotium rolfsii*, *S. cepivorum*, and *Sclerotinium* spp. but is ineffective against species of *Pythium*, *Fusarium*, and *Phytophthora*, a fact that has been utilized in the development of selective media for the isolation of these fungi. The high degree of selectivity of PCNB has led to an increase in the incidence and severity of diseases caused by *Fusarium* and *Pythium*, which are used for control of *Rhizoctonia*.[18] An increase in peanut pod rot[19] and damping-off of Pinus seedlings (*Pythium* spp.)[20] was observed after application of PCNB. PCNB reduced damping-off of the two pine species caused by *R. solani* and increased postemergence damping-off caused by a *Pythium* sp. Since PCNB was not fungitoxic to *Pythium*, its nontarget effect was presumed to be due to the suppression of the natural antagonist *Penicillium paxilli*.[20] Root rot of various hosts caused by *Fusarium* and *Helminthosporium* spp. is also reported to be increased by the application of PCNB.[21]

Katan and Lockwood[22] showed that the proportion of fungi which were tolerant of PCNB was greater in plant residues recovered from soil containing PCNB than from soil without the compound. Population-colonizing alfalfa particles increased in the presence of PCNB in soil, whereas those of *Penicillium oxalicum*, *Rhizopus stolonifer*, and *Streptomyces* spp. decreased. The decrease in the quantities of these microorganisms reduced competition and allowed the inoculum density of PCNB-insensitive pathogens to increase. A slightly more complex nontarget effect for PCNB was also reported.[23] An increase was observed in wilt of strawberry *(Verticillium albo-atrum)* and was attributed to an increase of the nematode by PCNB. A more complex reaction is exemplified by the action of chloropicrin and quintozene on *Phytophthora cinnamomi* root rot of pineapple. It was observed that chloropicrin controlled the disease for three consecutive years despite the fact that the pathogen could be detected in the soil only weeks after treatment. By contrast, quintozene, which was no less fungitoxic to *P. cinnamomi* than chloropicrin, increased disease.[24] These results were attributed to the fact that antagonists of *P. cinnamomi*, notably *Trichoderma viride*, were stimulated by chloropicrin but depressed by quintozene. Since *T. viride* is antagonistic to many root pathogens it would be expected that quintozene would exacerbate other root diseases. This seems to be the best evidence for the suppression of hyperparasitism by agricultural chemicals. Hence, work with soil-borne pathogens emphasizes the need for a better understanding of microbial interactions in the soil and the need for crop protection fungicides which reinforce rather than eliminate biological controls.

The exchange of *Rhizoctonia* for *Pythium* after the application of fenaminosulf and the increased severity of *Fusarium* and *Pythium* spp. following the application of the selective fungicide PCNB for the control of *Rhizoctonia bataticola* has been demonstrated.[24] In this situation, control of a sensitive pathogen is followed by the increase of a tolerant pathogen that initially was of minor importance; the effect is also due to different sensitivities among the pathogens themselves or between the pathogens and their specific antagonists.

Another nontarget effect was demonstrated after Sclerotinia blight, caused by *Sclerotinia sclerotiorum*, became a serious disease problem in Virginia peanuts.[25,26] Chlorothalonil, a fungicide currently recommended for control of early *(Cercospora arachidicola)* and late *(Cercosporidium personatum)* leaf spots of peanut, increased Sclerotinia blight and decreased pod yields of peanuts heavily infested with Sclerotinia blight as compared to controls. This effect was noticed with several different doses of chlorothalonil. Plants treated with benomyl had less Sclerotinia blight and greater pod yields than untreated plants. However, these differences were noticed only when excessive rates of benomyl, higher than normally recommended for leaf spot control, were used [4.48 and 6.72 kg active ingredient (a.i.) ha^{-1}].[26] Preliminary evidence provided by Porter[26] points to the development of strains of a third fungus tolerant of the fungicide. Tolerant strains may have created economic concern in areas where blight was not a major disease before the early 1970s.

In another study,[27] Porter reported an increased severity of Sclerotinia blight *(Sclerotinia minor)* in peanuts treated with captafol at the recommended rates for leaf spot control. The increases were similar to those brought about by chlorothalonil, another leaf spot fungicide. The disease severity was similar in all plots at the time of the first fungicide treatment, but at the end of the growing season disease was significantly more severe in plots treated with captafol and chlorothalonil than in untreated control plots or plots treated with benomyl. At the end of the growing season, two and four times more plants were dead in plots treated with chlorothalonil and captafol, respectively. Pod yields and values were significantly lower in plots receiving four applications of captafol and chlorothalonil than in untreated control and benomyl-treated plots.[27] Captafol and chlorothalonil applied to control Cercospora leaf spot may create an imbalance in the microbiological ecosystem under the dense peanut foliage canopy that would favor the growth of *S. minor*. Recently, *S. minor* mycelia insensitive to chlorothalonil developed on culture media treated with this fungicide.[28,29] If fungicide-insensitive mycelia develop after chlorothalonil application, the microbial balance at the soil surface in peanut fields could shift drastically in favor of *S. minor*. A similar phenomenon might occur when captafol is applied to peanuts and could partially account for the increased severity of Sclerotinia blight. The mechanism(s) responsible for these results may be due in part to the production of fast-growing mycelial sectors along the margins of *S. sclerotiorum* colonies growing on the chlorothalonil-treated medium[29] or to the different sensitivities of developmental stages of mycelium of *S. sclerotiorum* to chlorothalonil.[30] Also, the cause of blight enhancement may be related to the suppression of antagonists or competitors of *S. sclerotiorum*, as was recently reported for *Sclerotium rolfsii*.[31]

Application of nonsystemic fungicides on turfgrass increased several nontarget pathogens.[32] It was observed that application of chlorothalonil increased Helminthosporium leaf spot, collar rot, Fusarium blight and patch, and Typhula blight; similarly, Dyrene increased Fusarium blight. Dithanes (zineb and mancozeb) and thiram increased Helminthosporium, collar rot, and Rhizoctonia diseases.[32]

Although there is as yet far too little information available, these field results suggest that conventional broad-spectrum fungicides applied to control plant pathogens may occasionally bring about undesirable nontarget effects manifested by an increase of disease that had not been a problem previously. They can weaken the existing biological controls and stimulate nontarget pathogens. This is not surprising in view of the inherent capacity of conventional fungicides to eliminate beneficial nontarget microflora in soil.

III. SYSTEMIC FUNGICIDES

Several systemic fungicides have been very effective as chemotherapeutic agents for reducing or controlling certain plant diseases otherwise not controlled by conventional fungicides.

Among the systemic fungicides, benzimidazole has often effected an increase of disease by pathogens such as those belonging to the class of phycomycetes and colored spores members of ascomycetes and deuteromycetes. Most research, however, is concerned with foot and root rot diseases; a few concern the increase of leaf spot diseases. Although fungicides are not always directly applied to soil, large quantities of these chemicals reach the soil from the air. Systemic fungicides are pathogen-specific, but a wide variety of nonpathogenic saprophytic fungi are also eliminated, which affects the antagonistic effect that exists in the untreated soil (Tables 2 and 3).

Benzimidazole compounds increase the severity of soil-borne diseases caused by nontarget pythiaceous pathogens. This effect is often dramatically evident where pythiaceous pathogens are already present in soils receiving applications of benzimidazoles.[33] It was reported that cowpea wet stem rot *(Pythium aphanidermatum)* was significantly greater in plots treated with benzimidazole fungicides such as benomyl, carbendazim, and thiabendazole than in

Table 2
EXAMPLES OF IATROGENIC DISEASES DUE TO APPLICATION OF SYSTEMIC FUNGICIDES

Fungicide(s)	Target pathogens	Nontarget pathogens	Ref.
Benomyl, thiabendazole, carbendazim	*Pythium aphanidermatum*	Antagonists	33
	Pythium spp.	Antagonists	34
	P. splendens, P. myriotylum	Antagonists	35, 36
Benomyl			
Peanut	*Cercospora arachidicola, Phaeoisariopsis personata*	*Sclerotium rolfsii*	31
Wheat	*Septoria nodorum, Epicoccum* sp.	*Drechslera* sp. *Alternaria* sp.	44
Pears	*Penicillium digitatum*	*Alternaria* sp.	45, 46
Cotton	*Thielaviopsis basicola, Rhizoctonia solani*	*Pythium* spp.	66
Wheat	*Fusarium* spp., *Cercosporella herpotrichoides*	*Rhizoctonia cerealis*	69
Metalaxyl (brassica)	*Peronospora brassicae, Albugo vandida*	*Alternaria brassicae*	67
Metalaxyl	*Phytophthora citrophilla*	*Penicillium digitatum*	68
Ethirimol (wheat)	*Erysiphe graminis*	*Helminthosporium sativum*	70
			71

Table 3
IATROGENIC EFFECTS OF SOME SOIL FUNGICIDES

Fungicide	Pathogen controlled	Pathogen increased
Etridiazole	*Pythium, Rhizoctonia,* and *Fusarium* spp.	*Phytophthora* sp.
Benomyl	*Rhizoctonia* and *Fusarium* spp.	*Pythium* and *Phytophthora* spp.
Fenaminosulf	*Pythium* spp.	*Phytophthora* spp.
Ferbam	*Rhizoctonia* and *Sclerotium* spp.	*Phytophthora* and *Pythium* spp.

plots treated with nonbenzimidazoles and controls. Of the three benzimidazoles, benomyl and carbendazim produced the highest nontarget increase in stem rot. The growth of *P. aphanidermatum* in corn meal agar was unaffected by the addition of up to 250 μg mℓ^{-1} on benzimidazole fungicides; a positive correlation between the use of benzimidazoles and increased incidence of *P. aphanidermatum* was observed. In some bulb crops, application of benomyl increased damage by *Pythium* spp.[34] Application of thiabendazole and benomyl increased *Pythium* spp., such as *P. splendens* and *P. myriotylum* on caladium.[35,36] Increase of rot was probably due to increased activity of *Pythium* spp. brought about by the suppression of nontarget antagonistic microorganisms in the agrosystem.

Many diseases affecting aerial plant parts, especially perennial and annual plants, are apparently inhibited by competitors, antagonists, or hyperparasites.[8,35,36] Where such biological control is operative as in soil, it is a reasonable expectation that fungicides will, at least at times, create problems rather than solve them.[6] This is so in the examples discussed here. They are not particularly well known but they illustrate exceptionally well many important features of iatrogenic diseases. Botrytis rot of cyclamen corms was initially controlled effectively by benomyl (500 ppm dip); however, continuous use caused the disease to be more severe on treated plants than controls.[37] The reason for this was the development of resistance in a strain of *Botrytis cinerea*. This alone, however, could hardly account for the results.[6,37] Bollen[37] further observed that the disease on treated plants was no worse than on untreated, on which two strains of *Penicillium brevicompactum* were found which were not only resistant to benomyl but also antagonistic to the fungus. *P. brevicompactum* checked the infection of corms by *B. cinerea,* although it did not completely suppress it. Use of benomyl effectively controlled both fungi; however, emergence of benomyl-tolerant strains of *B. cinerea* in the absence of the naturally occurring antagonist resulted in more severe infection of Botrytis. Subsequent adaptation of *P. brevicompactum* to benomyl returned the plants to their original condition.

Fungicide applied for the control of foliar diseases may induce significant changes in the soil environment leading to an altered relationship between soil-borne pathogens and the host. Benomyl was used for the control of early and late leaf spots of groundnut. Its use resulted in increased losses in yield due to southern stem blight *(Sclerotium rolfsii).*[29] However, benomyl afforded complete control of leaf spots when applied at 14-day intervals at 0.39 kg ha^{-1} throughout the season. Maintenance of a virtually intact peanut canopy for this resupinate plant resulted in at least three major changes to the ecology of soil-borne fungi: (1) leaves on the soil surface served as organic food for *S. rolfsii,* (2) fungicide was prevented from reaching the soil by the umbrella effect of the intact canopy, and (3) an altered subcanopy environment was created which was favorable to the development of soil-borne pathogens.[29]

Effects on pathogen/pathogen and pathogen/saprobic interactions induced by benzimidazoles have been due to broad-spectrum yet selective activity that promotes the action of *P. aphanidermatum* by suppression of competition and antagonists. Tolerance of this fungus to benzimidazoles has been reported.[39,41] According to Siegel,[42] the cause of the stimulation is not the direct effect on the pythiaceous fungal pathogens themselves, but possibly the effect on other fungi and actinomycetes existing in the soil that may be antagonistic to either parasites or competition with the pythiaceous pathogens. Simultaneously, monitoring of leaf spot development by *Cochliobolus sativus* and colonization by saprophytic fungi on benomyl-treated and untreated rye plants in the field provided evidence on the antagonistic capacity of the naturally occurring saprophytic mycoflora; benomyl reduced the saprophytic population tenfold, which was reflected in an increase of infection of rye leaves when inoculated with benomyl-insensitive *C. sativus.*[43] Population of *Drechslera* and *Alternaria* increased on wheat kernals, whereas *Septoria* and *Epicoccum* decreased after benomyl treatments at anthesis.[44] In pears increased *Alternaria* decay occurred when benomyl and thiabendazole were used to control *Penicillium* infection.[45,46] Soil application of benomyl increased the severity of *Alternaria* leaf spot of carnations.[47]

Much information is available on the efficiency of specific fungicides for controlling diseases. The positive aspects of such research are made accessible to turfgrass workers. Moreover, the package labels on fungicides used for ornamental turf list nearly every efficient registered use, because residue and related problems are few compared to those for food crops. The negative aspects of disease-control studies, however, are communicated less frequently. Information on the specificity of disease control can be easily obtained from the label, but there remains a dearth of available information regarding instances in which fungicides have increased the prevalence of diseases.[30] Research workers periodically observe occasions on which a fungicide allows a particular disease to become more severe or a second disease occurs soon after a fungicide has been applied to control an initial, or target, disease. During the past decade, over 90 examples of fungicide-induced increases in turfgrass diseases have been listed and reviewed. Benomyl and thiophanates have been used extensively for the control of turfgrass diseases caused by fungi in the genera *Fusarium, Rhizoctonia,* and *Sclerotinia* for the last 10 years. However, these fungicides are not effective against oomycetes. Thus, the potential has existed for Pythium blight due to *Pythium aphanidermatum* and other *Pythium* spp. to become amplified where the benzimidazoles have been overly emphasized in a disease control program.[30] The possibility was confirmed in studies by Warren et al.[48] The occurrence of fungicide-insensitive nontarget pathogens or fungicide-resistant strains of target pathogens after fungicide application may be the result of disturbances in the balance among pathogens, between the pathogen and antagonistic saprophytes, or both. These fungicides are known to be nontoxic to most basidiomycetes and certain hyphomycetes and thus capable of amplifying diseases caused by fungi in these taxonomic groups.

Smiley[32] presented such documentation for Typhula blight *(Typhula incarnata)*, rust *(Puccinia graminis)*, red thread *(Corticium fusiformis)*, some Rhizoctonia diseases *(Pellicularia filamentosa* and *Ceratobasidium* spp.), and some Helminthosporium diseases *(Drechslera* and *Bipolaris* spp.). Smiley[32] provided ample evidence of severe occurrences of several new diseases in benomyl-treated turfgrass. Smith et al.[47] and Jackson[49] reported that turf diseases due to an unidentified basidiomycete and Helminthosporium leaf spot *(Helminthosporium vagans)*, respectively, were promoted when benomyl and thiabendazole were used to control other diseases. Chloroneb and cycloheximide predisposed creeping bent grass to a disease caused by *Rhizoctonia cerealis* when they were used for the control of *Fusarium nivale* and *Typhula incarnata,* although the disease was absent where fungicides had not been used.[30] It was further observed that cycloheximide increased Rhizoctonia diseases; triadimefon increased Helminthosporium leaf spot; etaconazole red thread; benomyl *Helminthosporium,* collar rot, red thread, and rust; and chloroneb Rhizoctonia diseases.[30]

Use of benomyl to control strawberry rot due to *Botrytis cinerea* has often been followed by an increased incidence of rot due to *Rhizopus* spp.,[50] which may reflect not only the lack of activity of benomyl against this pathogen but also a modifying effect in the natural flora of the host. In cyclamen, benomyl-resistant strains of *B. cinerea* more severely attack plants sprayed with benomyl than nontreated or control plants, probably because of a reduction of competition with antagonists *Penicillium* sp.[41,51] Application of benzimidazole compounds used for control of storage decay due to *Penicillium digitatum* on pears increased *Alternaria* decay.[45] The widespread application of benzimidazoles to control Gloeosporum rot in apples and Botrytis rot in soft fruit has been followed by a marked rise in the incidence of rot caused by the phycomycetes *Phytophthora syringae* in apples[52] and *Mucor* and *Rhizopus* spp. in strawberries.[53] These fungi are resistant to benzimidazole fungicides and probably benefit from reduced competition with benzimidazole-sensitive organisms.

The survival of many pathogens outside host plants is often strongly influenced by competitive or antagonistic microorganisms; treatments which selectively inhibit such organisms tend to increase both the abundance of the pathogen in the environment and the incidence of disease. This is particularly true for soil-borne pathogens. There are many examples of soil treatments designed to control root pathogens which have produced quite the opposite effect. Wilt of asters *(Phytophthora cryptogea)* increased after treatment of the soil with benomyl and the effect could be attributed to an increase in tolerant microorganisms which include *P. cryptogea.*[54] This is a normal result which, it may be argued, could have been predicted from the known insensitivity of phycomycetes to benomyl.

Sprays with carbendazim fungicides at dosages even up to 5.6 kg ha^{-1} of the a.i. did not appreciably control root and foot rot of cereals caused by *Fusarium culmorum*[55,56] and *Gaeumannomyces graminis.*[57,58] Fusarium root rot was suppressed, but in benomyl-treated plots there was ten times as much sharp eyespot *(Rhizoctonia cerealis)* than in untreated plots.[59] Increase of sharp eyespot as a result of treatment with benzimidazole fungicides was also observed in wheat.[58,60-63] The increase of sharp eyespot in benomyl-treated cereals is a case in point, as the pathogen is not resistant to the fungicide in vitro. Although mycelial growth of Pythiaceae and other resistant fungi was slightly inhibited on media containing 100 μg mℓ$^{-1}$ of the fungicide,[40,64] growth of *R. cerealis* was retarded at concentrations below 5 μg mℓ$^{-1}$. The difference in sensitivity between *R. cerealis* and the dominant pathogen in the untreated crop, *F. culmorum,* was relatively slight.[63]

Sharma and Kolte[67] reported that application of metalaxyl to foliage for the control of white rust and downy mildew of silique, *toria,* and yellow *sarson* has been observed to increase the severity of Alternaria leaf spot and blight. Similarly, postharvest application of metalaxyl reduced brown rot of citrus fruits *(Phytophthora citriphilla)* but promoted the development of *Penicillium digitatum* (another fruit-rotting fungus) in infected fruits.[68] The increased severity of disease caused by nonphycomycetes fungi may be either because of increased susceptibility of the host towards the pathogen due to the interference of metalaxyl with normal metabolism or because of decreased competition, particularly with respect to the availability of healthy tissues, due to specific inhibition of a group of pathogens.

These observations indicate that biological controls are not restricted to soil pathogens but extend to aerial pathogens also, which are apparently inhibited by competition, antagonists, or hyperparasites. Where such biological control is operative, it is a reasonable expectation that fungicides would be effective and would not create problems.

During cotton seed treatment to control seedling disease caused by several soil-borne pathogens, Papavizas and Lewis[65] noticed that benomyl consistently increased damping-off when *Pythium* spp. were present in the soil. They performed experiments with nonsterilized soil with a natural population of *Pythium* spp. (5 to 10 propagules per 10 g soil); in another set the soil was infected with *Rhizoctonia solani* or *Thielaviopsis basicola* 3 weeks before planting. For mixed pathogen studies, the *R. solani-* or *T. basicola*-infested soils were mixed in equal parts. The fungicide furmecyclox was applied to cotton seed at 2.5% in acetone using the organic solvent infusion technique (OSIT)[66] and metalaxyl (2.5% in acetone), which is effective against *Pythium* spp., was added in combination with furmecyclox using OSIT; benomyl was added directly to the seed after the solvent evaporated. The treated seeds were planted in the infested soil and the containers were kept at 70°C for 1 week and then at 20°C for 5 weeks to facilitate *T. basicola* infection. The observations indicated that benomyl significantly increased damping-off of cotton when *Pythium* spp. were not eliminated from the soil by metalaxyl. The increased disease associated with benomyl treatment was apparently the consequence of infections by *Pythium* spp., because when metalaxyl had been added to the seed, excellent stands were obtained.[65]

Phenomena such as these have also been observed in other pathosystems. Zadoks and Schein[69] present data which show that field applications of benomyl at 1.5 kg a.i. ha^{-1} on rye led to a decrease of symptoms caused by *Fusarium* spp. and *Cercosporella herpotrichoides.* However, sharp eyespot, caused by *Rhizoctonia* spp., increased from 2 to 35% incidence in the sprayed crop. In a similar example, ethirimol (0.15%), used for the control of barley powdery mildew *(Erysiphe graminis),* caused *Helminthosporium sativum* to increase significantly. This is apparently due to the fact that ethirimol promotes germination of *H. sativum* conidia while increasing host susceptibility to *H. sativum.*[70,71]

Early blight *(Alternaria solani)* and late blight *(Phytophthora infestans)* of potato *(Solanum tuberosum)* are serious diseases and require effective control measures for the successful protection of the crop. Copper-based fungicides and dithiocarbamates were once recommended for the control of these diseases.[72] With the introduction of systemic fungicides, metalaxyl, and other acylalanines, the strategies for the control of potato diseases have changed drastically.[73] The systemic fungicides are selectively effective for the control of late blight pathogens. That the continuous use of metalaxyl and other acylalanines allowed early blight to increase would seem to encourage the use of broad-spectrum fungicides.[74,75] Plant disease control strategies have previously concentrated on one disease or specific pathogen at a time,[69] so that their spectrum of activity could be determined. Epidemiological principles and the growing awareness of the possible deleterious consequences of iatrogenic diseases[69,76] suggest that a good case can be made for a reevaluation of the criteria used by agencies responsible for the introduction and registration of new fungicides. For example, in addition to residue and toxicity data, information should be gained concerning other harmful side effects that could leave the farming community more vulnerable to new pathogens or a more dominant strain of present pathogens. Thus, instead of a fungicide being registered for use against one or more specific organisms, it could be registered for safe use in a particular agroecological context.

IV. MECHANISMS OF IATROGENIC EFFECTS

Higher disease incidence in a fungicide-treated crop than in an untreated crop can be caused in different ways. The mechanisms involved have been discussed[77] and some of the pertinent points are as follows:

1. The appearance of fungicide-resistant strains in the pathogen populations that are more virulent than wild fungi may occasionally occur. An example is a benomyl-resistant strain of *Verticillium fungicola,* a major parasite of commercial mushrooms.[78] The reasons that there is little chance that these strains will maintain themselves in the long run in untreated crops have been discussed.[79]
2. The fungicide may inhibit host-resistant mechanisms.[80]
3. The microbial antagonism to which the pathogen is exposed in the soil or on the surface of roots and plant organs may decrease.

It must be expected that crop-protection fungicides are likely to have some effect on elements of the microflora associated with the plant other than target pathogens, and that they might compete with and suppress the target pathogens. In the simplest case, pathogens other than those directly controlled are allowed to develop, which may be straightforward opportunism on the part of a pathogen not within the activity spectrum of the fungicide. This may be the cause of the increase in brown rust of wheat *(Puccinia recondita)* following the use of the specific fungicide ethirimol for the control of spring barley powdery mildew *(Erysiphe graminis).*[81]

Fungicides reduce antagonistic microflora which can suppress low levels of a disease organism. Such reduction of the antagonists allows the disease to develop. This effect is more noticeable when broad-spectrum fungicides such as benzimidazoles, captafol, or dithiocarbamates are used. Fokkema et al.[43] found increased levels of *Cochliobolus sativus* on rye following spray treatment with benomyl. The increase in the disease was directly related to the suppression of natural antagonists such that they were unable to exert their natural buffering role. Similar disruption of antagonism may be responsible for increases in known pathogens such as *Typhula incarnata* and the widely reported development of sharp eyespot *(Rhizoctonia cerealis)* in Europe.[82] Triadimefon has a very pronounced side effect on *T. incarnata.*[83] A similar result is also effected by nuarimol.[84] This increase has been observed after extensive use of benomyl for the control of eyespot of wheat *(Pseudocercosporella herpotrichoides).*[58,85] The reduction of saprophytic mycoflora by fungicides may have less effect on the development of diseases caused by pathogens which are sensitive to the fungicides applied. However, pathogens which are not sensitive, or pathogens which have become resistant to the fungicides applied, will certainly profit from the reduction of the saprophytic mycoflora. Reported increases in the field of *Alternaria* and *Cochliobolus* diseases, mainly on other crops, after treatment with benzimidazole fungicides may be explained in this way. However, information about the role of saprophytes in the field is lacking.[86]

V. FUTURE OUTLOOK

Detailed microbiological studies are needed to elucidate the exact mechanisms involved in the phenomena discussed above. Such studies would help to clarify the role of pathogen/

saprobic interaction in plant health and disease. Benzimidazole fungicides offer an excellent opportunity for such studies because they have a wide spectrum of activity against many important plant pathogens. Previous research also highlights the necessity for awareness of the possible effects of the disruption of biological balances through the widespread use of broad-spectrum, yet selective, pesticides.

The development of integrated pest management systems for control of plant diseases is now of considerable interest; problems of environmental pollution will be minimized and the cost of control measures will be reduced.[4] Detailed studies on iatrogenic diseases can make an important contribution in this field. Earlier, such diseases were merely regarded as nontarget effects or unfortunate side effects of fungicidal application. This viewpoint is unfortunate because it tends to ignore the fact, well illustrated by the reviews, that iatrogenic diseases may offer exceptional opportunities for learning about host/pathogen interactions and for developing more effective and efficient control practices.[4]

REFERENCES

1. **Kreutzer, W. A.**, The reinfestation of treated soils, in *Ecology of Soil Borne Plant Pathogens*, Baker, K. F. and Sinder, W. C., Eds., University of California Press, Berkeley, 1965, 495.
2. **Horsfall, J. G.**, Selective chemicals for plant disease control, in *Pest Control Strategies for the Future*, National Academy of Sciences, Washington, D.C., 1972, 216.
3. **Yarwood, C. E.**, Man made diseases, *Science*, 168, 218, 1970.
4. **Griffiths, E. and Berrie, A. M.**, Iatrogenic diseases in plants, *Ann. Appl. Biol.*, 89, 722, 1978.
5. **Horsfall, J. G.**, Iatrogenic disease: mechanism of action, in *Plant Disease: An Advance Treatise*, Vol. 2, Horsfall, J. G. and Cobling, E. B., Eds., Academic Press, New York, 1979, 343.
6. **Griffiths, E.**, Iatrogenic plant diseases, *Annu. Rev. Phytopathol.*, 17, 69, 1981.
7. **Bliss, D.**, The destruction of *Armillaria mellea* in citrus soils, *Phytopathology*, 41, 665, 1951.
8. **Baker, K. F. and Cook, R. J.**, *Biological Control of Plant Pathogens*, W. H. Freeman, San Francisco, 1974.
9. **Ohr, H. D., Munnecke, D. E., and Bricker, J. L.**, The interaction of *Armillaria mellea* and *Trichoderma* spp. as modified by methyl bromide, *Phytopathology*, 63, 1352, 1973.
10. **Ohr, H. D. and Munnecke, D. E.**, Effects of methyl bromide on antibiotic production by *Armillaria mellea, Trans. Br. Mycol. Soc.*, 62, 65, 1974.
11. **Gibson, I. A. S.**, Trial of fungicides for control of damping-off in pine seedlings, *Phytopathology*, 46, 181, 1956.
12. **Carter, M. V.**, Biological control of *Eutypa armeniacae, Aust. J. Exp. Agric. Anim. Husb.*, 11, 687, 1971.
13. **Dimock, A. W. and Baker, K. F.**, Effect of climate on disease development, injuriousness and fungicidal control, as exemplified by snapdragon rust, *Phytopathology*, 41, 536, 1951.
14. **Griffiths, E., Gibbs, J. N., and Waller, J. M.**, Control of coffee berry disease, *Ann. Appl. Biol.*, 67, 45, 1971.
15. **Mulinge, S. K. and Griffiths, E.**, Effects of fungicides on leaf rust berry disease defoliation and yield of coffee, *Trans. Br. Mycol. Soc.*, 62, 495, 1974.
16. **Griffiths, E.**, Negative effects of fungicides in coffee, *Proc. 6th Br. Insectic. Fungic. Conf.*, 1, 817, 1971.
17. **Cox, R. S. and Hayslop, N. C.**, Progress in the control of grey mould of tomato in South Florida, *Plant Dis. Rep.*, 40, 718, 1956.
18. **Bird, L. S., Ranney, C. D., and Watkins, G. M.**, Evaluation of fungicides mixed with the covering soil at planting as control measures for the cotton seedling disease complex, *Plant Dis. Rep.*, 41, 165, 1957.
19. **Garren, K. H.**, The stem rot of peanuts and its control, *Va. Agric. Exp. Stn. Tech. Bull.*, 144, 1, 1959.
20. **Gibson, I. A. S., Ledger, M., and Boehm, E.**, An anomalous effect of pentachloronitrobenzene on the incidence of damping off caused by *Pythium* sp., *Phytopathology*, 51, 531, 1961.
21. **Farley, J. D. and Lockwood, J. L.**, Effect of pentachloronitrobenzene (PCNB) on the soil microflora, *Phytopathology*, 58 (Abstr.), 1050, 1968.
22. **Katan, J. and Lockwood, J. L.**, Effect of pentachloronitrobenzene on colonization of alfalfa residues by fungi and streptomycetes in soil, *Phytopathology*, 60, 1578, 1970.

23. **Rich, S. and Miller, P. M.,** Verticillium wilt of strawberry made worse by soil fungicides that stimulates meadow nematode populations, *Plant Dis. Rep.*, 48, 246, 1964.
24. **Baker, K. F. and Cook, R. J.,** *Biological Control of Plant Pathogens*, W. H. Freeman, San Francisco, 1974.
25. **Beute, M. K., Porter, D. M., and Hadley, B. A.,** Sclerotinia blight of peanut in North Carolina and Virginia and its chemical control, *Plant Dis. Rep.*, 59, 697, 1975.
26. **Porter, D. M.,** The effect of chlorothalonil and benomyl on the severity of Sclerotinia blight of peanuts, *Plant Dis. Rep.*, 61, 995, 1977.
27. **Porter, D. M.,** Increased severity of Sclerotinia blight of peanuts treated with captafol and chlorothalonil, *Plant Dis.*, 64, 394, 1980.
28. **Lankow, R. K.,** The differential sensitivity of developmental stages of *Sclerotinia sclerotiorum* to chlorothalonil, *Proc. Am. Phytopathol. Soc.*, 3, 253, 1976.
29. **Porter, D. M.,** The influence of chlorothalonil on the severity of Sclerotinia blight of peanuts, *Proc. Am. Phytopathol. Soc.*, 3, 253, 1976.
30. **Hartill, W. F. T. and Campbell, J. M.,** Control of Sclerotinia in tobacco seedbeds, *Plant Dis. Rep.*, 57, 932, 1973.
31. **Backman, P. A., Rodriguez-Kabana, R., and Williams, J. C.,** The effect of peanut leafspot fungicides on the nontarget pathogens, *Sclerotium rolfsii, Phytopathology*, 65, 773, 1975.
32. **Smiley, R. W.,** Nontarget effects of pesticides on turfgrasses, *Plant Dis.*, 65, 17, 1981.
33. **Williams, R. J. and Ayanabe, A.,** Increased incidence of Pythium stem rot in cowpeas treated with benomyl and related fungicides, *Phytopathology*, 65, 217, 1975.
34. **Dekker, J.,** Problems of selectivity in the field of systemic fungicides, *Acta Phytopathol. Acad. Sci. Hung.*, 6, 329, 1971.
35. **Knauss, J. F.,** Field evaluation of several soil fungicides for control of *Scindapsus aureus* cutting decay incited by *Pythium, Plant Dis. Rep.*, 56, 1074, 1972.
36. **Knauss, J. F.,** Soil fungicides for tropical foliage plants — efficacy and other considerations, *Proc. Fla. State Hortic. Soc.*, 90, 340, 1977.
37. **Bollen, G. J.,** Resistance to benomyl and some chemically related compounds in strains of *Penicillium* species, *Neth. J. Plant Pathol.*, 77, 187, 1971.
38. **Rodriguez-Kabana, R. and Curl, E. A.,** Nontarget effect of pesticides on soil-borne pathogens and disease, *Annu. Rev. Phytopathol.*, 18, 311, 1980.
39. **Bollen, G. J.,** A comparison of the in vitro antifungal spectrum of thiophanates and benomyl, *Neth. J. Plant Pathol.*, 78, 55, 1972.
40. **Bollen, G. J. and Fuchs, A.,** On the specificity of the *in vitro* and *in vivo* antifungal activity of benomyl, *Neth. J. Plant Pathol.*, 76, 299, 1970.
41. **Bollen, G. J. and Scholten, A.,** Acquired resistance to benomyl and some other systemic fungicides in a strain of *Botrytis cinerea* in cyclamen, *Neth. J. Plant Pathol.*, 77, 83, 1971.
42. **Siegel, M. R.,** Benomyl-soil microbial interactions, *Phytopathology*, 65, 219, 1975.
43. **Fokkema, M. J., van der Laar, J. A. J., Nelis-Blomberg, A. L., and Schippers, B.,** The buffering capacity of the natural flora of rye leaves to infection by *Cochliobolus sativus* and its susceptibility to benomyl, *Neth. J. Plant Pathol.*, 81, 176, 1975.
44. **Rapilly, F., Foucault, B., Poisson, J., and Cahagnier, B.,** Consequences au niveau du grain des traitements fongicides realises en vegetation sur bled'hiver, *Phytiatr. Phytopharm.*, 22, 281, 1973.
45. **Spalding, D. H.,** Post harvest use of benomyl and thiabendazole to control blue mold rot development in pears, *Plant Dis. Rep.*, 54, 655, 1970.
46. **Valdebenito, R. M. and Pinto de Torres, A.,** Control of *Penicillium expansum, Botrytis cinereay, Alternaria alternata* en manzanas y peras, *Agric. Tec., Santiago*, 32, 148, 1972.
47. **Smith, A. M., Stynes, B. A., and Moore, K. J.,** Benomyl stimulates growth of a basidiomycetes on turf, *Plant Dis. Rep.*, 54, 774, 1970.
48. **Warren, C. G., Sanders, P. L., and Cole, H., Jr.,** Increased severity of Pythium blight associated with use of benzimidazole fungicides on creeping bent grass, *Plant Dis. Rep.*, 60, 932, 1976.
49. **Jackson, N.,** Evaluation of some chemicals for control of stripe smut in Kentucky blue grass turf, *Plant Dis. Rep.*, 54, 168, 1970.
50. **Jordan, V. W. L.,** The effects of prophylactic spray program on the control of pre- and post-harvest disease of strawberry, *Plant Pathol.*, 22, 67, 1973.
51. **van Dommelen, L. and Bollen, G. J.,** Antagonism between benomyl-resistant fungi on cyclamen sprayed with benomyl, *Acta Bot. Neerl.*, 22, 169, 1973.
52. **Upstone, M.,** Evaluation of chemicals for control of Phytophthora fruit rot in stored apples, *Proc. 1977 Br. Crop. Prot. Conf. Pests and Diseases*, 1, 565, 1977.
53. **Dennis, C. and Davis, R. P.,** The selective effect of fungicides on postharvest spoilage fungi of strawberries, *Proc. 1977 Br. Crop. Prot. Conf. Pests and Diseases*, 1, 203, 1977.

54. **van den Berg, G. A. and Bollen, G. J.,** Effect of benomyl on incidence of wilting of *Callistephus chinensis* by *Phytophthora cryptogea, Acta Bot. Neerl.,* 20, 256, 1971.

55. **Bruchl, G. W. and Cunfer, B.,** Control of Cercosporella foot rot of wheat by benomyl, *Plant Dis. Rep.,* 56, 20, 1972.

56. **Duben, J.,** Untersuchungen zum Fusskrankheitskomplex an Winterweizen unter besonderer Berucksichtigung von Arten der Gattung *Fusarium* Lk., Thesis, University of Göttingen, West Germany, 1978, 147.

57. **Huber, D. M. and Mulanax, M. W.,** Benomyl rates and application time for wheat foot rot control, *Plant Dis. Rep.,* 56, 342, 1972.

58. **Prew, R. D. and McIntosh, A. H.,** Effects of benomyl and other fungicides on take all and eyespot diseases of winter wheat, *Plant Pathol.,* 24, 67, 1975.

59. **van der Hoeven, E. P. and Bollen, G. J.,** The effect of benomyl on antagonism towards fungi causing foot rot in rye, *Acta Bot. Neerl.,* 21, 107, 1972.

60. **Jenkyn, J. F. and Prew, R. D.,** Activity of six fungicides against cereal foliage and root diseases, *Ann. Appl. Biol.,* 75, 241, 1973.

61. **Obst, A., Graf, R., and Ruppold, H.,** Beobachtungen in einem Monokultur Weizen Sortiment zur Vikarianz der Fuss Krankheitserreger, *Nachrichtenbl. Dtsch. Pflanzenschutzdienstes (Braunschweig),* 29, 113, 1977.

62. **Reinecke, P.,** Untersuchungen zum Erregerspektrum des Fusskrankheitskomplexes an Getreide unter besonderer Berucksichtigung von *Rhizoctonia solani* Kuhn, Thesis, University of Göttingen, West Germany, 1977, 1.

63. **van der Hoeven, E. P. and Bollen, G. J.,** Effect of benomyl on soil fungi associated with rye. I. Effect on the incidence of sharp eyespot caused by *Rhizoctonia cerealis, Neth. J. Plant Pathol.,* 86, 163, 1980.

64. **Edgington, L. V., Khew, K. L., and Barron, G. L.,** Fungitoxic spectrum of benzimidazole compounds, *Phytopathology,* 61, 42, 1971.

65. **Papavizas, G. C. and Lewis, J. A.,** Side-effects of pesticides on soil-borne plant pathogens, in *Soil-Borne Plant Pathogens,* Schippers, B. and Gams, W., Eds., Academic Press, New York, 1979, 483.

66. **Papavizas, G. C. and Lewis, J. A.,** Acetone infusion of pyroxychlor into soybean seed for the control of *Phytophthora megasperma* var. *sojae, Plant Dis. Rep.,* 60, 484, 1976.

67. **Sharma, S. P. and Kolte, S. J.,** Control of downy mildew of *Brassicae* spp., *Pestology,* 3(5), 21, 1980.

68. **Cohen, E.,** Metalaxyl for post harvest control of brown rot of citrus fruit, *Plant Dis.,* 65, 830, 1981.

69. **Zadoks, J. C. and Schein, R. D.,** *Epidemiology and Plant Disease Management,* Oxford University Press, New York, 1979.

70. **Saur, R.,** Untersuchungen ueber den Einfluds von Ethirimol auf die Pathogenese einer Helminthosporiose *(H. sativum)* as Gertse, *Phytopathol. Z.,* 87, 304, 1976.

71. **Schuette, F. and Diercks, S.,** Moeglichkeiten und Granzen des integrierten Pflanzenschutzes im Ackerbau, *Mitt. Biol. Bundesanst. Land. Forstwirtsch. Berlin-Dahlem,* 165, 63, 1975.

72. **Nene, Y. L. and Thapliyal, P. N.,** *Fungicides in Plant Disease Control,* Oxford & IBH Publishing, New Delhi, 1979, 507.

73. **Vyas, S. C.,** *Systemic Fungicides,* Tata McGraw-Hill, New Delhi, 1984, 360.

74. **Putter, C. A. J.,** An Epidemiological Analysis of the *Phytophthora* and *Alternaria* Blight Pathosystem in the Natal Midlands, Ph.D. thesis, University of Natal, South Africa, 1980, 1.

75. **Nema, S. and Vyas, S. C.,** Control of early and late blight of potato, *Pestology,* 8(8), 31, 1985.

76. **Dekker, J.,** Acquired resistance to fungicides, *Annu. Rev. Phytopathol.,* 14, 405, 1976.

77. **Bollen, G. J.,** Side effects of pesticides on microbial interactions, in *Soil Borne Plant Pathogens,* Schippers, B. and Bollen, W. G., Eds., Academic Press, New York, 1979, 451.

78. **Bollen, G. J. and van Zaayen, A.,** Resistance to benzimidazole fungicides in pathogenic strains of *Verticillium fungicola, Neth. J. Plant Pathol.,* 81, 157, 1975.

79. **Dekker, J.,** Strategies for avoiding resistance to fungicides, in *Strategies for the Control of Cereal Disease,* Jenkyn, J. F. and Plumb, R. T., Eds., Blackwell Oxford, London, 1982, 123.

80. **Swinburne, T. R.,** The effect of benomyl and other fungicides on phenylalanine-ammonia lyase activity in *Hordeum vulgare* and *Phaseolus vulgaris, Physiol. Plant Pathol.,* 5, 81, 1975.

81. **Little, R. and Doodson, J. K.,** The comparison of yields of some spring barley varieties in the presence of mildew and when treated with a fungicide, *Proc. 6th Br. Insectic. Fungic. Conf.,* 1, 91, 1971.

82. **Cook, R. J.,** Unexpected effects of fungicides on cereal yields, *EPPO Bull.,* 11, 277, 1981.

83. **Fehrmann, H.,** Modern developments in fungicide use on cereals, *EPPO Bull.,* 11, 259, 1981.

84. **Ebenebe, C. and Fehrmann, H.,** Evaluation of a number of systemic fungicides for the control of *Typhula incarnata* in winter barley, *Z. Pflanzenkr. Pflanzenschultz,* 81, 711, 1974.

85. **van der Hoeven, E. P. and Bollen, G. J.,** Effect of benomyl on soil fungi associated with rye. I. Effect of the incidence of sharp eyespot caused by *Rhizoctonia cerealis, Neth. J. Plant Pathol.,* 86, 163, 1980.

86. **Hislop, E. C.,** Some effects of fungicides and other agrochemicals on the microbiology of the aerial surfaces of plants, in *Microbiology of Aerial Plant Surfaces,* Dickinson, C. H. and Preece, T. F., Eds., Academic Press, New York, 1976, 41.

Chapter 7

VIRUS AND VIRAL DISEASES

I. INTRODUCTION

The application of chemicals to crops to protect them from infection by fungal pathogens is now a standard treatment for the control of many plant diseases. The principal means of such protection until about 20 years ago was the use of protectant fungicides. Since then, however, the control of these diseases has been revolutionized by the discovery of systemic fungicides, the development of which has been fully discussed.[1-3] In contrast to these spectacular advances, no progress has been made on direct practical control or eradication of plant virus diseases by chemical treatment.[4,5] Ever since the identification of plant viral diseases, plant pathologists have devoted much of their time to finding the antiviral chemicals which could inhibit virus infection and multiplication to achieve direct protection of crops in the way that fungicides protect against fungal pathogens. However, this has not yet been successful. The chemicals which were tried include inorganic substances, organic substances, metabolites, antimetabolites, and substances isolated from the organisms. Although many of them inhibit viruses in glasshouse conditions, they have not been useful for disease control in the field, mainly because they do not meet the requirements of an effective drug for virus control: that the chemicals should be safe, systemic in plants, readily available, and not toxic to plants or human beings. Another limitation of these compounds is that they are toxic to the plant at concentrations only a little above those required to inhibit viral multiplication.

The systemic fungicides carboxin and oxycarboxin were introduced in 1966 by Schmeling and Kulka[6] for the control of smut and rust, respectively. Since then about 300 compounds have been introduced, and many of them are now employed commercially for the control of diseases which were hitherto not controllable by protectant (or nonsystemic) fungicides. Systemic fungicides are meant for fungal diseases but they also possess nontarget effects towards viruses and viral diseases of plants. Tomlinson et al.[7] demonstrated for the first time the nontarget effects of systemic fungicides towards the viruses. It was observed that the visible symptoms of tobacco mosaic virus (TMV) infection of tobacco plants could be suppressed by treating the plants with nonphytotoxic doses of carbendazim (methyl-benzimidazole-2-yl-carbamate). This compound is the water decomposition product and fungitoxic principal[8] of the systemic fungicide benomyl, which is already in wide use in agriculture for the control of a range of plant pathogens except for phycomycetous pathogens. The nontarget effects of systemic fungicides on the control of viral diseases are reviewed in Table 1.

II. CONTROL

A. TMV in Tobacco

Tobacco plants of uniform maturity were watered with 200 mℓ of a suspension of carbendazim twice at 10-day intervals and 2 days after the final application of the chemicals; two leaves of each treated and control plant were mechanically inoculated with a diluted purified preparation of TMV. By 30 days after inoculation, control plants had a conspicuous mosaic of yellow and dark green tissue, particularly evident in leaf numbers 5 to 7. By contrast, symptoms on the equivalent leaves of carbendazim-treated plants were either inconspicuous or absent. At 100 days after inoculation, the plants had attained a height of about a meter and had 28 to 30 leaves. The inflorescences of control plants were brittle with

Table 1
EFFECT OF SYSTEMIC FUNGICIDES ON CONTROL OF VIRAL DISEASE OF PLANTS

Systemic fungicide	Crop	Virus disease	Ref.
Benomyl	Urdbean *(Vigna radiata* var. *munga)*	Leaf crinkle virus	21
Carbendazim	Urdbean *(Vigna radiata* var. *munga)*	Leaf crinkle virus	23
Thiabendazole lactate	Beet	Beet yellow, beet yellowing virus	20
Carbendazim	Lettuce, tobacco	Beet western yellow, tobacco mosaic virus	9
	Rice	Rice tungro virus	25
Benomyl	Tomato, cucumber	Tobacco mosaic virus, cucumber mosaic virus	13
Carbendazim	Tobacco	Tobacco mosaic virus	11,12
Triadimefon, benomyl, prothiocarb, metalaxyl	Broad bean *(Vicia faba)*	Broad bean true mosaic virus	15
Carbendazim	Zinnia	Zinnia mosaic virus	19
Benomyl, carboxin, carbendazim, thiabendazole	Papaya	Papaya mosaic virus	17
Carbendazim	Lettuce	Lettuce big vein virus	30, 39
	Tobacco	Tomato spotted wilt virus	14

pronounced TMV symptoms, but carbendazim-treated plants were flexible, uniformly green, and free from symptoms. The treated plants contained approximately three times more chlorophyll than the control.[9] At specified intervals, the virus content of the inoculated leaves and the systemically infected leaves of both carbendazim-treated and control plants was determined. Discs (1 cm diameter) were cut from the leaves and homogenized with a pestle and mortar in 1% KH_2PO_4 (1 mℓ to 1 g tissue) and filtered sap assayed on *Nicotiana glutinosa*. Tomlinson[9] observed no significant difference in the infectivity in carbendazim-treated and control leaves, nor any loss of virus. The application of carbendazim to the roots caused a marked suppression of symptoms without affecting the virus content of the plant as measured by the infectivity assay.[7] Similar results have also been obtained with tobacco plants infected with five strains of TMV. Carbendazim caused suppression of normal symptoms irrespective of whether the dominant symptoms of the strain used were mosaic or severe leaf malformation.[10] Control of TMV carbendazim[11] and TMV in tomato and cucumber mosaic virus in cucumber by benomyl[13] has been observed. Tobacco plants were sprayed twice at 3-day intervals with an aqueous suspension of carbendazim at 0.375 to 1.75 g active ingredient (a.i.) ℓ^{-1} and inoculated with tomato spotted wilt virus (TSMV) just 2 days after the second spray.[14] The number of local lesions was increased by doses up to 0.75 g ℓ^{-1} and reduced by a dose of 1.5 g ℓ^{-1}. It was observed that systemic symptoms appeared earlier in plants with more local lesions. Plants given higher doses had a higher chlorophyll content. The activity of carbendazim, probably due to cytokinin, appears to be related to its physiological effect on the host which is maintained in a state unsuitable for virus multiplication.[14]

B. Broad Bean True Mosaic Virus (BBTMV)

Broad bean *(Vicia faba* L.) plants of cultivars Giza 1 and 2 were planted in clay and loamy and vermiculite soils and inoculated with BBTMV. The inoculated plants were drenched with the systemic fungicides triadimefon, benomyl, prothiocarb, and metalaxyl at the rate recommended for disease control before 8 and 4 and after 4 and 8 days of inoculation.

Triadimefon before inoculation afforded significant protection from BBTMV to the broad bean plant.[15] It was found that benomyl inactivated BBTMV when added to the infectious crude sap at the rate of 250 ppm. However, El-Banna[16] did not observe similar results with benomyl and thiophanate-methyl. These differences may be attributed to the mode of application and concentration used.

C. Papaya Mosaic Virus

Papaya (*Carica papaya* L.) suffers from two serious virus diseases in India. Mosaic is beginning to limit its cultivation. Spraying of thiabendazole at 0.6 and 0.8 g ℓ^{-1} and trunk injection at 0.5 g per tree reduced the viral strength of papaya mosaic virus. Carboxin spraying at the same concentration produced significant results but trunk injection was not effective. Carbendazim at 1.6 g ℓ^{-1} reduced viral strength significantly.[17] Therefore, carboxin and thiabendazole sprayings have been recommended for the control of papaya mosaic virus.

D. Common Bean Mosaic Virus

Moghe[18] reported that carboxin suppressed the common bean mosaic virus symptom by seed treatment at 0.15%.

E. Zinnia Mosaic Virus (ZMV)

Rafiq et al.[19] demonstrated that a 5-min immersion of roots in carbendazim solution of 1:10 to 1:1,000,000 inhibited ZMV when the viral and fungicidal solutions were inoculated on the indicator plants of *Chenopodium amaranticolor*. The reduction in virus was 87 to 16.6% in various dilutions. However, the concentration 1:10 was slightly phytotoxic.

F. Beet Western Yellow Virus of Lettuce (BWYV)

BWYV is an aphid-transmitted persistent virus that causes a widespread disease of lettuce of economic concern in Europe. Major symptoms are interveinal leaf yellowing and premature senescence. BWYV was cultured in *Glaytonia perfoliate* plants and transmitted to lettuce by feeding aphids *(Myzus persicae)* on infected *G. perfoliate* plants and then transferring them in groups of ten to 21-day-old lettuce plants grown in a glasshouse.[9] Inoculated and noninoculated plants were treated by applying 100 mℓ of 1 g/ℓ suspension of carbendazim to the soil, 7 days before and 5 days after inoculation with virus-bearing aphids. Each treated plant received 0.1 g carbendazim. The plants were transplanted in the field and symptoms were recorded on the plants 30 days after inoculation, by which time they were of marketable size.[9] The carbendazim-treated leaves were free from the BWYV symptoms and marketable, but control plants were diseased and unmarketable. Further experiments have demonstrated that benomyl applied to young lettuce (100 mℓ of a.i. g/ℓ suspension) before transplanting also prevents the disease symptoms. However, neither carbendazim nor benomyl, applied as weekly foliar sprays at 0.1 or 1.0% of the formulated products, suppressed the disease symptoms.[9]

The most important effect of some viruses in their hosts is not the decrease in yield but the change in their appearance. BWYV when it infects lettuce causes chlorosis, which makes it unmarketable. Carbendazim, although it does not decrease BWYV infection, multiplication, or spread in lettuce, seems to stabilize the host chloroplasts so that the chlorotic symptom is suppressed and the crop appears healthy.[7] One problem with symptom suppression is that unrevealed but possibly large reservoir of virus persist, which could infect untreated crops and other plant species.

Tomlinson[9] found that treatment of lettuce with carbendazim also rendered the plants somewhat less vulnerable to aphids, an effect similar to that noted by Russell[20] with sugar beet plants treated with thiabendazole. Although the test aphids were observed to settle and feed on all the receptor plants, it was concluded that on the carbendazim-treated plants, their feeding period had been insufficient to transmit the disease.[8]

G. Beet Yellow Virus (BYV)

BYV is transmitted by mechanical inoculation and by aphids, *Myzus persicae* and *Aphis fabae*. Of these two, *M. persicae* is the more problematic. The virus, a filamentous rod, is persistent in the vector and attacks a large number of plant species of different families. Russell,[20] in a field exposed to natural infection by virus-bearing aphids, observed fewer sugar beet plants infected with aphid-transmitted yellowing viruses in plots that had been sprayed with solutions of thiabendazole lactate than in water-sprayed plots. Subsequent glasshouse tests showed that foliar sprays of 0.01% thiabendazole lactate in water significantly reduced the portion of inoculated sugar beet plants which became infected with persistent BYV when inoculated with virus-bearing *M. persicae*. Treatment of test plants with the fungicide did not affect the transmission of the nonpersistent beet mosaic virus.

H. Urdbean Leaf Crinkle Virus (ULCV)

Benomyl suppressed symptom expression and inhibited virus acquisition, and inoculation of sap and aphids transmitted ULCV in the urdbean *(Vigna radiata* var. *munga)* plant.[21-23] The plants were drenched twice with 0.1 to 10.0% benomyl solutions, before and after inoculation; drenching with benomyl prevented development of the disease. However, preinoculation treatment was far superior as none of the plants receiving 1.0% or higher of the fungicide developed disease symptoms. Even at 0.5% concentration, only 20% of the plants were infected as compared to 80% in postinoculation treatment. Similar results were reported when pre- and postinoculation drenched plants were inoculated by virus-bearing *A. pisum* and *A. craccivora*. However, *A. craccivora* transmitted the virus to fewer plants drenched before inoculation at all the concentrations used; *A. craccivora* failed to transmit the virus in plants drenched with 1.0% benomyl whereas 11.11% of the plants inoculated by *A. pisum* produced symptoms. Neither of the aphids, however, could transmit the virus in plants supplied with 2 and 5% benomyl.[21]

In another experiment, Bhardwaj et al.[21] reported on the effect on virus acquisition by aphids in benomyl-treated plants at 1.1, 0.2, 0.5, 1.0, and 2.0% solution at an interval of 5 days; the transmission of the virus decreased drastically with increase in the benomyl concentration; at 1.0%, virus-bearing *A. pisum* and *A. craccivora* infected only 11.11 and 10.00% of the plants, respectively. Both aphids failed to acquire the virus from the infected plants drenched with 2.0% benomyl.[21-24]

I. Rice Tungro

Tungro disease of rice commonly occurs in India, the Philippines, and Indonesia. The disease is caused by rice tungro virus (RTV) which is transmitted by the green leaf hopper *(Nephotettix virescens)*. Symptoms of tungro virus include orange-yellow coloring of leaves, stunting, reduction in tiller number, and poor panicles. Suppression of tungro virus symptoms in rice plants after the application of carbendazim at concentrations of 450 and 600 ppm were reported. The symptom expressions were rated as 23.30 and 3.50 (preinoculation application) and 60.10 and 38.70 (postinoculation application), respectively. The lower concentrations had no effect. Carbendazim markedly alleviated leaf chlorosis, stunting, and decreased tillers in rice plants.[25] The virus was present in carbendazim-treated plants and could be easily transmitted to healthy rice seedling by the green hopper. The treated plants showed more vigorous growth and gave better yield than RTV-infected and nontreated plants. Low concentration of carbendazim as inoculation against RTV delayed the onset of symptoms.[25]

Benzimidazole derivatives such as carbendazim join with chloroplast membranes and prevent senescence.[26] BWYV infection of lettuce plants resulted in the destruction of chloroplast grana and stroma lamellae[27] and tungro infection caused similar breakdown of chloroplast lamellae resulting in premature senescence. Carbendazim may suppress symptoms by preventing or delaying chlorophyll destruction. The growth-promoting properties of carbendazim may be the reason that treated plants show greater tillering and plant height than tungro-infected plants. Leaf chlorosis was suppressed less effectively in plants that received carbendazim after inoculation. Possibly tungro infection in these plants had already resulted in an initiation of chloroplast destruction and metabolic alterations. Such plants may fail to bind carbendazim in their chloroplast lamellae.[25]

Systemic fungicides could not free diseased plants from the virus; however, carbendazim treatments may enable infected plants to maintain photosynthesis and thereby produce more green matter content and, subsequently, increase grain yield.

J. Lettuce Big Vein Agent (BVA)

Lettuce BVA is an infectious, graft-transmissible entity that causes a distinctive vein-banding symptom on lettuce leaves.[28] This causal agent may be a virus but has not been identified as such; hence it is termed the lettuce BVA.[29] The virus is harbored in the soil within the resting spores of the chytrid fungus *Olpidium brassicae* and transmitted to lettuce plants by zoospores of the fungus. The lettuce BVA is not mechanically transmissible and no morphological or chemical entity has been associated with it.[28]

Fungicides were tested at 100 μg a.i. mℓ^{-1} for effects on zoospore motility and infectivity and on growth and maturation of thalli in vivo. Fenaminosulf was generally ineffective. Metalaxyl stopped zoospore motility but not infection or reproduction in vivo. Pyroxychlor, captan, and ethazole stopped motility and infection by zoospores, but were ineffective in vivo. Triadimefon or benomyl did not prevent motility, but did prevent infection or reproduction. Benomyl was systemic in the roots and prevented reproduction in seedlings for more than 7 days after 1 day of uptake.[30]

Olpidium brassicae has a distinctly different response to fungicides than do pathogens in the class Oomycetes. *Olpidium* is slightly affected by fungicides (fenaminosulf, pyroxychlor, ethazole, and metalaxyl) with specificity for pathogens in Peronosporales.[31-34] *Olpidium,* like *Plasmodiophora brassicae,*[35] is sensitive to benomyl, which is active against the higher pathogens, but not the Oomycetes.[36] Motility and infectivity of *Olpidium* zoospores, however, were not affected by benomyl at 100 μg mℓ^{-1} whereas the *Olpidium* thallus is killed after infection of the root.[28] These observations are consistent with the report that the active ingredient methyl-2-yl-benzimidazole carbamate carbendazim is absorbed by the plant and is effective in vivo[37] by interfering with fungal mitosis.[38] Campbell et al.[30] confirmed the systemic movement of methyl-2-yl-benzimidazole carbamate in the root system because new roots formed 1 week after a 24-hr application of benomyl did not support growth of *Olpidium.* The sensitivity of *Olpidium* to benomyl in vivo has been discussed.[29] Carbendazim has recently been reported to be nontoxic to zoospores, as it was the surfactants in the formulation that killed them.[39] However, Campbell et al.[30] reported that this report is incorrect for two reasons: (1) the tests evaluated only zoospore motility and (2) they neglected the in vivo activity of systemic fungicides. Nonmotile zoospores were infective in tests with heat-shocked zoospores and in tests with zoospores incubated in soil without host plants, as well as with metalaxyl at 50 μg mℓ^{-1}.[30] Although captan, copper sulfate, and arasan stopped zoospore movement, their ability to kill either zoospores or thalli in vivo is unknown, since they caused moderate infection to the lettuce roots, and healthy roots are required for multiplication of the obligate parasitic *Olipidium.* Captan was less phytotoxic, but even so the roots were discolored and only a few zoospores were recovered from in vivo tests.[30] When *Olpidium* was killed by benomyl treatment during its first vegetative generation in lettuce seedlings, the frequency of infection by BVA was reduced.[29]

In field trials, more transplanted lettuce had big vein than did direct-seeded lettuce.

Benomyl drench prior to transplanting reduced the incidence of the disease. This allowed better uptake and distribution of benomyl in the roots which resulted in better control of the disease.[30] Recently, White[40] also confirmed that lettuce BVA is controlled by the application of carbendazim.

K. Tomato Black Ring and Raspberry and Strawberry Ringspot Viruses

Reduced incidence of soil-borne viruses in raspberries and strawberries after treatment with quintozene (pentachloronitrobenzene) at 20 kg ha^{-1} was reported as early as 1965 by Murant and Taylor.[41] These viruses are transmitted by the nematode *Longidorus elongatus* and quintozene controls nematodes in the field.

III. MECHANISM OF VIRAL INHIBITION IN PLANTS

There are two types of mechanisms by which carbendazim might inhibit TMV RNA in plants after inoculation.

A. Direct Inhibition

The means by which carbendazim treatment restricts TMV multiplication have been studied. TMV RNA synthesis was measurable in systemically infected leaves at the same point in time after the inoculation of lower leaves in control and carbendazim-treated plants. It is therefore unlikely that carbendazim slows the systemic spread of the virus through the host;[11] carbendazim might inhibit or reduce viral RNA synthesis directly by inhibiting the replicase. Other benzimidazole derivatives have been shown to inhibit certain animal viral RNA polymerase[42] and heterogeneous nuclear RNA synthesis by RNA polymerase II in HeLa cells.[43] It is conceivable that carbendazim inhibits TMV RNA synthesis directly. Experiments conducted by Fraser and Whenham[11,12] on the inhibition of multiplication of TMV by carbendazim showed that in the nucleic acids extracted from comparable systemically infected upper leaves of control and carbendazim-treated plants, 28 days after primary infection, the amount of TMV RNA in the carbendazim-treated leaves was only about 10% of the control. Carbendazim therefore strongly inhibited virus multiplication. In contrast, there was no significant effect of carbendazim on the accumulation of host nucleic acids; the peak areas of DNA and 25S rRNA were similar in gels of nucleic acid from control and carbendazim-treated leaves.[11]

The extent of inhibition of TMV multiplication by carbendazim treatment depended on the height of the leaf of the plant.[11] It was reported that 90% inhibition of TMV RNA synthesis occurred in the ninth leaf above the upper inoculated leaf. In the fifth leaf above the inoculated leaf, carbendazim treatment inhibited TMV multiplication by about 60%. In the inoculated leaf itself, TMV multiplication was inhibited only 20 to 30% by carbendazim treatment. The different levels of inhibition of TMV RNA accumulation in upper and lower leaves were probably not caused by differences in carbendazim concentration in different parts of the plants. The inhibition of TMV RNA accumulation in the upper leaves by carbendazim treatment was long-lasting. However, the rate of accumulation in treated leaves decreased earlier than in control leaves. The TMV RNA content remained fairly constant for a period of at least 10 days following the phase of virus RNA accumulation in carbendazim-treated leaves. In later studies Fraser and Whenham[12] also showed that in the lower leaves, if inoculated when they were almost fully expanded, carbendazim treatment had little or no effect on the pattern or amount of TMV RNA accumulation. However, in the higher systemically infected leaves, carbendazim treatment reduced the level of TMV RNA accumulation. They further observed that in leaves in which the accumulation of TMV RNA was reduced by carbendazim treatment, there was also a reduced level of infectivity.[12] Similarly, Tomlinson et al.[7] reported a lower level of infectivity in leaves of carbendazim-treated plants than in nondark green areas of infected control plant leaves. Fraser and Whenham[11] confirmed the results of Tomlinson et al.[7] that 100 days after inoculation, the levels of infectivity in the uppermost systemically infected leaves of carbendazim-treated and control plants were reduced. Bailiss et al.[13] found a slightly lower level of infectivity in TMV-infected tomato leaf discs floated on low concentration solutions of benomyl.

Fraser and Whenham[12] tested the variability in inhibition of TMV RNA in upper and lower leaves. First they thought that there might be uneven distribution of carbendazim in different parts of the plant and hence uneven effects on viral multiplication. Thus, the plants were exposed to a wide range of doses of carbendazim in an attempt to alter the internal carbendazim levels, and effects on leaf growth and TMV RNA accumulation in lower and upper leaves were tabulated. Tobacco plants were treated with a dosage of carbendazim with from 0.1 to 20 g and these plants were inoculated with TMV strains. The growth of the lower inoculated leaves was severely inhibited when plants were treated with 5 g or more carbendazim; in plants treated with 10 g carbendazim, no leaf growth occurred after the first application. Leaves of plants treated with 20 g carbendazim actually lost weight during the period of treatment, due to severe wilting. It was observed that the amount of carbendazim which will inhibit leaf growth is not constant, but depends on the size of the plant treated.

The effects of carbendazim on accumulation of TMV RNA in the inoculated leaves were complex. In plants treated with 1 or 2 g carbendazim, TMV RNA concentration equaled or was slightly lower than in untreated control plant leaves. With higher levels of carbendazim treatment, the concentration of TMV RNA was higher than in control plant leaves. This elevated concentration of the virus was due to the severe inhibition of leaf fresh weight increase by high levels of carbendazim. In plants treated with low doses of carbendazim, in the range of 0.1 to 0.5 g, the TMV RNA concentration in inoculated lower leaves was higher than in comparable leaves of untreated control plants. This increased concentration could result from stimulation by some means of TMV RNA synthesis in the inoculated leaves by low levels of carbendazim. Alternatively, the increased concentration of RMV RNA in inoculated leaves might occur as a consequence of the effect of low levels of carbendazim on the systemically infected upper leaves. It has been demonstrated by Fraser and Whenham[12] that one factor limiting the production of TMV RNA in lower, older leaves is the competition for precursors by synthetic activity in the upper leaves.[44] Carbendazim treatment of 0.1 to 0.5 g per plant greatly reduced the level of TMV RNA synthesis in upper leaves. This reduction in competition may allow increased TMV RNA accumulation in lower leaves.[12] In contrast to these results, Bailiss et al.[13] found that low levels of benomyl slightly reduced the level of infectivity reached in excised, TMV-infected tomato leaf discs, while high concentrations of benomyl caused increased infectivity. The different experimental systems used, and the different methods used to assay TMV, may have contributed to this apparent contradiction.

While conducting experiments for testing the direct inhibition of TMV RNA by carbendazim, Fraser and Whenham[11] examined rates of ^{32}P incorporation into TMV RNA in infected leaf discs newly exposed to carbendazim. The rate of incorporation of ^{32}P into S rRNA was followed by an internal control to detect any depression of RNA synthesis from nonspecific causes. It was found that 2 μg/mℓ carbendazim had little or no effect on ^{32}P incorporation into TMV RNA or 25S rRNA; 20 μg/mℓ carbendazim reduced incorporation into TMV RNA by 40% and into 25S rRNA by 60%. There was no further reduction in incorporation at 200 μg/mℓ but carbendazim is not completely soluble at this concentration. Total uptake of ^{32}P by the discs was actually slightly stimulated by 2 or 20 μg/mℓ carbendazim, so the inhibition of incorporation of ^{32}P into TMV RNA and 25S rRNA was likely to be an effect of carbendazim on RNA synthesis, not merely on precursor uptake. Fraser and Whenham[11] concluded from the experiments that there is no direct evidence for a general depression of RNA synthesis at higher levels of carbendazim; there is no evidence for any selective inhibition of TMV RNA synthesis by carbendazim. However, they observed the effects of the dosage level of carbendazim on growth and found significant inhibition of TMV RNA accumulation without any effect on 25S rRNA content; in another experiment on effects of carbendazim on RNA synthesis, negative effects were observed; however, by comparing both the experiments, it was concluded that carbendazim promotes inhibition of TMV RNA accumulation in old leaves and that carbendazim supplied early in the life of the leaf inhibits accumulation, while carbendazim supplied later, during the period of viral multiplication, does not.[12]

B. Indirect Inhibition

Systemic fungicides indirectly alter leaf metabolism in certain ways so as to make conditions in plants unfavorable for viral multiplication. Thomas[45] and Skene[46] proposed that a possible mechanism for the indirect suppression of TMV RNA accumulation by carbendazim is its cytokinin-like activity. Cytokinins have been reported to influence local lesion formation and activity.[43,47] Furthermore, cytokinins are known to delay leaf senescence. It has been shown that the major accumulation of TMV RNA in leaves coincides with the onset of the senescent phase of leaf development. Thus carbendazim, if applied to young leaves, may prevent viral multiplication by maintaining those leaves in a juvenile state unsuitable for viral multiplication. Carbendazim-treated plants have more leaves of a greater fresh weight than control plants.[11] This conclusion is consistent with the demonstrated failure of carbendazim to inhibit viral multiplication in old leaves unless applied at extremely high concentrations or very early in the life of the leaf.[11]

IV. PERSISTENT EFFECTS OF CARBENDAZIM ON GROWTH AND TMV RNA CONTENT

Fraser and Whenham[12] traced the TMV RNA content up to 30 days after inoculation, by which time it had reached fairly stable levels in most treatments. The absence of any apparent direct viricidal effect of carbendazim indicates a possibility that it may inhibit TMV multiplication by maintaining the leaf in some juvenile state. This may have a persistent effect on growth and TMV RNA content. Therefore, Fraser and Whenham[12] examined tobacco plants for 100 days after inoculation. These plants had completed vegetative growth and were well into the flowering phase. In these plants, there was still a very striking difference in growth between carbendazim-treated and control plants. Carbendazim-treated plants had more leaves of a greater fresh weight than control plants. The mean shoot weight of carbendazim-treated plants was over twice that of control plants. The concentration of chlorophyll a and b in treated plants remained higher than in control plants;[11,12] thus plants receiving carbendazim treatments 9 and 2 days before inoculation showed significant beneficial effects on plant growth which persisted for a long time. Differences in TMV RNA concentrations in carbendazim-treated and untreated plants 100 days after inoculation were observed. In lower leaves, TMV per gram was lower in carbendazim-treated plants than controls; the concentration of a these leaves corresponds to that of upper leaves 30 days after inoculation. The inhibition of TMV RNA accumulation in these leaves by carbendazim treatment thus appears to be highly persistent. In the upper leaves of carbendazim-treated plants, the concentration of TMV RNA at 100 days after inoculation was much higher than in controls. There is therefore an apparent paradox: that treated plants grew much better than control plants, but also contained much higher concentrations of virus. This result was possibly the consequence of two separate processes. During the vegetative growth phase of the plant, suppression of TMV RNA synthesis by carbendazim treatment prevented inhibition of leaf growth by the virus; thus, treated plants grew better than untreated control plants. However, once the plants had finished vegetative growth and entered the flowering phase, the enhanced growth resulting from carbendazim treatment was available as a source of substrates for viral synthesis.[11,12]

The enhanced growth of TMV-infected plants as a result of carbendazim treatment is accompanied by a reduction in the severity of visible symptoms[7] and reduction in viral RNA accumulation in actively growing parts of the plant. Fraser and Whenham[11] concluded that leaf growth, symptoms, and TMV multiplication are related in tobacco. They measured the expression of visible symptoms by the loss of chlorophyll. In young infected leaves, carbendazim dosages of more than 1 to 2 g per plant were required to maintain chlorophyll content at the same levels as in healthy leaves, i.e., 1 mg/g. Strong inhibition of viral multiplication was observed at dosages as low as 0.1 g carbendazim per plant. Thus it seems that there is no direct correlation between the amount of viral synthesis and expression of visible symptoms as measured by the loss of chlorophyll. In young infected leaves the levels of carbendazim treatment which prevented inhibition of leaf growth by TMV infection were much lower than those required to prevent chlorophyll loss.[11] Thus the beneficial effect of carbendazim treatment on growth of infected leaves was probably not brought about by prevention of chlorophyll loss or any resulting reduction in photosynthetic capacity.

Viral RNA accumulation was inhibited by levels of carbendazim treatment similar to those preventing inhibition of leaf growth. TMV reaches very high concentrations in tobacco leaves. Its multiplication requires the activity of a large proportion of the host nucleic acid and protein synthesizing capacity. Thus TMV multiplication is likely to act as a competitive inhibitor of the host growth. Fraser and Whenham[11] suggested that the enhanced growth of carbendazim-treated infected plants compared to control plants may result from the indirect inhibition of TMV multiplication by carbendazim treatment during the active growth phase of the plant.

Carbendazim at high levels inhibits growth of the host. Fraser and Whenham[11] compared the dosages required to inhibit viral multiplication and host growth. In old leaves treated late in their development with carbendazim, the dosages inhibiting viral multiplication and host growth were similar. It appears, however, that from the standpoint of preventing viral inhibition of leaf growth by carbendazim, this finding is not important, as these leaves had partly completed growth in any case. More significant is the finding that in young, actively growing leaves, the carbendazim dosage required to inhibit leaf growth was 50 to 100 times the dosage required to inhibit viral multiplication and viral inhibition of leaf growth. Carbendazim thus displays low phytotoxicity and permits significantly improved growth and appearance of the host.[7] One possible drawback to its use as a chemotherapeutic agent is that treated plants do eventually contain more virus than other plants.[13]

V. FUTURE OUTLOOK

Even since their identification, plant virus diseases have posed a great problem to plant pathologists. Since there was no direct control, indirect controls such as insect control, plant resistance, growing disease-free seeds or cutting, etc. were adopted. The introduction of systemic fungicides for the control of fungal diseases with nontarget effects on the viruses may prove to be a boon to agriculture for the control of these diseases.

Although there is no full explanation of the nontarget effects of systemic fungicides on viruses, fundamental and developmental work has now reached a relatively advanced stage. Further research is required on the mechanism of action of systemic fungicides as symptom suppressants and on the structure-activity relationships of the chemical and other related compounds to establish functional groups. Fraser and Whenham[11,12] have made some progress in establishing the effect of carbendazim on TMV. Such research is needed on other fungicides which demonstrate antiviral activity. Many other systemically translocated fungicides will not require testing as possible disease therapeutic agents, but in which direction the search should be directed has yet to be determined. That the search should include compounds already commercially available and used for other purposes is clear, not only from the results of the present review but also from the fact that the fungicides benomyl[48] and thiabendazole[49] were originally introduced as a mite ovicide and as an antihelminthic substance, respectively.

REFERENCES

1. **Erwin, D. C.,** Systemic fungicides: disease control, translocation and mode of action, *Annu. Rev. Phytopathol.*, 11, 389, 1973.
2. **Wain, R. L. and Carter, G. A.,** Historical aspects, in *Systemic Fungicides,* March, R. W., Ed., Longman, London, 1977, 6.
3. **Vyas, S. C.,** *Systemic Fungicides,* Tata McGraw-Hill Publishing, New Delhi, 1984.
4. **Mathews, R. E. F.,** *Plant Virology,* Academic Press, London, 1970.
5. **Gibbs, A. and Harrison, B.,** *Plant Virology, The Principle,* Edward Arnold, London, 1976.
6. **von Schmeling, B. and Kulka, M.,** Systemic fungicidal activity of 1,4-oxathiin derivatives, *Science,* 152, 659, 1966.
7. **Tomlinson, J. A., Faithfull, E. M., and Ward, C. M.,** Chemical suppression of symptoms of two virus diseases, *Ann. Appl. Biol.,* 84, 31, 1976.
8. **Clemons, G. P. and Sisler, H. D.,** Formation of fungitoxic derivatives from Benlate, *Phytopathology,* 59, 705, 1969.
9. **Tomlinson, J. A.,** Chemotherapy of plant virus diseases, *Proc. 1977 Br. Crop. Prot. Conf. Pests and Diseases,* 1, 807, 1977.
10. **Tomlinson, J. A., Ward, C. M., Webb, M. J., and Faithfull, E. M.,** Chemotherapy of plant virus diseases, *Natl. Veg. Res. Stn. Rep. for 1976,* p. 105, 1977.
11. **Fraser, R. S. S. and Whenham, R. J.,** Inhibition of multiplication of tobacco mosaic virus by methyl-benzimidazole-2-yl-carbamate, *J. Gen. Virol.,* 39, 191, 1978.
12. **Fraser, R. S. S. and Whenham, R. J.,** Chemotherapy of plant virus disease with methyl-benzimidazole-2yl-carbamate: effects on plant growth and multiplication of tobacco mosaic virus, *Physiol. Plant Pathol.,* 13, 51, 1978.
13. **Bailiss, K. W., Cocker, F. M., and Cassells, A. C.,** The effect of Benlate and cytokinins on the content of tobacco mosaic virus in tomato leaf discs and cucumber virus in cucumber cotyledon discs and seedlings, *Ann. Appl. Biol.,* 87, 383, 1977.
14. **de Fazio, G. and Kudamaatsu, M.,** Action of carbendazim on tobacco plant infected by tomato spotted wilt virus, *Biologico,* 50(3), 51, 1984.
15. **Osman, A. and El-Shaieb, M. K. Z.,** Suppression of broad bean true mosaic virus by fungicides, Proc. 5th Congr. Un. Phytopathol., Patras, Greece, 1981, 169.
16. **El-Banna, O. M. I.,** Studies on Broad Bean True Mosaic Virus (BBTMV) Disease, M.Sc. thesis, Cairo University, Egypt, 1979, 123.
17. **Raut, B. T., Somani, R. B., and Wangikar, P. D.,** Spectrophotometric assay of some systemic fungicides against papaya mosaic virus, *Pesticides,* 17(7), 22, 1983.
18. **Moghe, P. G.,** Influence of Heat and Systemic Chemicals on the Multiplication of Bean Common Mosaic Virus, Ph.D. thesis, Pennsylvania State University, University Park, 1974, 1.
19. **Rafiq, M., Jabri, A., Siddique, K. A., and Mohmood, K.,** Effect of Bavistin on the infectivity of zinnia mosaic virus, *Indian Bot. Rep.,* 3(1), 81, 1984.
20. **Russell, G. E.,** Some effect of spraying with thiabendazole on susceptibility of sugar beet to yellowing viruses and their vector, *Mycus persicae, Ann. Appl. Biol.,* 62, 265, 1968.
21. **Bhardwaj, S. V., Dubey, G. S., and Sharma, I.,** Effect of Benlate on infection and transmission of urdbean *(Vigna radiata* var. *munga)* leaf crinkle virus, *Phytopathol. Z.,* 105, 87, 1982.
22. **Bhardwaj, S. V.,** Virus-Vector-Host Relationship of Urdbean Leaf Crinkle Virus (ULCV), Ph.D. thesis, Himachal Pradesh Agricultural University, Solan, India, 1981, 126.
23. **Sharma, I.,** Studies on Leaf Crinkle Diseases of Urdbean *(Vigna radiata* var. *munga),* Ph.D. thesis, Himachal Pradesh Agricultural University, Solan, India, 1980, 1.
24. **Sharma, I. and Dubey, G. S.,** Control of urdbean leaf crinkle virus: heat treatment and chemotherapy, *Indian J. Mycol. Plant Pathol.,* 10(22), 1, 1980.
25. **Tomas, J. and John, V. T.,** Suppression of symptoms of tungro virus diseases by carbendazim, *Plant Dis.,* 64, 402, 1980.
26. **Waygood, E. R.,** Benzimidazole effect in chloroplast of wheat leaves, *Plant Physiol.,* 40, 1242, 1965.

27. **Tomlinson, J. A. and Webb, M. J. W.,** Ultrastructural changes in chloroplasts in lettuce infected with beet western yellow virus, *Physiol. Plant Pathol.,* 12, 13, 1977.
28. **Campbell, R. N.,** Effects of benomyl and ribavirin on the lettuce big vein agent and its transmission, *Phytopathology,* 70, 1190, 1980.
29. **Westerlund, F. V., Campbell, R. N., and Grogan, R. G.,** Effect of temperature on transmission, translocation and persistence of the lettuce big vein agent and big vein symptom expression, *Phytopathology,* 68, 921, 1978.
30. **Campbell, R. N., Greathead, A. S., and Westerlund, F. V.,** Big vein of lettuce: infection and methods of control, *Phytopathology,* 70, 741, 1980.
31. **Hills, F. J. and Leach, L. D.,** Photochemical decomposition and biological activity of *p*-dimethylaminobenzenediazo sodium sulfates (Dexon), *Phytopathology,* 52, 51, 1962.
32. **Noveroske, R. L.,** Dowco 269, a new systemic fungicide for control of *Phytophthora parasitica* of tobacco, *Phytopathology,* 65, 22, 1975.
33. **Paulus, A. O., Nelson, J., Snyder, M., and Gafney, J.,** Downy mildew of lettuce controlled by systemic fungicide, *Calif. Agric.,* 31(12), 9, 1977.
34. **Wheller, J. E., Hine, R. B., and Boyle, A. M.,** Comparative activity of Dexon and Terrazole against *Phytophthora* and *Pythium, Phytopathology,* 60, 561, 1970.
35. **Jacobson, B. J. and Williams, P. H.,** Control of cabbage clubroot using benomyl fungicide, *Plant Dis. Rep.,* 54, 456, 1970.
36. **Edgington, L. V., Khew, K. L., and Barron, G. L.,** Fungitoxic spectrum of benzimidazole compounds, *Phytopathology,* 61, 42, 1971.
37. **Sims, J. J., Mee, H., and Erwin, D. C.,** Methyl-2-benzimidazolecarbamate, a fungitoxic compound isolated from cotton plants treated with methyl-1-(butylcarbamoyl)-2-benzimidazole-carbamate(benomyl), *Phytopathology,* 59, 1775, 1969.
38. **Davidse, L. C.,** Mode of action, selectivity and mutagenicity of benzimidazole compounds, *Neth. J. Plant Pathol.,* 83(S), 135, 1977.
39. **Tomlinson, J. A. and Faithfull, E. M.,** Effects of fungicides and surfactants on the zoospores of *Olpidium brassicae, Ann. Appl. Biol.,* 93, 13, 1979.
40. **White, J. G.,** Control of lettuce big vein disease by soil sterilization, *Plant Pathol.,* 29, 124, 1983.
41. **Murant, A. F. and Taylor, C. E.,** Treatment of soil with chemicals to prevent transmission of tomato black ring and raspberry ring spot viruses by *Longidorus elongatus, Ann. Appl. Biol.,* 55, 227, 1965.
42. **Tamm, G. J. and Eggers, H. J.,** Specific inhibition of replication of animal viruses, *Science,* 142, 24, 1963.
43. **Sehgal, P. B., Derman, E., Molley, G. R., Tamm, I., and Darnell, J. E.,** 5,6-Dichloro-1-β-D-ribofuranosynlbenzimidazole inhibits initiation of nuclear heterogenous RNA chains in HeLa cells, *Science,* 194, 431, 1976.
44. **Fraser, R. S. S.,** Effects of two strains of tobacco mosaic on growth and RNA content of tobacco leaves, *Virology,* 47, 261, 1972.
45. **Thomas, T. H.,** Investigations into the cytokinin-like properties of benzimidazole-derived fungicides, *Ann. Appl. Biol.,* 76, 237, 1974.
46. **Skene, K. G. M.,** Cytokinin-like properties of the systemic fungicide benomyl, *J. Hortic. Sci.,* 47, 179, 1972.
47. **Aldwinckle, H. S.,** Stimulation and inhibition of plant virus replication in vivo by 6-benzylaminopurine, *Virology,* 66, 341, 1975.
48. **Delp, C. J. and Klopping, H. L.,** Performance attributes of a new fungicide candidate, *Plant Dis. Rep.,* 52, 95, 1968.
49. **Staron, T. and Allard, C.,** Properties antifongiques du 2-(4-thiazolyl) benzimidazole ou thiabendazole, *Phytiatr. Phytopharm.,* 13, 163, 1964.

Chapter 8

AIR POLLUTANTS

I. INTRODUCTION

Photooxidants, the principal component of which is ozone, have been described as the most damaging air pollutants to vegetation.[1] They induce premature senescence and degradation of the chloroplast in leaves. Ozone increases permeability to the membrane and induces the conversion of free sterol (FS), causing a net decrease in the FS content of the membrane and an increase in sterol glycoside (SG) and acylated sterol glycoside (ASG) in ozonated leaves over the tissue and chloroplast, which are free of ozone.

Various attempts have been made to prevent damage by air pollutants. In an effort to find an interim solution until adequate emission controls are established, researchers have looked at a number of possible ways to reduce ozone damage in plants. Protecting plants from ozone by the use of chemicals has been suggested as a method of reducing injury until emissions are lowered. However, progress has been limited by a lack of basic knowledge concerning the physiology of ozone injury.

Chemical protective methods, in which ozone is adsorbed or destroyed on the plant surface before entering the tissue, have been established.[2] Incorporation of chemicals into plant systems has been shown to reduce ozone injury. Compounds such as diphenylamine (DPA), phenothiazine, and 1,4-naphthoquinone have been sprayed on leaf surfaces while indole acetic acid, hydrazine, octodecene-1, ascorbic acid, and disphosphopyridine nucleotide have been applied to roots.[3] Many of these materials have no other commercial uses in or on plants, however; hence there is little likelihood that they will ever be marketed for ozone protection alone. If chemical inhibition of ozone injury is to become standard, it is obvious that effective chemicals must be found among those materials used for other types of plant protection. This would minimize the investment in research on toxicological and environmental effects normally associated with the development of new compounds.

II. NONSYSTEMIC FUNGICIDES

Rich and Taylor[4] made the first attempt to control weather fleck of tobacco by using protectant manganous 1,2-naphthoquinone-2-oxime and found that the compound protected against ozone damage. It proved to be an effective antiozonant at 100 ppm concentration. They also observed that cobaltous and manganous chelates of 8-quinolinol were also effective. The materials were applied to cloth of the type used to make field tents for shade-grown tobacco. Tomato plants also were protected against otherwise damaging concentrations of ozone. According to these authors the fungicides and method may afford a useful way to reduce weather fleck of tobacco and other plant injuries caused by excessive atmospheric ozone. Later attempts by Walker[3] with dithiocarbamates and by Jones[5] with other particular materials that would react with ozone before entering the plants met with only partial success. In fact, the actual breakthrough came with the introduction of systemic fungicides, which incidentally also possess antisenescent properties and alter the sensitivity of the plant at the biochemical and physiological level. These fungicides have also been applied to plants for the control of phytooxidant damage.

DPA at the rate of 1% in 0.1 Triton X-100 gave protection to strapbean cv. Tendercrop and tobacco cv. Bel W-3 seedlings when these plants where exposed to 25 ppm O_3.[6,7] Koiwai[8] reported that piperonyl butonide and piperonyl butoxide at 500 ppm reduced the weather fleck of tobacco and that synergistic activity was observed with pyrethrium insecticide; the application checked the reduction of free amino acid in affected plants.

III. SYSTEMIC FUNGICIDES

Ozone injury was expressed as yellow to tan necrotic flecks and varying degrees of chlorosis. Ozone injury appeared to predispose plants to increased drought damage under dry conditions and such plants become more susceptible to fungal diseases. Benomyl provided a high level of reduction in ozone injury to *Poa anua* at 20 to 200 μg active ingredient (a.i.) per cm^3 soil and a consistent trend of injury reduction was observed when it was applied as a soil drench.[9] Spraying at 500 to 1000 ppm benomyl reduced ozone damage on soybean exposed to ambient levels but had no effect on seed weight or yield.[10,11]

Pinto beans are susceptible to ozone at lower levels (0.05 to 0.10 $\mu\ell$ ℓ^{-1}) which occur in ambient air for longer periods. Manning et al.[12] conducted experiments on this subject and reported that benomyl soil amendments provided short-term protection to Pinto bean plants repeatedly exposed to ozone at 0.06 $\mu\ell$ ℓ^{-1} (6 parts per hundred million [pphm]) for 5-day periods. However, benomyl was not effective in overcoming the long-term deleterious effects of repeated exposures to a low level of ozone in the growth, reproduction, and nodulation of Pinto beans. The authors concluded that compounds being evaluated for the suppression of ozone injury should be used with plants grown in nonsterilized mineral soils and exposed to low levels of ozone for long periods of time.

An increase in weather flecking of tobacco following the application of nematicides was observed. This is due to the fact that nematicides applied for root injury will permit the roots to supply adequate water to the leaves. The leaves then remain turgid and stomata remain open; when the stomata are open, ozone can enter the leaf to cause fleck. Soil application of benomyl at the rate of 60 kg ha^{-1} reduced *Pratylenchus penetrans* populations over 50% and reduced the severity of fleck until late harvest in tobacco; however, benomyl appeared to stunt plant growth during the preceding 3 weeks, for the new leaves were small in benomyl-treated plots.[13,14]

Benomyl as a soil amendment or drench protected Pinto bean seedlings from ozone damage.[15,16] The hypothesis that the protection that benomyl provides against ozone injury is due to benzimidazole concentration was tested. Three concentrations of benzimidazole, benomyl, and thiabendazole were incorporated into a soil-peat-perlite (2:1:1) growth mixture; plants grown in treated soil mixtures were fumigated for 4 hr with 25 pphm ozone. Benzimidazole and benomyl protected the plants from ozone injury; thiabendazole did not.[16] Through a bioassay technique it was established that the failure of thiabendazole to protect against ozone injury may have been caused by a low uptake of the chemical. The technique involved measurement of the zone of *Penicillium cyclopium* growth inhibition surrounding a leaf disc placed on an inoculated agar plate.

Benomyl reduces ozone injury to Pinto bean, annual bluegrass, and petunias in vitro.[9,12] In conjunction with these studies, a number of field trials were established. Ozone-sensitive species — bean (cvs. Tempo and Pinto III), azalea (cv. Snow), tobacco (cv. Conn 7272), and grape (cvs. Ires and Concord) — were planted at a number of locations with acknowledged air pollution problems. Untreated plants exhibited typical ozone injury symptoms whereas benomyl-treated plants were protected.[12,17-20]

Thiophanate has also been shown to provide adequate protection to Pinto beans under greenhouse conditions.[2] When applied as a foliar spray to azaleas, 500 to 750 μg mℓ^{-1} was an effective rate of application under field conditions.[20] Kitano et al.[21] reported that thiophanate-methyl at 1.5 g per plant or a 0.5% solution of benzimidazole prevented chlorophyll degradation and ozone damage in mature tobacco leaves.

Most of the experiments reported so far relate to the control of ozone injury of tobacco and beans under controlled conditions with known dosages of a single pollutant. It has been shown that a synergistic interaction exists when ozone is present in combination with other pollutants such as sulfur dioxide and nitrogen dioxide, resulting in greater damage than that induced by a single pollutant. Hence, there is need for much more information concerning the effectiveness of chemical protectants under outdoor conditions if practical use is to be made of such protection.

Foliar spray of azalea plants with benomyl, thiophanate-methyl, or carboxin reduced oxidant injury, when applied at 14-day intervals.[20] Soil application, drench, and foliar application were made at the rate of 60 to 100 $\mu g/cm^3$, 1.12 to 100 $\mu g/cm^3$, and 500 μg to 750 μg $m\ell^{-1}$, respectively. Benomyl applied as a soil drench at 10 lb a.i. acre^{-1} and as a spray at 1.5 lb a.i. acre^{-1} (three sprays) and foliate, a hydrocarbon emulsion acting as a transpiration suppressant, applied as a spray at 1 or 5% both failed to control weather fleck of tobacco.[22]

Lewis and Brennan[23] reported that an ozone and sulfur dioxide mixture caused a paracetyl nitrate (PAN)-type injury to petunias. The undersurface of newly matured leaves became glazed, especially at the apex of the youngest susceptible leaf; they reported that foliar treatment of petunia plants by benomyl afforded some protection against the mixture of ozone and sulfur dioxide.

Potato plants (cv. Norchip) were given soil treatment of benomyl and DPX-4891 at 3-week intervals at field rates. DPX-4891 significantly reduced the amount of foliage damage and injury resulting from elevated ozone levels, but benomyl had no effect. Application of antioxidants had no effect on yield irrespective of the degree of control achieved.[24] Although certain systemic fungicides are effective in reducing ozone injury in vitro, it is difficult to achieve the same results in vivo. In laboratory studies, plants are usually exposed to short durations of an acute ozone level, whereas in the field vegetation is exposed to low ozone concentrations for many days during the summer; when ozone damage does occur, ozone levels are usually much lower than the levels used in the laboratory. Therefore, field studies are necessary to determine the economic feasibility of spray programs for the control of ozone injury. Curtis[25] reported that carboxin and carboxin-monooxide sprays at the rate of 2.2 kg ha^{-1} protected white beans from ozone bronzing for 6 to 9 days under laboratory conditions, but after 15 days they had no effect. Similar results were also reported by Rich et al.[26] for beans, soybeans, cotton, tomatoes, and tobacco grown in carboxin-treated soil (9.5 ppm) and subsequently drenched twice with carboxin at the same concentration. Since it is not economically feasible to spray agricultural crops at 1- to 2-week intervals with a chemical to control ozone injury, timing becomes very critical for successful use in field conditions. Curtis et al.[27] and Manning and Vardaro[28] observed that carboxin controls ozone injury on white beans under field conditions. Application of carboxin (50%) diluted 500 times and oxycarboxin (50%) diluted 1000 times to tobacco reduced ozone injury, but higher concentrations were toxic in young leaves. Oxycarboxin diluted 1500 times protected tobacco leaves from oxidant injury without any phytotoxicity; however, an oxycarboxin and thio-phanate-methyl mixture was toxic.[29,30] Similar results were also reported by Fukuda et al.[31]

Carboxin significantly reduces oxidant injury on white beans as a foliar application[25] and when applied to Pinto beans as a soil drench.[32] However, as a soil drench it is not effective on azaleas.[9] The lack of protection could possibly have been due to retarded uptake caused by peat amendments in the planting and/or a greater lignin concentration in the azalea plant than would be found in an herbaceous plant. Carboxin has been shown to be inhibited by lignin.[33] This suggests that the use of carboxin may be limited to herbaceous plants and planting beds which do not contain lignaceous materials. Curtis et al.[27] reported that carboxin is an effective protectant against ozone injury because it readily oxidizes to its sulfoxide in the plant. Pyracarbolid, a carboximide with a carbon substituted for the sulfur in the fourth position of the oxathiin ring of carboxin, was ineffective in reducing ozone injury. Curtis et al.[27] also observed the toxicity of pyracarbolid sprays to white beans. Successful ozone control with carboxin was dependent upon the timing of the spray. The chemical must be applied 1 to 2 days before the incidence of high ozone levels to ensure sufficient time for the chemical to become active in the plant and reduce plant sensitivity to ozone. Depending on the ozone concentration,[25] the chemical must remain active in the plant for at least 4 days, since ozone levels may be high for 3 to 4 consecutive days. Since carboxin moves in the transpiration stream of the leaves and accumulates in the leaf margins and tips, chemical protection will last for only 6 to 9 days, depending on the transpiration rate. According to Curtis et al.,[27] the timing of a single spray of carboxin is critical if increased yields are expected. White beans are particularly sensitive to ozone injury from full bloom until seed set. Therefore, if ozone levels can be predicted during this sensitive period, ozone damage on white beans may be reduced by one single spray of carboxin.[27]

Ethylenediurea (EDU) showed more antioxidant properties than carboxin and benomyl in suppressing ozone injury of three *Phaseolus vulgaris* cultivars at four sites in Ontario, Canada. While seed yields at two sites were unaffected by the application of any of the three chemicals, yields at the other two sites were increased from 1.57 to 1.69 t ha^{-1} to 2.04 to 2.31 t ha^{-1} with the application of EDU. Carboxin and benomyl caused similar increases in yield; part of the yield increase was due to large seed size. EDU was almost effective in reducing bronzing and delaying leaf drop. The effectiveness of the chemical in reducing ozone injury was influenced by the date of application, amount of bronzing on the crop, and the sensitivity of the cultivars to ozone.[34]

Triarimol and EL-279 were shown to suppress ozone injury to 11-day-old greenhouse- and chamber-grown *Phaseolus vulgaris* Pinto III bean seedlings fumigated with air containing 25 to 250 ppm ozone for 4 hr; treatments included soil amendments at 2, 5, and 10 μg g^{-1} soil and foliar sprays of 20, 50, 100, and 200 μg mℓ$^{-1}$ spray solution. Both the fungicides and treatments significantly reduced ozone injury.[2] Triarimol appears to provide greater ozone protection and greater greening and internode shortening than benomyl when tested on bean plants. EL-279 appeared to be more active in this respect than triarimol. Treated plants were not only severely inhibited in internode growth but also had markedly smaller leaf areas. The growth reduction may be related to an age factor wherein immature plants are most affected by this particular effect of the chemical and, due to its continuing systemic effect, the plants are never able to recover. Covey[35] noted similar effects of increased greening and shortened internodal length on apple seedlings sprayed with triarimol. On mature plants such as bearing apple trees, however, stunting or phytotoxic effects have never been observed and triarimol is considered to be extremely safe. On mature turfgrass on field stands, greening of treated areas is commonly observed, although stunting or any other evidence of phyto- toxicity is not. In creeping bent grass 10 μg g^{-1} soil resulted in stunting of seedlings; plants grown from seed in soil treated with 80 μg mℓ$^{-1}$ water were not subjected to stunting or dry weight reduction.[36]

Field plots of bean (cvs. Pinto III and Tempo) were used to determine effectiveness of benomyl, carboxin, and triarimol in the suppression of foliar oxidant injury under natural conditions. They were applied at 0.5, 1, 2, and 3 lb/100 gal, 8 and 24 ppm, and 50, 70, and 100 ppm, respectively, once a week for 4 weeks as a foliar spray beginning on the fifth day after seedling emergence. Higher doses of benomyl prevented ozone injury and others were not effective. Carboxin as granules at 2.7, 5.3, and 7.9 g/15-ft row was applied over seeds at planting time; again higher doses provided complete control. Carboxin as granules gave protection for 40 days but later the plant became susceptible to ozone.[28] Higher doses of carboxin and triarimol were toxic; the phytotoxic symptoms appeared as yellowing, burning, and cupping of leaves. Similar symptoms have been observed with tobacco grown in carboxin-treated soil.[19] Leaves treated with triarimol showed as much injury as untreated controls and those treated at 75 to 100 μg mℓ$^{-1}$, which exhibited cupping, dwarfing, and greening. The adverse effect of triarimol on bean leaves and internodes has also been previously noted.[36]

The ability of seed treatment of benomyl to protect Pinto bean plants from injury by PAN or combinations of PAN and ozone was tested by Pell.[37] After a 7-day treatment with 0, 60, 80, and 100 μg benomyl per gram of soil in 15-day-old Pinto bean plants which were exposed to 745 μg/m³ PAN or 745 μg/m³ PAN + 492 μg/m³ ozone for 3 hours, he observed that benomyl did not protect the young primary leaves from either pollutant. However, when 20-day-old plants were exposed to the same dosage of ozone and PAN following a 7-day benomyl treatment, the older primary leaves were protected.

The response of petunia *(Petunia hybrida)* cvs. White Cascada and Coral Magic to the pollutant PAN was tested 7 days after application of soil drench of benomyl at the rate of 0 to 120 μg g^{-1} dry weight of soil; susceptible cultivars showed increased sensitivity to PAN at 60 μg g^{-1} or higher concentrations. Foliage of White Cascada treated with lower concentrations of benomyl responded to PAN similarly to plants which were not treated with fungicides. The PAN sensitivity of tolerent cv. Coral Magic was unaffected by benomyl application.[38]

IV. MODE OF ACTION

Benomyl is effective in reducing ozone damage to Pinto bean and tobacco plants. Benomyl is converted to methyl benzimidazole-2-yl carbamate (carbendazim).[39,40] The benzimidazole concentration of the benomyl molecule is reported to be responsible for the reduction in ozone damage in Pinto beans[41] and spinach.[42] When protection of ozone-sensitive species was observed in the field, it was speculated that benomyl prevented oxidant damage per se.[17,18] There are two hypotheses which could explain prevention of air pollution injury to ozone-sensitive plants by benomyl treatments:[37]

1. Ozone was the only oxidant present in phytotoxic concentrations at the time these studies were conducted.
2. Benomyl protected all phytotoxic oxidants that were present.

The effect of ozone on plant membranes resulting in subsequent changes in ion flux and water relations has been observed. Sublethal ozone concentrations caused a significant decrease in permeability of potato tuber disks and tobacco leaves to phosphate and rubidium ions.[41] Similar observations have been made in pea chloroplasts. Ozone reduced the reflection coefficient of glycerol, indicating a marked increase in permeability of the chloroplast-limiting membrane. Thus, ozone altered the membrane permeability of several plant species under various exposure conditions and a difference in effect between sublethal ozone levels and toxic levels was observed. The effects of benzimidazole on ion flux and water relations, which may be related to alteration of plant membranes, have been studied. Benzimidazole increases K^+, Na^+ and Ca^{2+} uptake in excised barley roots and K^+ uptake in intact barley roots. It was effective in decreasing the loss of previously absorbed K^+ when roots were subjected to low pH. A similar effect of benzimidazole on K^+ uptake in tobacco was observed.[43]

Benzimidazole treatment of *Phaseolus vulgaris* plants increased water efflux of normal plants and also O_3-treated plants and reduced electrolytic leakage after O_3 treatment, but had no effect on osmotic potential. It is hypothesized that benzimidazole changes some structural components of plant cell membranes and that these membranes are then less subject to O_3-induced damage; O_3 at 0.25 μℓ ℓ$^{-1}$ caused a decrease in the cholesterol content of *P. vulgaris* leaves but had no effect on campesterol, stigmasterol, or β-sitosterol. Treatment with cholesterol maintained increased susceptibility, indicating that cholesterol may be a factor in the O_3 resistance of *P. vulgaris*. The cholesterol content of the leaves was increased by benzimidazole.[44]

Spotts et al.[45] studied the mechanism of benomyl-induced resistance against ozone damage using bean (cv. Pinto III) plants grown under controlled conditions. Water flux, water potential, electrolyte leakage, and osmotic potential measurements were made immediately and 48 hr after ozonation. Benzimidazole treatment alone increased water efflux of plants and water influx of ozonated plants and appeared to increase the calculated hydraulic conductivity for these fluxes; benomyl also decreased electrolytic leakage and had no effect on osmotic potential or water potential. Benomyl prevented the immediate ozone-induced electrolytic leakage, increased water potential, and decreased osmotic potential. It is proposed that benomyl changes some structural components of plant cell membranes.[45]

Antisenescent compounds[2,16] have the ability to inhibit yellowing and chloroplast degradation and thus protect the plants from phytooxidant damage. Benzimidazole stimulates chlorophyll formation,[46] increases the number of intergrana and grana chloroplast lamellae,[47] and conserves chlorophyll and protein.[48] Benzimidazole *n*-benzylaldehyde, and kinetin retard senescence of bean leaves and inhibit ozone injury to chloroplasts exposed to ozone. Carbendazim also possesses cytokinin-like properties on host plants. Since cytokinins are antisenescent, they can reduce ozone injury and associated biochemical changes.[49-51] There is evidence that these compounds inhibit senescence by preventing changes in membrane permeability.[52] It may be suggested that they also inhibit ozone injury by inhibiting changes in membrane permeability. One way that ozone may increase permeability is to induce the conversion of FS to SG and ASG, causing a net decrease in the FS content of membranes.[53,54] Tomlinson and Rich[51] reported that whole tissue and chloroplasts from leaves exposed to ozone ($0.5\ \mu\ell\ \ell^{-1}$) for 1 hr contained less FS and more SG and ASG than did tissue and chloroplasts from nonozonated leaves. Chloroplasts from bean leaves contained 40% of the total FS in while leaf tissue and ozone decreased the FS content of leaves and chloroplasts by 25 and 21%, respectively.

Ozone damage altered Hill's reaction in the photosynthesis of lipid peroxide formation and the chemical pitromyl bentoxide and systemic fungicides take effect at that point and prevent ozone damage. Similarly, the spinach leaves contained 37% of the FS and ozone decreased in the whole tissue and chloroplast by 44 and 39%, respectively, after ozonation. Tomlinson and Rich[51] reported that benzimidazole, *N*-benzyladenine, and kinetin inhibited ozone injury and also inhibited senescence in bean and spinach leaves. They further observed that these resistant plants did not lose FS from their membranes or show symptoms of cell leakage during ozonation. Therefore, the factor in common among chemicals that have both antisenescent and ozone-resistant properties may be their ability to maintain normal membrane structure and functions.

Experiments were performed to measure the effects of ozone and benzimidazole compounds on the cholesterol, campesterol, stigmasterol, and β-sitosterol content of Pinto bean leaves. Pretreatment with benzimidazole did not prevent cholesterol decrease but benzimidazole did increase the cholesterol content of nonozonated plants and of ozonated plants 48 hr after exposure. Pretreatment with cholesterol protected plants against ozone damage; pretreatment with steroid inhibitor increased the susceptibility of plants to ozone. This evidence suggests that cholesterol may be a factor in the ozone resistance of beans.[45]

REFERENCES

1. **Heggestad, E. H.,** Consideration of air quality standard for vegetation with respects to ozone, *J. Air Pollut. Control Assoc.,* 19, 424, 1969.
2. **Seem, P. C., Cole, H., and Lacasse, N. L.,** Suppression of ozone injury to *Phaseolus vulgaris* Pinto bean III with triarimol and its monochlorophenyl cyclohexyl analogue, *Plant Dis. Rep.,* 56, 386, 1972.
3. **Walker, E. K.,** Evaluation of foliar sprays for control of weather fleck of flue cured tobacco, *Can. J. Plant Sci.,* 47, 99, 1967.

4. **Rich, S. and Taylor, G. S.,** Antiozonants to protect plants from ozone damage, *Science,* 132, 150, 1960.
5. **Jones, J. L.,** Ozone damage: protection for plants, *Science,* 140, 1317, 1963.
6. **Lisk, D. J.,** Protecting plants against injury from air pollution, *N.Y. Food Life Sci. Q.,* 8(4), 3, 1975.
7. **Gilbert, H. D., Maylin, G. A., Elfring, D. C., Edgerlon, J. J., Gutermann, W. H., and Lisk, D. J.,** The use of diphenylamine to protect plant against ozone injury, *Hortic. Sci.,* 10, 228, 1975.
8. **Koiwai, A.,** Studies on protection chemicals against ozone injury to tobacco leaves, *Bull. Okayama Tob. Exp. Stn. (Okayama Tab. Shikenjo Hokoku),* 38, 1, 1977.
9. **Moyer, J. W., Cole, H., Jr., and Lacasse, N. L.,** Suppression of azalea plants by drench or foliar spray treatment with benzimidazole or oxathiin compounds, *Plant Dis. Rep.,* 58, 136, 1974.
10. **Walker, J. T. and Melin, J.,** Effectiveness of benomyl and oxamyl to reduce oxidant damage under greenhouse conditions, *Proc. Am. Phytopathol. Soc.,* 1 (Abstr.), 167, 1976.
11. **Walker, J. T. and Melin, J.,** Response of soybean to antioxidant sprays at ambient ozone levels in Georgia, *Plant Dis. Rep.,* 62, 400, 1978.
12. **Manning, W. J., Feder, W. A., and Vardaro, P. M.,** Benomyl in soil and response of Pinto bean plants to repeated exposures to a low level of ozone, *Phytopathology,* 63, 1539, 1973.
13. **Taylor, G. S.,** Tobacco protected against fleck by benomyl and other fungicides, *Phytopathology,* 60 (Abstr.), 60, 578, 1970.
14. **Miller, P. M. and Taylor, G. S.,** Effects of several nematicides and benomyl on the incidence of weather fleck of tobacco, *Plant Dis. Rep.,* 54, 672, 1970.
15. **Pellisser, M.,** Effect of Foliar and Root Treatments of Benomyl in Reducing Ozone Injury to Pinto Bean and Cucumber, M.S. thesis, Pennsylvania State University, University Park, 1971, 1.
16. **Pellisser, M., Lacasse, N. L., and Cole, H.,** Effectiveness of benzimidazoles benomyl and thiabendazole in reducing ozone injury to pinto bean, *Phytopathology,* 62, 580, 1972.
17. **Kender, W. L.,** Benomyl protection of grapevines from air pollution injury, *Hortic. Sci.,* 8, 396, 1973.
18. **Manning, W. J., Feder, W. A., and Vardaro, P. M.,** Suppression of oxidant injury by benomyl: effects on yield of bean cultivars in the field, *J. Environ. Qual.,* 3, 1, 1974.
19. **Taylor, G. S. and Rich, S.,** Ozone injury to tobacco in field influence by soil treatment with benomyl and carboxin, *Phytopathology,* 64, 814, 1974.
20. **Moyer, J. W., Cole, J., Jr., and Lacasse, N. L.,** Reduction of ozone injury on *Poa annua* by benomyl and thiophanate, *Plant Dis. Rep.,* 58, 41, 1974.
21. **Kitano, H., Shinohara, T., Yamamoto, Y., Kimura, T. I., Koiwai, A., Kisaki, T., and Fukuch, M.,** Studies on weather fleck on tobacco leaves. XX. Effects of some protective chemicals for weather fleck on the production of tobacco leaves, *Bull. Okayama Tob. Exp. Stn. (Okayama Tab. Shikenjo Hokoku),* 35, 87, 1975.
22. **Rhodes, F. M. and Jappan, W. B.,** Influence of weather fleck control chemical on yield and quality of cigar wrapper tobacco, *Tob. Int. (N.Y.),* 177(16), 37, 1975.
23. **Lewis, E. and Brennan, E.,** Ozone and sulphur dioxide mixture cause a PAN-type injury to petunia, *Phytopathology,* 68, 1011, 1978.
24. **Clarke, B., Henniger, M., and Brennan, E.,** The effect of two antioxidants on foliar injury and tuber production in Norchip potato plants exposed to ambient oxidant, *Plant Dis. Rep.,* 62, 715, 1978.
25. **Curtis, L. R.,** Control of Ozone Injury on White Beans with Oxathiin Analogues, M.Sc. thesis, University of Guelph, Ontario, 1973, 1.
26. **Rich, S., Ames, R., and Zukel, J. W.,** Oxthiin derivatives protect plants against ozone, *Plant Dis. Rep.,* 58, 162, 1974.
27. **Curtis, L. R., Edgington, L. V., and Little-Johns, D. J.,** Oxathiin chemicals for the control of bronzing of white beans, *Can. J. Plant Sci.,* 55, 151, 1975.
28. **Manning, W. J. and Vardaro, P. M.,** Suppression of oxidant air pollution on bean plants by systemic fungicides under field conditions, *Phytopathology,* 63, 204, 1973.
29. **Shoda, M., Imaizumi, S., Wada, Y., and Suyama, I.,** Effect of oxathiin compounds spray on oxidant injury to tobacco plants, *Proc. Crop Sci. Soc. Jpn.,* 44, 178, 1975.
30. **Shoda, M., Imaizumi, S., Wada, Y., and Suyama, I.,** Studies on weather fleck on tobacco leaves. XXI. Effect of oxathiin compound spray on reducing oxidant injury to tobacco leaves, *Bull. Okayama Tob. Exp. Stn. (Okayama Tab. Shikenjo Hokoku),* 35, 97, 1975.
31. **Fukuda, M., Kisaki, T., Koiwai, A., Kitano, H., Shinohara, T., and Tanaka, Y.,** Studies on weather fleck on tobacco leaves. XIX. Screening of chemicals for control of ozone injury to tobacco, *Bull. Okayama Tob. Exp. Stn. (Okayama Tab. Shikenjo Hokoku),* 35, 75, 1975.
32. **Anon.,** Experimental Plant Smog Protectants, Uniroyal, Bethway, Conn., 1972, 1.
33. **Chin, W. T., Stone, G. M., Smith, A. E., and von Schmeling, B.,** Fate of carboxin in soil, plants and animals, *Proc. 5th Br. Insectic. Fungic. Conf.,* 3, 322, 1969.
34. **Hopstra, G., Little-Johns, D. A., and Wukasch, R. J.,** The efficacy of antioxidant ethylene diurea (EDU) compared to carboxin and benomyl in reducing yield losses from ozone in navy bean, *Plant Dis. Rep.,* 61, 350, 1978.

35. **Covey, R. P.,** Orchard evaluation of two new fungicides for the control of apple powdery mildew, *Plant Dis. Rep.,* 55, 514, 1971.

36. **Seem, R. C., Cole, H., Jr., and Lacasse, N. L.,** Suppression of ozone injury to *Phaseolus vulgaris* L. with thiophanate-ethyl and its methyl analogue, *J. Environ. Qual.,* 2, 314, 1973.

37. **Pell, E. J.,** Influence of benomyl soil treatment on Pinto bean plants exposed to peroxyacetyl nitrate and ozone, *Phytopathology,* 66, 731, 1976.

38. **Pell, E. J. and Gardner, W.,** Enhancement of peroxyacetyl nitrate injury to petunia foliage by benomyl, *Hortic. Sci.,* 14(1), 61, 1979.

39. **Peterson, C. A. and Edgington, L. V.,** Entry of pesticides into the plant symplast as measured by their loss from an ambient solution, *Pestic. Sci.,* 7, 482, 1976.

40. **Clemons, G. P. and Sisler, H. D.,** Formation of fungitoxic derivative from Benlate, *Phytopathology,* 59, 705, 1969.

41. **Spotts, R. A., Lukezic, F. L., and Hamilton, R. H.,** The effect of benzimidazole on some membrane properties of ozonated pinto bean, *Phytopathology,* 65, 39, 1975.

42. **Tomlinson, J. A.,** Chemotherapy of plant virus diseases, *Proc. 1977 Br. Crop Prot. Conf. Pests and Diseases,* 1, 807, 1977.

43. **Dyar, J. J.,** Certain effects of benzimidazole on young tobacco plants, *Plant Physiol.,* 43, 477, 1968.

44. **Spotts, R. A.,** The effects of benzimidazole on some membrane properties, leaf sterols and ozone resistance of *Phaseolus vulgaris, Diss. Abstr. Int. B,* 35, 5224, 1975.

45. **Spotts, R. A., Lukezic, F. L., and Lacasse, N. L.,** The effect of benzimidazole, cholesterol and steroid inhibitor on leaf sterols and ozone resistance of bean, *Phytopathology,* 65, 45, 1975.

46. **Wang, D. and Waygood, E. R.,** Effect of benzimidazole and nickel on the chlorophyll metabolism of detached leaves of Khapli wheat, *Can. J. Bot.,* 37, 743, 1959.

47. **Waygood, E. R.,** Benzimidazole effect in chloroplast of wheat leaves, *Plant Physiol.,* 40, 1242, 1965.

48. **Person, C., Samborski, D. J., and Forsyth, F. R.,** Effect of benzimidazole on detached wheat leaves, *Nature (London),* 180, 1294, 1957.

49. **Skene, K. G. M.,** Cytokinin-like properties of the systemic fungicide benomyl, *J. Hortic. Sci.,* 47, 179, 1972.

50. **Thomas, T. H.,** Investigation into the cytokinin-like properties of benzimidazole-derived fungicides, *Ann. Appl. Biol.,* 76, 237, 1974.

51. **Tomlinson, H. and Rich, S.,** Effects of ozone on the sterols and sterol derivatives in bean leaves, *Phytopathology,* 61, 1401, 1971.

52. **Sacher, J. A.,** Permeability characteristics and amino acid incorporation during senescence (ripening) of banana tissue, *Plant Physiol.,* 41, 701, 1965.

53. **Grunwald, C.,** Effects of free sterols, sterol esters and sterol glycoside on membrane permeability, *Plant Physiol.,* 48, 653, 1971.

54. **Grunwald, C.,** Effects of free sterols on the permeability of alcohol treated red beet tissue, *Plant Physiol.,* 32, 484, 1968.

Chapter 9

PHYTOTOXICITY

I. INTRODUCTION

Phytotoxicity or injury to the host plant is often a serious problem with experimental and commercial fungicides. This occurs even when the fungicides are used at lower or recommended doses, because most fungicides are relatively nonspecific in their modes of action. They function by disrupting cell membranes through surface effects, precipitating enzymes and other macromolecules, or reacting indiscriminately with amino acids, peptides, and other intermediary metabolites. This is because fungi are relatively undifferentiated organisms which do not possess specialized structures, such as chloroplasts, nerve tissue, or erythrocytes, which can be acted upon by the fungicides selectively. Hence, most of the fungicides are general protoplasmic poisons which, given the opportunity, can be as injurious to the host as to the invader. Phytotoxicity to the host plant is often a limiting factor in the use of fungicides. This must be expected, since most chemicals used as plant protectants can react with vital biological processes and interfaces. However, they are less specific in their action with other pesticides and are thus likely to be phytotoxic if they can penetrate the cuticularized tissue protecting the host.

Phytotoxicity is a phenomenon of adverse interaction and reaction of the host and fungicides, which is manifested in the form of a range of symptoms on treated plants such as necrosis, defoliation, stunting, etc. This becomes even more evident when fungicides are sprayed at recommended or lower rates. Although phytotoxicity may seem to be a simple effect, it is not, since it is influenced by the concentration, configuration, dosage, formulation, and method of application of the compound, as well as the host and its cultivar and atmospheric conditions. Compounds such as dichlone, which otherwise can be used safely on many plants, will cause severe burning when formulated in soils. It is possible that the fungicide dissolves in the hydrocarbon, which enables it to permeate the leaf tissue more efficiently. The literature available on the subject analyzed below in order to categorize the phytotoxic effects of fungicides for agricultural scientists and farmers.

II. NONSYSTEMIC FUNGICIDES

A. Sulfur

The phytotoxic effect of sulfur when used as a fungicide is a very serious problem. It is phytotoxic to apples, pears, gooseberries, currants and cucurbits. The symptoms are scorching, burning, dwarfing of foliage, a reduction in yield, and sometimes premature defoliation. The most severe symptoms are usually indicative of lime-sulfur injury. Elemental sulfur more frequently results in slight dwarfing and burning. The degree of phytotoxicity varies with varieties of the crop. The varieties which are more susceptible to the phytotoxic action of fungicides have been commonly called "sulfur shy" or "copper shy".

Sulfur dust was less phytotoxic to young apple trees than lime-sulfur.[1] While comparing two different mixtures of sulfur it was found that the more dilute mixture of lime-sulfur was less phytotoxic than the mixture containing more lime; however, when effects of lime-sulfur were compared to flotation sulfur, less growth and yield and more leaf injury were observed.

Lime-sulfur is more phytotoxic than elemental sulfur. The phytotoxic effects of lime-sulfur are more apparent before the material had dried on the leaf.[2] Low-volume sprays of lime-sulfur on apple are less toxic than high volume. The material applied with the low-volume application was dry almost immediately upon contact with the leaves. The reason that the elemental form of sulfur causes phytotoxicity is uncertain and the subject of much speculation. In order to reduce phytotoxicity, lime-sulfur was precipitated with ferrous sulfur; this removed the more soluble sulfide which was believed to be responsible for plant injury. Phytotoxic effects of hydrogen sulfide and lime-sulfur were investigated on several plant species.[3] It was found that symptoms produced by lime-sulfur and hydrogen sulfide were similar in many respects. In both cases increased temperature caused increased phytotoxicity.

Acute injury by sulfur is rare in temperate climates, but in warmer climates (above 80°F) severe burning is sometimes caused on cucurbits when sulfur is used for powdery mildew control. Apple treated with sulfur in semiarid areas may develop lesions on the sun-exposed side of the fruit, an injury attributed to sulfur sun scald.[4] Some fruit cultivars are sulfur sensitive, such as apple varieties Stirling Castle and Lane's Prince Albert. Toxicity by sulfur-containing fungicides has also been noticed when they are applied during the blossoming period.[5] Sulfur placed on the stigma of apple blossoms prevents germination of the pollen and occasionally reduces fruit set to a considerable degree. It was shown that sulfur exerted its inhibitory action on the pollen grains only when they were in actual contact with or in very close proximity to it, and that fertilization took place when the pollen tube had entered the stylet. The inhibitory action of sulfur on the germination of apple pollen is consistent with its fungicidal action on hop powdery mildew. It has been observed that lime-sulfur when applied to green leaves considerably reduces photosynthesis by these green leaves.[6]

The effect of fungicides on the host may be modified by environmental factors and a compound may cause injury under some conditions, but not others. For example, damage by sulfur fungicides is exacerbated by heat,[7] probably due to vapor phase action. Similarly, the fungicide dichlofluanid, safe on strawberry plants in the open, can cause damage when used under closed conditions.[8]

The mechanism of sulfur phytotoxicity has not yet been identified, but a reduction in carbon dioxide assimilation appears to be a fundamental factor underlying the phytotoxic effects.[7] Sulfur results in a decrease in photosynthesis, but there is also the possibility that it might increase respiratory activity rather than cause a decrease in carbon dioxide assimilation.[9] Christopher[1] also reported that lime and flotation on sulfur reduced carbon dioxide assimilation and that the reduced assimilation was much more marked with lime-sulfur than with elemental sulfur. The reduced photosynthetic activity resulting from lime-sulfur application was much more marked at high temperatures than moderate ones.

B. Copper

Until World War II, copper-based fungicides, especially Bordeaux mixture (BM), were used widely, but their use has decreased since the introduction of dithiocarbamates. These can be used on seed, soil, and foliage. The copper in the fungicides is present as copper sulfate, which is highly soluble in water and results in phytotoxicity to plants, especially at high temperatures. The plants show selective toxicity and are called copper shy. The injurious effect is characterized by chlorosis followed by brownish purple flecks on the leaves or fruit. In leaves, the leaf veins become purple and leaves fall off. On fruits such as apple, a russeting effect may be seen. Lime is added to BM to reduce phytotoxic symptoms, but sometimes the lime itself may be injurious.[10] In some high-yield cultivars of paddy (IR-8) and maize (Ganga Hybrid Makka No. 3) copper toxicity is characterized by the appearance of chlorotic patches somewhere in the middle of the leaf, particularly where the upper leaf bends, which is followed by necrosis.[4] Widespread phytotoxicity due to wrongly recommended copper sprays for the control of paddy diseases was observed. The phytotoxicity of solutions of copper salts has been known for over a century, yet by formulating copper as a compound of low water solubility (such as copper oxychloride), with no loss of fungitoxicity plant damage is largely avoided. When Burgundy mixture labeled with radioactive copper was applied to bean leaves, Sommers and Richmond[11] found that only a negligible amount of copper was absorbed by the leaf surface of broad bean.

An interesting relationship between copper sprays and increased frost damage in tomatoes and potatoes has been observed by Wilson:[12] the effect of frost damage in these crops when sprayed with various copper fungicides was severe as compared to those plants sprayed with noncopper fungicides such as zineb.

A very interesting observation has been made by van der Vassen:[13] the application of copper fungicides (0.25%) to resistant cultivars of *Coffee arabica* decreased the disease and increased the quality and yield. On the other hand, in susceptible cultivars the application drastically reduced the quality and yield. Hence, it is suggested that indiscriminate use of fungicides be avoided.

Sometimes the different salts of the source compounds enhance or reduce phytotoxicity. Copper sulfate and copper hydroxide, which are the copper sources for many copper fungicides, were evaluated at various rates between 0 and 486 kg Cu per hectare for several years in snapbeans.[14] It was observed that hydroxide was more toxic than sulfate in the reduction of yield in snapbeans.

C. Mercurial Compounds

Mercurial compounds have been known for their fungicidal value for many years. They have been used mostly for treating seeds to protect them from fungal attacks. Because of their extreme toxicity, their use as foliage sprays could not be encouraged, though there have been some mercurial compounds used as fruit and foliage fungicides.

Seed injury is considered to be one of the hazards in the use of organic mercurials. In general, the effect of treatment on germination capacity is favorable when recommended dosages are used. However, even at recommended rates, seed injury has been observed.[15,16] The injury to the seed, however, is reported to be more pronounced at higher rates and the degree of seed injury increases in direct proportion to increasing rates of application.[15,16]

Injury by organic mercurials to cereal seed has been characterized by abnormal germination. The effect is a characteristic hypertrophy of the roots and coleoptile of cereal seedlings, when higher dosages of fungicides are used or the storage conditions are inadequate.[16-19] Sass[20] made an anatomical study of the hypertrophy induced in maize by New Improved Ceresan; the primordia were greatly thickened and developed irregular carnations and lobes. In the apical meristem of the plumule, cell division was inhibited and extreme enlargement of the existing cells took place. The cells of the hypertrophied organs became multinucleate, containing nuclei ranging from micronuclei to polyploid giant nuclei, and the number of chromosomes in these often exceeded 200. Sass[20] considered the malformation to be due to incomplete meiosis.

The hazards which organomercurials pose are too well known to be discussed here. Through their persistence, they are introduced into the food chain. Several countries have now banned the use of organomercurials. A ban is one approach to reducing the amount of the material that comes in contact with the food chain. Huisingh[21] has argued that another approach could be to explore the concept of "tolerable concentration" instead of "zero concentration". A program using tolerable concentration would be more useful and would take into account the fact that all elements, essential as well as nonessential, are potentially toxic, depending upon their form, concentration, ratio with other elements, and duration and route of exposure. It must also be remembered that heavy metals have always been present naturally in the biological world and thus all organisms have developed a degree of tolerance. Hazards can be reduced by diligent enforcement of reasonable regulations designed to reduce emission of metal wastes into the environment.

D. Carbamates

Carbamates form a very important group among fungicides. Most are foliage fungicides, while some are used for soil and seed treatments. All carbamate fungicides presently available commercially are derivatives of dithiocarbamic acid. Organic acid does not occur in the free state and was synthesized in the 1920s to accelerate the action of sulfur in vulcanizing rubber.

Prasad and Pramer[22] reported the mutagenic activity of ferbam, a member of the carbamate group, when applied to growing *Allium cepa* roots. They observed unusually high chromosomal aberrations in ferbam-treated roots; this indicates the possibility of an injurious effect on certain host plants. A report on nitrogen deficiency disease of sugarcane induced by a large number of applications of zineb or maneb to the preceding tomato crop was made by Dubey.[23] Maneb is phytotoxic to tobacco seedlings, cauliflower, pepper, tomato, and lettuce.[4] Maneb seems to adversely affect pollen germination.[24]

Application of captan and ferbam at disease-control rates against apple scab led to considerably more leaf spot than thiram, mancozeb, and propioneb; metiram had an intermediate effect. A similar classification of these compounds was observed with regard to influence on premature leaf drop, although the effect of fungicides could only be observed under conditions of heavy leaf drop.[25] Maneb, captan, and benomyl at a rate of 0.1% depressed plant growth, the bain concentration, and the bain yield per plant when applied repeatedly (i.e., over 22 weeks rather than 10 weeks) to *Papaver bracteatum*. This repeated application did not affect the bain concentration.[26] Ziram, thiram, and captan were not phytotoxic to apple but mancozeb and wettable sulfur showed toxicity to apple trees at the rate used for the control of plant disease.[27]

E. Other Organic Compounds

Dodine, a fungicide and bactericide, is used extensively for the control of peach and apple diseases. Daines[28] and Diener and Carlton[29] reported that dodine at recommended rates causes shot-holes in peach foliage. The fungicide produces spots of discoloration on fruit of Sunlight cultivar of peach.[30] Lukens[31] also observed dodine injury to peach foliage. The injury increased with concentrations of the active ingredient (a.i.) and with the temperature of the leaf following application. Addition of captan also sharply increases phytotoxic response with an increase in temperature. The phytotoxic response to captan alone increases with temperature just as sharply as the response to the dodine-captan mixture; apparently captan is the phytotoxic agent in the mixture.[31] The phytotoxic action of dodine may require moisture because injury by dodine was associated with the wetness of the residue of the fungicide. The defoliation of some cultivars of peach by a spray residue of dodine-captan was attributed by Chandler[32] to prolonged moistness of the spray residue on the foliage, due to a light rain. Miller[33] reported that toxicity of dodine to rose foliage can be avoided if the plants dry quickly after spraying. Lukens[31] observed that captan alone apparently does not cause injury at any concentration or temperature, but in dodine-captan mixtures toxicity is observed. This was because dodine as a surfactant increases the amount of captan entering the leaves; when captan decomposes inside the leaf, phytotoxicity results. However, when captan is applied alone, it does not enter the leaf in any appreciable amount, it decomposes on the leaf surface, and it does not injure the underlying cells.

Burchfield and Storrs[34] reported that Dyrene causes necrotic flecking and defoliation of pepper plants when they are incubated in a moist chamber for 1 day before being placed in a greenhouse. Plants sprayed with Dyrene and transferred directly to the greenhouse bench were unaffected; the leaves dropped while still turgid because of the disintegration of the abscission layer. Evidently the compounds were translocated to conditions of extreme humidity. It was concluded that solubility may be a factor in this toxicity. However, this cannot be the only factor governing toxicity, since captan is at least as soluble as Dyrene in water and it is considerably safer for use on apples. The reason for this is that the half-life of captan in an aqueous buffer at pH 7 is only about 2 hr, compared to about 22 days for Dyrene, so that their longevities in the aqueous phase differ by a factor of more than 200.

While captan might diffuse far enough to reach fungus spores of localized regions within the plant system, it would not have the range of penetration of Dyrene because of its shorter life. Thus, the interrelated effects of several physical factors, in combination with intrinsic biological activity, might help to explain why some compounds are phytotoxic and others are not.[34]

Captan, a widely used fungicide, is recommended for the control of brown rot of stone fruits and several apple and peach diseases during bloom and prior to harvest. Captan has a tolerance of 50 μg g^{-1} on peaches, nectarines, and apricots, with no time limitation on its application to these fruit crops.[35] Captan, in addition to combating diseases, has also been effective in improving the finish of apple fruits; however, the foliage of several apple and peach cultivars may develop phytotoxic symptoms such as small circular spots followed by necrotic burning symptoms.[36,37]

These spots most frequently develop during warm periods of low light intensity; although they usually appear near the blooming period, they may occur any time during the growing season. Peaches sprayed with this fungicide frequently developed a leaf-lacing effect from the loss of injured tissue. In bean, severe leaf scorching was occasionally encountered.[36] Daines et al.[38] studied the effect of formulation, environment, and plant factors responsible for phytotoxicity of captan. Laboratory and field studies indicated that injury increased as the concentrations of captan and wetting agent increased. The addition of kaolin as a filler also increased phytotoxicity, but the presence of calcium carbonate and magnesium oxide lessened this effect. Toxicity to bean increased with increasing temperature and with decreasing light intensity. It was further observed that captan decomposition is accelerated by increasing temperature and time, with a resulting increase in hydrogen ion concentration, chloride concentration, and hydrogen accumulation.

Phytotoxicity of fungicides on plant leaf and fruit surfaces is influenced by physical and chemical factors operative in the fungicide film. This was indicated in the study of the influence of light on susceptibility of bean plants to injury by captan.[38] In these studies, shading was found to increase phytotoxicity. Captan is appreciably phytotoxic at a concentration of 1 ppm when applied to plant cells lacking a protective cutin coating, i.e., roots of bean and tomato in liquid culture.[39] Under conditions of low light intensity, plants develop thin cuticles and cell fluids with increased hydrogen ion concentrations. Under such conditions, resistance to penetration into the leaf may be decreased and injury from fungicide degradation products would be expected. Phytotoxicity also resulted because of the penetration of fungicides into the stomata. There is some evidence that phytotoxicity may result within the cell. Reaction with sulfhydryl groups in the cell may provide an additional mechanism of toxic action whereby the fungicide is decomposed with the production of hydrogen and other ions and hydrogen sulfide. Daines et al.[38] also found that as the pH of a captan solution was increased above 7.0 the rate of captan decomposition increased. The pH of the wax solution used in the postharvest treatment was about 8.0; however, Chastagner and Ogawa[41] were unable to detect pH drop when mixtures of captan, wax, and fruit juices were stored under conditions similar to those used for the fruits in these experiments.

After harvest, fresh-market stone fruits often receive a postharvest fungicide-in-wax treatment to reduce decay during shipment and storage. Injury to peach fruit (cv. Suncrest) has been observed during and after storage when preharvest spray of captan was followed by commercial postharvest treatment.[41] Similarly, apricots (cv. Blenheim) that received a captan spray 20 days before harvest and a subsequent commercial postharvest treatment developed surface discoloration during storage. Subsequently, studies were conducted to determine if postharvest sprays of captan followed by commercial postharvest treatment resulted in the development of fruit surface discoloration and, if so, to determine the factors that contributed to it. Chastagner and Ogawa[41] reported that a preharvest spray of 50% captan was applied on nectarine, apricot, and peach cultivars. Peaches (cv. Suncrest) developed surface streaking ranging from brown to black after 4 days of storage and nectarines (cv. Le Grand) developed irregular or circular surface discoloration after 7 days. Apricots (cv. Blenheim) field-sprayed with captan 20 days before harvest and treated postharvest with benomyl in wax developed diffused brown surface discoloration after 3 days at 0.5°C and more distinct lesions after 3 days at 20°C. Only the fruit treated with captan and Botran and/or benomyl in wax after harvest developed surface discoloration. Laboratory studies in cv. Tilton apricots show that similar discoloration developed only when fruit with a captan residue was brushed; more severe symptoms occurred when brushed fruits were treated with wax containing captan. However, peach fruits (cvs. Fay Elberta, Carnival, and Halloween) did not develop discoloration, regardless of pre- and postharvest treatment.[41] The exact nature of captan phytotoxicity has not been determined, but conditions that enhance the development of injury are related to postharvest brushing and waxing.

III. SYSTEMIC FUNGICIDES

Systemic fungicides, by virtue of their systemicity, have a profound influence on the physiology and metabolism of plants. Borderline differences have been reported between the fungitoxicity and phytotoxicity of systemic fungicides. Several reports are available that indicate a range of symptoms, and the subject has also been recently reviewed.[42]

Carboxin and oxycarboxin do not appear to be phytotoxic except at very high rates.[45,46] Carboxin and oxycarboxin at 20 or 30 lb a.i. acre^{-1} caused yellowing of leaf tips in *Bromus willdenowii* and 40 lb caused stunting.[43] In greenhouse trials both killed tobacco plants. However, at the same rates, i.e., 50 mℓ of 0.5 g ℓ^{-1} per pot, benomyl was used with no adverse effects.[44] Foliar sprays of oxycarboxin (15 kg a.i. ha^{-1}) on black gram and urdbean reduced the nitrogen content of the leaves, but seed treatment with fungicide did not.[47] However, the latter treatment in black gram resulted in less phosphorus and more sulfur and magnesium in leaves, although in urdbean seed treatment reduced photosynthetic activity but increased sulfur and magnesium. Soil application of carboxin at 4 g/m^2 was phytotoxic to carnations.[48] Carboxin and oxycarboxin were phytotoxic to soybean seedlings at 100 ppm when incorporated into soil.[49] The effects of systemic fungicides such as carboxin and benomyl on uredo metabolism in germinating seeds of groundnut, soybean, green gram, and black gram have been studied. Whereas all these fungicides delayed germination, inhibited radical elongation, allantoin and allantoic acid production, and allantoinase activity in seeds, maximum effect was seen with carboxin followed by benomyl. Carboxin and oxycarboxin caused chromosomal aberration in onion root tips similar to those caused by radiation and other mutagenic treatments.[50]

Carboxin, oxycarboxin, and diclobutrazole inhibit growth of clover and were most phytotoxic, even with the lowest concentration of the fungicides. There was a significant difference between plants grown in treated and untreated soil.[51,52] Plant weight was significantly reduced by a concentration in the soil greater than that which might result from a single application during the season.[51] Fisher et al.[53] showed that much greater concentrations of triadimefon in soil have similar effects; benodonil, pyracarbolid, and tridemorph were also phytotoxic at the highest concentration.[51] Triarimol also retards plant growth. Clover plants are evidently much more sensitive to diclobutrazole.[51,52] The effects of both triadimefon and triarimol have been ascribed to interference with gibberellin biosynthesis.[54,55] Diclobutrazole is structurally very similar to triadimefon and probably acts in the same way. However, none of the other systemic fungicides arrested clover growth.[51,52] The observed reduction in plant size and toxicity was reflected in a significant decrease in nitrogen fixation per plant. Fisher and Hayes[51,52] reported that symbiotic nitrogen fixation was decreased by carboxin, oxycarboxin, tridemorph, and diclobutrazole; diclobutrazole halved this concentration. Root nodulation was reduced by diclobutrazole and tridemorph; however none of the other fungicides reduced nitrogen fixation and nodulation.

Carlson[56] found that carboxin almost completely inhibits photosynthesis in barley. However, the physiological mechanism was not evident from the study. A nonspecific action of the chemical is suggested from the observation on inhibition of respiration. McMillan and Riedhart[57] observed that inhibition of photosynthesis by hydrocarbon oils is thought to be a mechanical interference of gaseous exchange. Other chemicals, for instance, thiram and nabam, actually enter the cells and interrupt photosynthesis.[58] Studies on the uptake and binding of carboxin by *Ustilago maydis* and *Rhizoctonia solani* have shown the chemical to be bound in the ribosomal-soluble fraction of the cell.[59] Studies indicated that carboxin is also absorbed by the plant cells and, once the fungicide is in the cell, is bound to the chloroplast, thus interfering with photosynthesis.[60] Mathre[59] studied the effect of carboxin on photosynthetic activity in barley, Pinto beans, and corn and found that whereas carboxin is inhibitory, oxycarboxin is less effective. However, he observed inhibition only at high concentrations. The likelihood that nontarget organisms would have contact with high concentrations of these compounds in nature would appear to be extremely small except where materials are misused, such as in the feeding of treated seed to livestock. The reason for the low sensitivity of oxathiin compounds to nontarget organisms is not known, but may be related to the uptake of the compounds into cells and/or mitochondria.[59]

Sometimes application of benzimidazole fungicides produces undesirable effects on plant parts. Benomyl causes superficial damage to apples and pears.[61] Spraying of benomyl causes russeting of the skin of certain apple cultivars.[61-64] The damage takes two forms: there may be increased russeting of the fruits of apple cultivars such as Golden Delicious, Rome Beauty, Cox's Orange Pippin, or Worcester Pearmain or of pear cultivar Williams Bon Chretien. However, there may also be another effect on the appearance of the fruit: an opalescent finish that makes the fruit look unripe or of an unusual color. This has been reported repeatedly in Rome Beauty, Stayman Winesap, Sunset, and Worcester Pearmain. Kirby[65] reported that benomyl causes undesirable side effects, inclusing russeting and poor fruit color and a reduction of fruit size on some cultivars of apple and pear. He was of the view that confining its use to postharvest applications to eliminate perithecial production or (if it is necessary to use it postblossom) the use of extended intervals or perhaps low-volume applications may serve to reduce these side effects to acceptable levels. Similar results were also reported for triarimol. Attempts to mitigate the russeting effects of benomyl with captan, or organoarsenical, known to act in this way in some instances, led to results that varied with the cultivars used.[65] The possibility was considered that increased russet was related to the use of high-volume sprays (200 gal acre^{-1} = 2280 ℓ ha^{-1} or above). According to Byrde et al.,[66] all trials were carried out at this volume and often produced russet on susceptible cultivars; however, the use of 100 gal acre^{-1} did reduce the incidence of russet on Cox's Orange Pippin.[66] The beneficial influence of these compounds clearly varies with the cultivar as well as with season.[67] It is possible that a higher rate might give more satisfactory results, although this might cause some problems in ripening, especially in storage. Preblossom application of benomyl can advance the flowering date of plum and somewhat increase the effective pollination period.[68] Schreiber and Hock[69] found that benomyl and thiabendazole soil application reduced the growth of American elm, marigold, buckthorn, and silver maple seedlings and were phytotoxic to sycamore. They observed that pepper, tomato, and beans were less sensitive to benomyl whereas peas and corn were more sensitive.

Adele et al.[70] found that soil application of 100 mℓ benomyl (125 and 250 ppm) to potgrown wheat in a glasshouse reduced the protein level in treated plants after 18 days to about half the value of the untreated plants. Seed treatment inhibited the DNA-transcription process, thus decreasing RNA synthesis, which led to a decrease of protein synthesis and cell division. However, it was found that 10 ppm carbendazim initially depressed slightly ribonuclease and deoxyribonuclease activity, although pretreatment led to an apparent decrease in RNA and DNA synthesis.[70] One of the primary toxic effects of benomyl is an inhibition of protein and RNA synthesis by 50% at 15 and 10 ppm, respectively.[71] Direct effects of carbendazim on RNA synthesis have also been reported in tobacco mosaic virus.[72]

On Kentucky bluegrass and bent grass, thiabendazole has been reported to be phytotoxic at levels which were quite safe with benomyl.[73,74] Cabbage seedlings in a 1:1 sphagnum peat-organic mulch soil were stunted by 0.49 benomyl per 10 cm.[75]

Benomyl and carbendazim are antimitotic in their mode of action on fungi. Benomyl and carbendazim interfere with mitosis in cells of onion root tip.[76] Pandit[77] reported that carbendazim application on the root cells of *Allium cepa* var. *cepa* produces cytological effects in the form of mitotic chromosome aberrations at low concentration (0.05%) and showed the least clastogenic or turbogenic effects, whereas higher concentrations (0.2%) given for 24 hr or more increased both the number of cell divisions and chromosomal aberration. Wheat seed treatment or foliar spray at 0.15% with benomyl and thiabendazole slightly reduced the development of endotrophic mycorrhizae in the roots.[78]

It is possible, experimentally, to produce phytotoxic symptoms such as stunting and leaf necrosis with drench applications of benzimidazole derivatives at high concentrations. There have been reports of other physiological effects of some of these compounds. Cole et al.[79] found that benomyl applied to the soil at levels above 20 ppm caused an increase in fresh weight and dry weight of cucumber plants. Lettuce growers using benomyl for the control of *Botrytis cinerea* can produce plants which have a more intense green coloration and a mature appearance when young. Such plants may be heavier than those not treated with the fungicide. Benomyl-treated celery has been found to be crisp and the green parts still fresh after winter storage, whereas the untreated crop becomes yellow in storage. Tomato plants which have been treated with benomyl, thiabendazole, or thiophanate derivatives usually have a darker shade of green, particularly in younger leaves, than the untreated controls. Benomyl used for the control of *Verticillium albo-atrum* in chrysanthemum cuttings caused the treated plants to remain dark green.[80]

Metalaxyl provided complete control to the plants of *Brassica campestris* var. *toria* and *B. campestris* var. yellow *sarson* against white rust *(Albugo candida)* and downy mildew *(Peronospora parasitica)*. However, the fungicide considerably reduced the overall seed yield and 1000 seed weight as compared to the control.[81] There was reduced seed germination and seedling growth when the seeds obtained from metalaxyl-treated yellow *sarson* plants were sown in soil. However, it has been indicated that, apart from phytotoxicity, the reduced yield might be partly or solely due to the increased severity of *Alternaria brassicae* in metalaxyl-treated plants. A similar observation has also been made in silique, another cruciferous plant, by Sharma and Kolte.[81] When a mixture of metalaxyl and captafol was sprayed on yellow *sarson,* a reduction in Alternaria blight severity occurred and yield increased instead of decreasing. Phytotoxic symptoms were observed on *toria* seedlings raised from metalaxyl-treated seeds. The fungicide at comparatively high concentrations damaged young pepper plants.[82] Soil drench of metalaxyl was phytotoxic to rhododendron but only at high dosages.[83] In spite of some adverse effects of metalaxyl on the mycorrhizal fungus *Glomus etunicatus,* the growth of plants with mycorrhizal association was not adversely affected by the fungicidal treatment.[84] Triforine is a good mildew fungicide. Sometimes its use resulted in greener foliage,[85] but an early spray sometimes caused increased fruit russet.[62,85] Church et al.[86] reported that spraying of triforine (125 to 250 ppm) on undehisced anthers of apple reduced the viability of the pollen, discolored anthers, and impaired pollen release. On dehiscence, death of the anthers was also observed. Jalali and Domsch[78] reported that soil and foliar application of triforine (125 to 250 ppm) reduced the total amino acid content of root exudates of wheat, although liberation of some individual acids was enhanced. Exudation of aromatic amino acids (phenylalanine and tyrosine) was totally or partially suppressed.

A noticeable effect of formulation on phytotoxicity was observed in greenhouse tests with triforine reported by Fuchs et al.[87] A 10% emulsifiable concentrate (EC) containing a nonionic emulsifier damaged barley, bean, and tomato by sprays and barley and tobacco by root application. A 20% EC formulated with an ionic emulsifier caused no damage by foliar sprays and less than the other EC by root application.

The organophosphorus compound Hoe 2873 was reported to be phytotoxic to rose and aquilegia.[88] Triadimefon caused various side effects in plants. Following seed treatment of cereals, the germination rate was generally not affected, but the compound slightly delayed emergence and treated plants were a deeper green than the control plants.[89,90] Triadimefon[91] and tridemorph[92] delayed senescence in wheat and barley, respectively. Triarimol was reported to cause russet on Cox's Orange Pippin apples[63,66] and on Doyenne du Connie pears.[65] Dodemorph acetate russeted apples, both Cox's Orange Pippin and Golden Delicious.[93] Soil application of chloroneb at 0.96 g per pot was phytotoxic to cotton seedlings.[94] Similarly, seed treatment at 0.15% to durum wheat produced phytotoxic symptoms which resembled those of colchicine.[95] Foliar application of 0.5 to 1.0% tridemorph on Agra local variety of wheat was phytotoxic.[96] Triarimol and pyracarbolid inhibited respiration at concentrations which control plant disease and did not produce visible symptoms. Pyracarbolid caused 40% inhibition of photosynthesis 10 days after treatment in one of the cultivars tested, at a concentration considerably below the field application rate.[97]

Pepin and co-workers[98] applied triadimefon biweekly to actively growing Willamette red raspberry primocanes at the rates of 0, 0.28 and 0, 0.50 kg ha^{-1}. The treated canes were thinner and shorter than controls and had shortened petioles and reduced leaf area; the total number of nodes per cane was unaffected by the sprays, but after cane heading the treated plants had a greater number of nodes and fruiting laterals the following season. The decreased production of laterals balanced the increase in lateral number to produce no net effect of triadimefon on overall yield. However treated plants ripened earlier.

Spotts[99] recorded severe foliar stunting and deformation of treated grape plants in growth chambers. The same symptoms were much less severe when treatments were applied in the vineyard. Sanders et al.[100] found that growth reduction on bent grass and bluegrass with triadimefon varied with genus and cultivar. It may be possible, through the use of gibberellic acid (GA$_3$) and gibberellin inhibitors such as triadimefon, to produce a plant with all its fruiting laterals concentrated at the top of the cane for more efficient machine harvesting. In addition to its commercial possibilities, triadimefon is a useful research tool in the study of raspberry yield components or in the study of the role of GA$_3$ in raspberry development.

Studies also indicated that cultivars react differently to fungicidal application, particularly in cereals where they were extensively used.[101,102] Wheat cultivars Indus 66, Nayab, C 591, and C 42 and mutants derived from these cultivars were sprayed 18 days after sowing with 0.06% tridemorph to control powdery mildew *(Erysiphe graminis);* the treatment caused 25% of the leaf area to become chlorotic in Indus 66 and Nayab and 37.5% in C 591 and 42, but mutants developed no chlorosis.[101] Later studies by Siddique and Haahr[102] with rye, oats, and barley indicated that they were resistant to leaf chlorosis after the application of tridemorph; of the 11 hexaploid wheat tested for chlorosis only mutant 11 and somara were resistant to phytotoxicity. These observations indicate that among cereals phytotoxicity is associated with wheat, although some mutants showed resistance to tridemorph. Hence, further studies are needed to explain the genetic behavior of the crop, its cultivars, and their mutants before the application of fungicides is made.[101,102]

IV. FUTURE OUTLOOK

The application of fungicides on the plant surface is made with the sole intent of controlling plant pathogens. Sometimes this results in an array of effects which may or may not be discernible. Phytotonic effects were discussed in Chapters 3 and 4. Phytotoxic effects, produced by fungicides used at the rates recommended for disease control, occasionally baffle scientists. It has been hypothesized that phytotoxic effects depend upon the genetic makeup of the plants, although many other theories have been proposed. Hence, it is suggested that a coordinated effort be made to study the behavior pattern of phytotoxicity before conclusions are drawn.

REFERENCES

1. **Christopher, E. P.,** A comparison of lime-sulphur and flotation and sulphur spray on apple trees, *Proc. Am. Soc. Hortic. Sci.,* 40, 63, 1942.
2. **Wallace, E.,** Spray injury induced by lime-sulphur preparations, *Cornell Univ. Agric. Exp. Stn. Bull.,* p. 288, 1910.
3. **McCallan, S. E. A., Hortsell, A., and Wilson, F.,** Hydrogen sulphide injury to plants, *Contrib. Boyce Thompson Inst.,* 3, 13, 1936.
4. **Nene, Y. L. and Thapliyal, P. N.,** *Fungicides in Plant Disease Control,* Oxford & IBH Publishing, New Delhi, 1979.
5. **McDaniels, L. H. and Furr, J. R.,** The effect of dusting sulphur upon the germination of pollen and the set of fruit of the apple, *Cornell Univ. Agric. Exp. Stn. Bull.,* 499, 13, 1930.
6. **Kozlowski, T. T. and Keller, T.,** A short review of the effects of fungicides on photosynthesis in plants, *PANS (Pest Artic News Summ.),* 14, 31, 1968.
7. **Torgeson, D. C.,** *Fungicides — An Advanced Treatise,* Vol. 2, Academic Press, New York, 1969, 119.
8. **Kavanagh, T. and O'Callaghan, T. F.,** A comparison of fungicides for the control of Botrytis of outdoor and protected strawberries, *Proc. 5th Br. Insectic. Fungic. Conf.,* 1, 139, 1969.
9. **Hoffman, M. B.,** The effect of lime-sulphur spray on the respiration rate of apple leaves, *Proc. Am. Soc. Hortic. Sci.,* 33, 173, 1936.
10. **Horsfall, J. G., Magie, J. O., and Suit, R. F.,** Bordeaux injury to tomatoes and its effect on ripening, *N.Y. Agric. Exp. Stn. Geneva Tech. Bull.,* 251, 1, 1938.
11. **Sommers, E. and Richmond, D. V.,** Translocation of copper by broad bean plants, *Nature (London),* 194, 1194, 1962.
12. **Wilson, J. D.,** Relation between spray treatments and frost damage to potatoes, *Bull. Ontario Agric. Exp. Stn.,* 32, 77, 1947.
13. **van der Vassen, H. A.,** Consequences of phytotoxic effects of fungicides to breeding for disease resistance, yield and quality in *Coffee arabica, J. Hortic. Sci.,* 57, 321, 1982.
14. **Walsh, L. M., Erhardt, W. H., and Seibelh, D.,** Copper toxicity in snapbeans *(Phaseolus vulgaris), J. Environ. Qual.,* 1, 197, 1972.
15. **Koehler, B. and Beaver, W. M.,** Oat and wheat seed treatment methods, injuries and gains, *Plant Dis. Rep.,* 38, 762, 1954.
16. **Purdy, L. H.,** The nature and extent of mercury injury and associated anatomical responses in wheat seedlings, *Res. Stud. Wash. State Univ.,* 24, 343, 1956.
17. **Crosier, C. and Keitt, G. W.,** Abnormal germination in dusted wheat, *Phytopathology,* 24, 544, 1934.
18. **Noll, W.,** Deformities induced in wheat seeds by treatment of the grains, *Arch. Fitotec. Uraguay,* 3, 86, 1938.
19. **Weston, W. A. R. and Booer, J. R.,** Seed disinfection — an outline of an investigation of disinfectant dust containing mercury, *J. Agric. Sci.,* 25, 628, 1935.
20. **Sass, J. E.,** Histological and cytological studies of ethylmercury phosphate injury to corn seedlings, *Phytopathology,* 27, 95, 1937.
21. **Huisingh, D.,** Heavy metals: implications for agriculture, *Annu. Rev. Phytopathol.,* 12, 375, 1974.
22. **Prasad, I. and Pramer, D.,** Genetic effects of ferbam on *Aspergillus niger* and *Allium cepa, Phytopathology,* 58, 1488, 1968.
23. **Dubey, H. D.,** A nitrogen deficiency disease of sugarcane probably caused by repeated pesticide applications, *Phytopathology,* 60, 485, 1970.
24. **Karov, S.,** Effect of some fungicides on the pollen germination of peach, apricot and plum, *Loz. i Ovosht.,* 18, 43, 1969.
25. **Bremmer, H. and Bunemann, G.,** Side effects of organic scab fungicide. I. Leaf spot and leaf drop of Golden Delicious trees, *Gartenbauwissenschaft,* 47(2), 56, 1982.
26. **Wu, F. F., Dobbersteen, R. H., and Morris, R. W.,** The effect of selected fungicides and insecticides on growth and the bain production of *Papaver bracteatum, Lloydia,* 41, 355, 1978.

27. **Burth, U. and Ramson, A.**, On the effect of organic fungicide on the vegetative performance of apple trees, *Arch. Gartenbacu*, 21, 567, 1973.

28. **Daines, R. H.**, Copper-zinc chromate and dodine show promise in the control of bacterial stop of peach, *Plant Dis. Rep.*, 48, 152, 1964.

29. **Diener, U. L. and Carlton, C. C.**, Dodine-captan combination controls bacterial spot of peach, *Plant Dis. Rep.*, 44, 136, 1960.

30. **Daines, R. H.**, Control of bacterial spot of peach with dodine, *Plant Dis. Rep.*, 46, 251, 1962.

31. **Lukens, R. J.**, Injury to peach foliage by dodine and dodine-captan spray, *Plant Dis. Rep.*, 54, 152, 1970.

32. **Chandler, W. A.**, Effects of certain mixtures of fungicides and spray additives on the control of peach scab and brown rot, *Plant Dis. Rep.*, 46, 476, 1962.

33. **Miller, H. N.**, Control of rose leaf disease in Florida, *Proc. Fla. State Hortic. Soc.*, 74, 400, 1968.

34. **Burchfield, H. P. and Storrs, E. F.**, Effects of chlorine substitution and isomerism on the interactions of s-triazine derivatives with conidia *Neurospora sitophilla*, *Contrib. Boyce Thompson Inst.*, 18, 429, 1957.

35. **Daines, R. H.**, Captan, the new discovery in fruit fungicides, *Am. Fruit Grow.*, 73, 16, 1953.

36. **Anon.**, Compendium of Registered Pesticides, Vol. 2. Part I. Fungicides and Nematicides, U.S. Environmental Protection Agency, Washington, D.C., 1973, 10-00.01.

37. **Miller, P. M.**, Heat decomposition products of captan as phytotoxic agents, *Phytopathology*, 47(Abstr.), 245, 1957.

38. **Daines, R. H., Luckens, R. J., Brennan, E., and Leone, I. A.**, Phytotoxicity of captan as influenced by formulation, environment and plant factors, *Phytopathology*, 47, 567, 1957.

39. **Sommers, E.**, Fungicide selectivity and specificity, *World Rev. Pest Control*, 8, 95, 1969.

40. **Daines, R. H.**, Foliage fungicides and combination sprays, *Agric. Chem.*, 12, 32, 1957.

41. **Chastagner, G. A. and Ogawa, J. M.**, Injury of stone fruits by preharvest captan sprays followed by postharvest treatments, *Phytopathology*, 66, 924, 1976.

42. **Vyas, S. C.**, *Systemic Fungicides*, Tate McGraw-Hill, New Delhi, 1984.

43. **Latch, G. C. M.**, Chemotherapy of head smut *(Ustilago bullata)* in *Bromus willdenowii*, *Plant Dis. Rep.*, 55, 284, 1969.

44. **Saenger, H. L.**, Control of Thielaviopsis root rot of tobacco with benomyl fungicide drenches, *Plant Dis. Rep.*, 54, 136, 1970.

45. **Reinbergs, E. and Edgington, L. V.**, Field control of loose smut in barley with systemic fungicides Vitavax and Plantvax, *Can. J. Plant Sci.*, 48, 131, 1968.

46. **Jones, J. P. and Collins, F. C.**, Control of loose smut of wheat with carboxin (Vitavax) and Benomyl (Benlate), *Plant Dis. Rep.*, 55, 1053, 1971.

47. **Chopra, S. L. and Purthi, S. P.**, Effect of plantvax on the chemical composition of black gram *(Phaseolus mungo* Roxb.)*, Indian J. Agric. Sci.*, 41, 345, 1971.

48. **Pergola, G. and Garibald, A.**, Results of control trials against collar rot *(Rhizoctonia solani* Kuhn.) of carnation, *Inf. Fitopatol.*, 25, 70, 1975.

49. **Gray, L. E. and Sinclair, J. B.**, Uptake and translocation of systemic fungicides by soybean seedlings, *Phytopathology*, 60, 523, 1970.

50. **Mohan, S. T.**, Cytological effects of fungicides Plantvax and Vitavax on somatic cells of *Allium cepa*, *Curr. Sci.*, 44, 813, 1975.

51. **Fisher, D. J. and Hayes, A. L.**, Effects of some fungicides used against cereal pathogens on the growth of *Rhizobium trifolii* and its capacity to fix nitrogen in white clover, *Ann. Appl. Biol.*, 98, 101, 1981.

52. **Fisher, D. J. and Hayes, A. L.**, Effects of some systemic imidazole and triazole fungicides on white clover and symbiotic nitrogen fixation by *Rhizobium trifolii*, *Ann. Appl. Biol.*, 101, 19, 1982.

53. **Fisher, D. J., Pickard, J. A., and McKenzie, C. M.**, Uptake of the systemic fungicide triadimefon by clover and its effect on symbiotic nitrogen fixation, *Pestic. Sci.*, 10, 75, 1979.

54. **Buchenauer, H. and Rohner, E.**, Effect of triadimefon and triadimenol on growth of various plant species as well as on gibberellin content and steroid metabolism in shoots of barley seedlings, *Pestic. Biochem. Physiol.*, 15, 58, 1981.

55. **Shive, J. B. and Sisler, H. D.**, Effect of ancymidol (a growth retardant) and triarimol (a fungicide) on the growth, sterols and gibberellins of *Phaseolus vulgaris* (L)., *Plant Physiol.*, 57, 40, 1976.

56. **Carlson, L. W.**, Effects of vitavax on chlorophyll content, photosynthesis and respiration of barley leaves, *Can. J. Plant. Sci.*, 50, 627, 1970.

57. **McMillan, R. T. and Riedhart, J. M.**, The influence of hydrocarbons on photosynthesis of citrus leaves, *Proc. Fla. State Hortic. Soc.*, 77, 15, 1964.

58. **Lindahl, P. E. B.**, The effect of sodium dimethyldithiocarbamate and tetramethylthiuram-disulphide on photosynthesis and endogenous respiration in *Enteromorpha linza* (L), *Plant Physiol.*, 16, 644, 1963.

59. **Mathre, D. E.**, Effects of oxathiin systemic fungicides on various biological systems, *Bull. Environ. Contamin. Toxicol.*, 8, 311, 1972.

60. **Sinclair, J. B. and Darrag, I.**, Systemic activity of 1,4-dichloro2,5-dimethoxybenzene against *Rhizoctonia solani* in cotton seedlings, *Dis. Counc. Proc.*, 26, 108, 1966.

61. **Kirby, A. H. M.**, Progress towards systemic fungicides, *PANS (Pest Artic. News Summ.)*, 18, 11, 1972.
62. **Butt, D. J., Kirby, A. H. M., and Williamson, C. J.**, Fungitoxic and phytotoxic effects of fungicides controlling powdery mildew on apple, *Ann. Appl. Biol.*, 75, 1973.
63. **Byrde, R. J. W., Harper, C. W., and Holgate, M. E.**, Spraying trials against apple mildew and scab at Long Ashton, *Long Ashton Res. Stn. Univ. Bristol Rep. for 1968*, p. 146, 1969.
64. **Kirby, A. H. M., Butt, D. J., and Williamson, C. J.**, Trials in 1968 of new fungicides on apple and pear, *East Malling Res. Stn. Maidstone Engl. Rep. for 1969*, p. 171, 1970.
65. **Kirby, A. H. M.**, Phytotoxic effects of some new fungicides for apple mildew control, *Meded. Fac. Landbouwwet. Rijksuniv. Gent*, 36, 355, 1971.
66. **Byrde, R. J. W., Harper, C. W., and Holgate, M. E.**, Spraying trials against apple mildew and apple scab at Long Ashton, *Long Ashton Res. Stn. Univ. Bristol Rep. for 1970*, p.194, 1971.
67. **Kremer, W.**, Influence of pesticides on russeting of Golden Delicious apples, *Bayer Pflanzenschultz-Nachr.*, 20, 629, 1967.
68. **Scott, K. and Jefferies, C. J.**, Fruit set, *Long Ashton Res. Stn. Univ. Bristol Rep. for 1971*, p. 149, 1972.
69. **Schreiber, L. R. and Hock, W. K.**, Effect of benomyl and thiabendazole on growth of several plant species, *Proc. Am. Soc. Hortic. Sci.*, 100, 309, 1975.
70. **Adele, S., Baku, T., and Ceausescu, S.**, Effects of some new fungicides on the amino acid, protein and sugar levels in wheat seedlings, *An. Univ. Bucuresti Biol. Veg.*, 22, 189, 1973.
71. **Rankin, P. W., Surale, J. G., and Thompson, N. P.**, Effect of benomyl and benomyl hydrolysis products on *Tetrahymena pyriformis*, *Food Cosmet. Toxicol.*, 15, 187, 1977.
72. **Fraser, R. S. S. and Whenham, R. J.**, Inhibition of the multiplication of tobacco mosaic virus by methyl-benzimidazole-2yl-carbamate, *J. Gen. Virol.*, 39, 191, 1978.
73. **Goldberg, C. W., Cole, H., and Duich, J.**, Comparative effectiveness of thiabendazole and benomyl for control of Helminthosporium leaf spot and crown rot, red thread, Sclerotinia dollar spot and Rhizoctonia brown patch of turf grass, *Plant Dis. Rep.*, 54, 1080, 1970.
74. **Vargas, J. M., Jr. and Laughlin, C. W.**, Benomyl for the control of Fusarium blight of merian Kentucky blue grass, *Plant Dis. Rep.*, 55, 167, 1970.
75. **Jacobson, B. J. and Williams, P. H.**, Control of cabbage clubroot using benomyl fungicide, *Plant Dis. Rep.*, 54, 456, 1970.
76. **Richmonds, D. V. and Phillips, A.**, The effect of benomyl and carbendazim on mitosis in hyphae of *Botrytis cinerea* and *Allium cepa*, *Pestic. Biochem. Physiol.*, 5, 367, 1975.
77. **Pandit, T. K.**, Cytological effects of Bavistin, *Nucleus (Calcutta)*, 24(3), 106, 1981.
78. **Jalali, B. L. and Domsch, K. H.**, Effect of some fungitoxicants on the amino acid spectrum of wheat root exudates, *Phytopathol. Z.*, 90, 22, 1977.
79. **Cole, H., Boyle, J. S., and Smith, C. B.**, Effect of benomyl and certain cucumber viruses on growth, powdery mildew and element accumulation by cucumber plant in the green house, *Plant Dis. Rep.*, 54, 141, 1970.
80. **Spencer, D. M.**, Result in practice-glasshouse crops, in *Systemic Fungicides*, Marsh, R. W., Ed., Longman, London, 1977, 240.
81. **Sharma, K. D. and Kolte, S. J.**, Metalaxyl in the control of downy mildew and white rust of rapeseed and mustard, *Pestology*, 9(1), 31, 1984.
82. **Aleksic, D., Aleksic, N. M., and Sutic, D.**, On the possibility of control of *Phytophthora capsici*, the pathogen of pepper mildew, *Zast. Bilja*, 29, 317, 1979.
83. **Benson, D. M.**, Efficacy and in vitro activity of two systemic fungicides acylalanine and ethazol for control of *Phytophthora cinnamomi* root rot azalea, *Phytopathology*, 69, 174, 1979.
84. **Nemec, S.**, Effects of 11 fungicides on endo-mycorrhizal development in sour orange, *Can. J. Bot.*, 58, 522, 1980.
85. **Adlung, K. G. and Drandarevski, C. A.**, The evaluation of Cela W 524, a systemic fungicide for the control of powdery mildew and apple scab, *Proc. 6th Br. Insectic. Fungic. Conf.*, 2, 577, 1971.
86. **Church, R. M., Williams, R. R., Morgan, N. G., and Cooke, B. K.**, Toxicity of fungicides to apple pollen in the orchard, *Long Ashton Res. Stn. Rep. for 1977*, p. 22, 1978.
87. **Fuchs, A., Doma, S., and Voros, J.**, Laboratory and green-house evaluation of a new systemic fungicide, *N,N*-bis-(1-formamido-2,2,2-trichloroethyl)-piperazine (Cela W 524), *Neth. J. Plant Pathol.*, 77, 42, 1971.
88. **Smit, F. M.**, Diaethyl-methyl-ethoxycarbomyl pyrazolopirimidineyl fosforothioaat eeen nieuw systemisch werkzaan meeldauwbestrijdingsmiddel, *Meded. Rijksfac. Landbouwwet. Gent*, 34, 763, 1969.
89. **Buchenauer, H.**, Untersuchungen zur systemischen Wirkung sowir zum Einfluß von Bayleton (triadimefon) auf verschiedene pilzliche Getreidekrankheiten, *Bayer Pflanzenschutz-Nachr.*, 29, 267, 1976.
90. **Frohberger, P. E.**, New approaches to the control of cereal diseases with triadimefon (MEB 6447), *Proc. 8th Int. Plant Congr. Moscow*, 3, 247, 1975.
91. **Rowley, N. K., Wainwright, A., and Chipper, M. E.**, The development of the systemic fungicide triadimefon for the control of foliar diseases of wheat and oats in the UK, *Proc. 1977 Br. Crop Prot. Conf. Pests and Diseases*, 1, 17, 1977.

92. **Jenkyn, J. F.,** Effects of chemical treatments for mildew control at different times on the growth and yield of spring barley, *Ann. Appl. Biol.*, 88, 369, 1978.
93. **Pommer, E. H. and Kradel, J.,** Substituert Dimethyl morpholinderivate als neue Fungicide aur Bekenfung echler Mehteaupile. *Meded. Rijksfac. Landbouwwet. Gent,* 33, 735, 1967.
94. **Kappleman, A. J., Jr.,** Fungicidal control of disease in cotton, *Plant Dis. Rep.,* 53, 161, 1969.
95. **Miayoukou, J. F. and Albertini, L.,** Study of the cytotoxic action of chloroneb (dichloro-1,4-dimethoxy-2,4-benzene) at the root level in *Triticum durum, C. R. Acad. Sci. Sec. D,* 281, 779, 1975.
96. **Bahadur, P., Sinha, V. C., and Upadhyaya, Y. M.,** Calixin, a new systemic fungicide for the control of powdery mildew of wheat, *Indian Phytopathol.,* 27, 410, 1975.
97. **Prusky, D., Dinoor, A., and Eshel, Y.,** Symptomless effects of experimental fungicides on wheat, *Phytopathology,* 64, 812, 1974.
98. **Pepin, H. S., MacPherson, E. A., and Clements, S. L.,** Effects of triadimefon on the growth of Willamette red raspberry, *Can. J. Plant Sci.,* 60, 1203, 1980.
99. **Spotts, R. A.,** Use of Bay Meb 6447 for eradication of grape black rot caused by *Guignardia bidwelli, Plant Dis. Rep.,* 63, 967, 1979.
100. **Sanders, P. L., Burpee, L. L., Cole, H., and Duich, J. M.,** Uptake, translocation and efficacy of triadimefon in control of turfgrass, *Phytopathology,* 68, 1482, 1978.
101. **Siddique, K. A.,** Resistance in some wild and cultivated grasses to the phytotoxicity of a systemic fungicide, *Experientia,* 28, 1239, 1972.
102. **Siddique, K. A. and Haahr, V.,** Different reactions of wheat mutants to a systemic fungicides, *Naturwissenschaften,* 58, 415, 1971.

Chapter 10

SOIL MICROORGANISMS AND THEIR ACTIVITIES

I. INTRODUCTION

The fertility of soil, i.e., its capacity to produce crops, is a function of a range of biological processes that occur within it. The activity of the microflora of the soil is generally favorable to vegetation, e.g., the fixation of atmospheric nitrogen; the production of nitrates, sulfates, and carbonic anhydride; the breakdown of plant residue and waste into compounds more easily utilized by plants; and the removal from the soil of diverse products that may be added to it, such as fungicides. Biologically, the soil system is in equilibrium, which is very sensitive. Each disturbance of the environment is likely to modify the activity of the microflora and consequently soil fertility. The application of organic matter and lime modifies biological processes. The universal application of fungicides to agricultural crops often exposes the environment to pollution hazards. Although certain fungicides are not directly applied to the soil, they eventually sink in one way or another. The influence of fungicides on soil microorganisms and microbial transformations in soil is dependent on physical, chemical, and biochemical conditions in the soil, in addition to the nature and concentration of the fungicides.

Application of fungicides has become an integral and economically essential part of agriculture. Although intended to protect crops from plant pathogens, they may alter the microflora of soil ecosystems[1,2] because their action is rarely limited to the specific organisms they are intended to control.[3] If the microflora are altered, the interactions among their component groups, including phytopathogenic fungi, may change. If antagonists to a pathogenic fungus are inhibited by a fungicide, the activity of that fungus could increase,[4,5] by direct or indirect action, after short, average, or longer periods of time, depending on whether the product acts quickly or persists longer in its initial state or in its metabolic form. It is conceivable that some fungicides could effectively inhibit the growth of a pathogenic fungus and also predispose the host to diseases caused by nonsensitive pathogens. Such interactions could explain the dependence upon fungicides to control disease where toxicants are applied repeatedly. The study of nontarget effects of fungicides on soil microflora is a reflection of the increasing interest in this subject, which is becoming more pronounced as the range of fungicidal products increases every year. Treatments are more often carried out at short intervals and are sometimes superimposed. Because of the continuous use of fungicides to keep soil and its crops sanitary, there is an element of risk for the soil microflora. This risk cannot be underestimated; it requires attention to the possible nontarget effects of fungicides on soil microflora. The subject, however, is vast and it is virtually impossible to establish its limits.

II. SOIL MICROORGANISMS

A. Fungi

Fungi form a large part of the total biomass of microflora in soil. These are usually abundant in the upper layers of soil where aerobic conditions prevail.[6-8] Their role in soil aggregation and degradation of complex molecules of fungicides in soil is important. Saprophytes, such as *Aspergillus* sp., *Penicillium* sp., *Trichoderma* sp., etc., have a marked effect on the population and/or pathogenicity of soil-borne pathogens such as *Pythium* sp., *Phytophthora* sp., *Rhizoctonia* sp., *Sclerotium* sp., *Fusarium* sp., and other species through competition, antagonism, and hyperparasitism.[9] While studying the nontarget effects of fungicides on soil fungi, a serious problem that is continually encountered is the difficulty of obtaining representative specimens from the soil due to the preponderance of spores on genera such as *Aspergillus* sp., *Penicillium* sp., and *Trichoderma* sp. This tends to reduce the results of such studies to approximations. In recent years, selective media have been developed for several genera, such as *Fusarium* sp., *Pythium* sp., *Phytophthora* sp., *Rhizoctonia* sp., and *Thielaviopsis* sp., but not for saprophytic fungi which are responsible for so many activities in the soil. It is desirable to develop a technique which could clarify the effect of fungicides on these saprophytic microbes.

The changes in the fungal population of soils treated with fungicides may be due to the direct effect of the chemicals on soil fungi, because most cellular processes are common to all fungi. Fungicides kill or adversely affect both the target fungi and a wide range of nontarget fungal species.[10-12] Changes in the fungal population of treated soils may also, however, be due to the alteration of the physiochemical characteristics of soil fungicides.[13] The application of Verdasan was shown to reduce decomposition of organic matter;[14] this in turn could result in waterlogging and anaerobic conditions which could further reduce microbial activity. The objective of studying the nontarget effects of fungicides is to explore the changes brought about by their successive application in soil. Their long-term effects on soil microflora were monitored to determine patterns and to distinguish between tolerant and recolonizing species.

1. Nonsystemic Fungicides

While studying the effects of several soil fungicides, Domsch[15] demonstrated that at concentrations of 40 to 400 ppm several microorganisms are affected differently. Vapam increased actinomycetes but decreased the growth of other microorganisms; captan inhibited all except bacteria while nabam inhibited only fungi. Contrary to these findings that captan reduced the population of fungi, Vaartaja and Agnihotri[16] observed that captan applied to the rhizosphere soil stimulated the growth of saprophytes of *Picea glauca* seedlings growing in forest seedbeds, particularly *Penicillium* and *Glicocladium* spp. Further Agnihotri[17] observed that captan at 62.5, 125, and 250 ppm acre^{-1} active incredient (a.i.) encouraged the growth of three saprophytic fungi, *Penicillium, Trichoderma,* and *Fusarium* spp. In respect to their reaction to soil fungicides such as captan, nabam, vapam, and thiram, Domsch[18] found the following fungi to be particularly sensitive to these fungicides: *Rhizoctonia solani, Doratomyces* sp., *Microsporus* spp., *Monilia pruinosa, Rhizopus nigricans, Morticerella alpina, Fusarium solani,* and *Aspergillus fumigatus,* whereas *Penicillium nigricans, Pullularia pullulans, Chaetomium* sp., and *Aspergillus versicolor* are tolerant.

Selective influence on microbes by different soil fungicides has been illustrated by Cordon and Young;[19] mylon soil treatment at 30 ppm decreased fungal population in the first 30 days but the soil could be easily reinfested with normal fungal flora during this period. The total bacteria increased but the actinomycetes remained unchanged. Nabam (20 ppm) had similar effects. However, they observed an increase in the total fungi because of the development of *Trichoderma* sp. Another soil fungicide, vapam (15 ppm), caused a decrease in fungi; certain penicillia contributed to a build-up after 20 to 30 days. However, bacteria and actinomycetes were not affected. On soil treatment with methylmercury oxinate (20 ppm), fungal count was reduced by 100% in 24 hr, but *Penicillium* spp. developed tremendously (ten times)[19] and no effect was observed on bacteria. Antinomycetes were drastically reduced. Application of Metasol resulted in rapid reduction of the fungal population and a subsequent increase in the recolonizing population was primarily due to *Penicillium* spp. and a *Chrysosporium* sp.[20] Chandra and Bollen[21] reported that mylon and nabam at 375 and 250 kg ha^{-1}, respectively, caused a depression of bacteria for 30 days and of fungi for 45 days.

Quintozene (pentachloronitrobenzene, PCNB) occupies a unique position among soil fungicides and is specifically effective for the control of *Rhizoctonia* spp. and *Sclerotium rolfsii*. PCNB at rates up to 20 kg ha^{-1} in loam soils produced no change in number of fungi or actinomycetes and no obvious shift in the species.[22,23] The effect of fungicides was significantly reduced in soils which are devoid of available nutrients. Farley and Lockwood[22] failed to show any effects of quintozene in unamended soil. Similarly, the addition of benomyl to an unamended soil did not affect microflora; however, in soils which are amended with easily available organic sources, some fungi resistant to benomyl increased.[24] Hence, it is apparent that the response of soil microflora to benomyl depends highly on the soil used. *Trichoderma viride* and *T. harzianum* in which no benomyl resistance has been observed could only successfully compete for the extra substrate in the absence of the fungicide. In a similar experiment in which a sandy soil was used, benomyl at 10 μg g^{-1} only slightly affected the microflora in nutrient-amended soil.[24]

In the competition for substrates, suppression of one group of organisms will result in an increase of less sensitive ones.[25] Reduced competition for nutrients is an important factor in the increase of resistant organisms after the use of selective and specific fungicides. This was probably the mechanism for the increase of *Fusarium* spp. and bacteria when quintozene was added to glucose-amended soil;[4] in soil amended with lucern residues, quintozene promoted *Fusarium* spp. and *Pythium ultimum* at the expense of *Penicillium oxalicum* and *Rhizopus stolinifer*. Similarly, in soil amended with other various energy sources, chitin in particular, PCNB strongly suppressed the number of actinomycetes; with glucose, fungi usually increased but some decreases were recorded in *Fusarium* spp. in the presence of quintozene. The fungicide suppressed competition from other fungi and actinomycetes and thus promoted germination of *Helminthosporium victoriae* conidia and subsequent elongation. PCNB affected the rate of colonization of alfalfa particles at 1.0 kg ha^{-1}.[26] PCNB at 1.2 and 2.5 kg ha^{-1} did not increase the frequency of reisolation of *Rhizoctonia solani*.[27] PCNB (10 kg ha^{-1}) added to soil which was amended earlier with glucose delayed the time of peak oxygen uptake by 1 day, compared with amended soil without the fungicide and soil amended with chitin. However, respiration was adversely affected by PCNB at 5 kg ha^{-1}.[28] In silt loam soil PCNB and its degradation products pentachloroaniline and pentachlorotetranaline inhibited all bacteria and actinomycetes but stimulated soil fungi at the operational dose, while thiabendazole had the opposite effect. Microbial activity was stimulated by PCNB in glucose-amended soil but was inhibited after amendment with wheat straw and yeast extract. Thiabendazole inhibited it under all conditions.[29]

Pugh and Williams[14] found that the fungal numbers in a golf green that had received approximately 20 years treatment with Verdasan (12.5 kg ha^{-1}) were significantly reduced compared with those in the untreated fairways. They found that dilution plate and soil plate counts in the treated soils were reduced by 97 and 67%, respectively, compared with the control soil. Comparable figures after 1 year of treatment were 38 and 37%, although immediately after spraying the reduction was 50 and 55%, respectively. These studies indicate that long-term effects of Verdasan on soil fungal population are cumulative. The toxicity of Verdasan is enhanced by its degradation products. Organic mercurial compounds applied to soil were found to be in the organomercury form after a lapse of 1$^{1}/_{2}$ months.[30]

Captan, dicloran, thiram, and Verdasan were applied at 9, 2, 6.7, and 12.5 kg ha^{-1}, respectively, at monthly intervals for 12 months consecutively. These fungicides reduced the number of fungal propagules in soil by 23, 11, 36, and 50%, respectively, as compared to controls.[13] Captan- and dicloran-treated soils were rapidly recolonized within 1 week of the application whereas the effects of thiram and Verdasan were more persistent. However, thiram at 94, 188, and 375 ppm decreased fungal populations of a soil.[13,31] Thiram caused a significant reduction of soil biomass at 5 ppm, but after 1 week, it was restored to the normal level.[32] At 50 ppm, the fungicide resulted in long-term reduction and increased the ratio of bacterial to fungal population. However, the fungal numbers in soils treated with these fungicides did not recover sufficiently to reach control levels. The descending order of toxicity was Verdasan > thiram > captan > dicloran. Wainwright and Pugh[33] showed that nitrification in soil was inhibited by these fungicides in a similar order. Although thiram and Verdasan caused the most deleterious effects on soil fungal numbers, cellulolytic species such as *Apiosordaria verruculosa, Chaetomium* spp., *Gliocladium roseum, Penicillium* spp., and *Trichoderma* spp. were not completely controlled by these fungicides.[13] They reported that the reason that *Chrysopermum pannomum* is predominant in Verdasan-treated soil may be because of the elimination of its faster growing competitors by the fungicides. However, its abundance in Verdasan-treated soil and its conspicuous absence in thiram-treated soil suggests that the fungus may detoxify and utilize Verdasan rather than promote its proliferation in the absence of competitors.[13] The fungus was capable of detoxifying the fungicide and utilizing it as a carbon source.[34] While *C. pannorum* occurred in increased frequencies in Verdasan-amended soil on the first day, *T. viride* and *Botryotrichum piluliferum* were sensitive to Verdasan and were eliminated on the first day. The relative abundance of these species can be used to monitor the levels of pollution of the soil by mercurials.[13]

Fenaminosulf (Dexon) has shown extraordinary promise in combating diseases caused by soil-borne phycomycetes fungi, e.g., *Pythium* and *Phytophthora* spp.[35,36] Tammen et al.[36] reported that fenaminosulf limited the colonization of steamed soil by *Pythium spinosum* for 92 days, whereas Alconero and Hagedorn[37] observed that the frequency of isolation of *Fusarium* spp., *Actinomucor elegans,* and other mucorales was not affected by fenaminosulf treatment. Agnihotri[38] incorporated this fungicide in potato dextrose agar (PDA) at the rate of 300 ppm and produced no effect on *Mortierella* spp. Field studies indicated that after application at 1.1 to 4.4 or 8.8 to 35 kg ha^{-1} in clay loam soil with pH 6.4 and organic matter 3.5%, the low range had no effect but the higher range decreased fungi and the highest rate decreased actinomycetes.[38] *Pythium* spp. are sensitive to fenaminosulf but *Mortierella* spp. are resistant to it. This was unexpected because *Mortierella* is related to *Pythium* taxonomically and also metabolically, as is evident by their tolerance to carbohydrates and amino acids. Furthermore, the addition of fenaminosulf to soil stimulated the population of *Mortierella* spp. This effect may have great significance because *Mortierella* spp. are known antagonists of *Pythium.*[38]

Several investigators have studied the effects of fungicides on various fungi developing in soil. Waksman and Starkey[39] observed that the types of fungal colonies developing on plates represented relatively few species compared to those developing in untreated soil. *T. viride, Penicillium* sp., and other fungal species became dominant following fungicidal application of thiram in sandy but not in compost soil.[40] Soil treatment with several fumigants resulted in the development of numerous fungal species.[41,42] The colonization of *T. viride* is due to its resistance to the toxicity of the fungicide and the less competitive environment. The other pertinent point is that sufficient fungicidal residues are left in the soil which are toxic to other fungal species. The changes in the microbial population of the soil following fungicide application may exert a biological control of soil-borne pathogens. *T. viride* is the best known example: it has demonstrated antagonistic influence on *Phytophthora, Pythium, Armillaria, Rhizoctonia,* and other pathogenic species. TCMTB (Busan) at 5 to 30 kg ha^{-1} caused a species shift resulting in the displacement of cellulolytic fungi such as *Fusarium, Penicillium, Aspergillus,* and *Streptomyces.*[43] *Trichoderma* spp. have been consistently observed to be major recolonizers in fungicide-treated soil.[31,38,44] *T. viride* is more a recolonizer than a survivor that invades and proliferates on the removal of the toxicant.[13] It was also observed that in soil where broad-spectrum fungicides such as thiram and carboxin were applied, *T. viride* was introduced from the outside to the treated area.

Sinha et al.[45] studied the effect of aeratan on soil bioecosystems and their related biochemical activity. They found that aeratan markedly reduced the fungal propagules in the early stage of the experiments and thereafter fungal propagules gradually increased. Low rates of aeratan reduced the numbers of Fusaria, while higher concentrations were lethal. They also observed the effect of the fungicide on the growth and reproduction of 14 fungi and found that the effect is more pronounced on pathogenic than on saprophytic fungi. The decrease in the fungal population by organomercurials was selective[46] and the cellulolytic fungi were less tolerant than other fungi.

Actually, the decrease in the number of fungi in soil, which occurs even with small rates of the fungicide, was only temporary. It was maximal after 2 days with captan in the laboratory assay and after 3 to 15 days with captan, dicloran, and drazoxolon in the field.[34] The return to a normal population occurred after 28 days[47] with a treatment using captan, thiram, dicloran, and quintozene and after 60 days[21] with nabam and dazomet. The inhibitory effect of captan in all the fungi lasted longer than that of thiram.[48] Captan, dicloran, and drazoxolon, at normal field rates, led to a reduction of fungi 15 days after treatment. After 157 days, the fungal population increased, sometimes beyond the normal level.[34]

Captafol and folpet (1 to 10 ppm) inhibit fungi and some soil microbial metabolic activities presumably associated with fungi. The greatest effect was seen on fungal growth measured at higher concentrations in vitro. At this concentration, high activity should be expected since adsorption effects that normally occur in soil are absent.[49]

2. Systemic Fungicides

Systemic fungicides are being used extensively through foliar and fruit applications for the control of several plant pathogens in vegetable and fruit crops. Sooner or later, however, they reach soil and persist there for a long time. Some of the fungicides exudate through the roots after foliar application. The presence of these fungicides affects the soil microflora and their activity.

Sinha and co-workers[50] studied the effect of carbendazim on the soil microflora, their physiological activity, and subsequently the growth of sugarbeet seedlings. They observed that at 5 to 40 ppm carbendazim did not impair the germination of sugarbeet and their emergence and stand, but affected different bioecosystems of soil and their physiological activity. The toxicant markedly reduced the population of fungi, bacteria, *Azotobacter*, *Rhizobia*, and actinomycetes for 45 days. The toxic effects were observed at 20 ppm and above. Fusaria were not affected significantly in carbendazim-treated soil during the early stage, but their number was drastically altered 30 days later. The reduction in the length of roots and the height of the top was perhaps due to the accumulation of ammonia to a toxic level in the soil. Since sugarbeet is a nitrate-preferring plant, it is suggested that seed should be sown 10 to 15 days after applying the chemical, because by the time seedlings emerge, the toxic effect of ammonia would be considerably reduced.[50]

The effect of benomyl on the growth of fungi has also been studied.[51,52] Hofer et al.[51] reported that benomyl at 25 ppm had a fungistatic effect on fungal properties in humus-sand soil in vitro. The fungicide at 4 and 20 kg ha^{-1} applied to field soil of pH 5.3 and organic matter 3.1% caused an increase in fungal propagules 4 weeks after treatment.[47] *Trichoderma* sp. was usually isolated from the soils treated with this fungicide at the rate of 2.2 to 89.6 kg ha^{-1}.[44] At 2.0 kg ha^{-1} the fungicide had no effect on the cellulolytic activity of organisms in sandy soils.[53] Wainwright and Pugh,[47] who used benomyl in vitro and in vivo, found that it caused increases in the amounts of exchangeable elements such as Mn, Zn, Cu, K, and Na in some cases, but had no effect or caused decreases in other cases; ammonification, however, was increased. Wainwright and Pugh[34] reported that benomyl at 5 or 25 kg ha^{-1} in grass soil caused qualitative and quantitative changes in the free amino acid content of the soil and thus the change in fungal flora. After 2 weeks with 300 kg ha^{-1} benomyl, 74% of the fungi disappeared, and only three species survived. The recovery to a normal population took place after 60 days with benomyl.[21]

Benzimidazole derivatives such as benomyl, thiophanate-methyl, and thiophanate-ethyl suppressed the number of fungi more than other risk fungicides, and this was shown to be expecially true for *Trichoderma, Penicillium,* and *Fusarium* spp.[54] but not for *Aspergillus* and *Mortierella.* This type of relationship has also been reported by other workers.[55-57] Benomyl and related fungicides are fungistatic in action and not fungicidal.[51,58,59] Dycloheximide is toxic towards *Mortierella, Trichoderma,* and *Aspergillus* spp.[25] and the fungicide inhibited various similar fungi.[60]

Evidence of the interrelationship between pathogenic fungi and fungicides has been established.[61] Benomyl was effective on *Cercosporidium personatum* and *Cercospora arachidicola* at 0.15%, but in unsprayed plots *Sclerotium rolfsii* damage was lower than in sprayed. This may be due to the direct effect of the fungicide on this pathogen or to an indirect effect, that of killing its antagonist, *T. viride.* The highest infestation of *S. rolfsii* was associated with plots receiving a fungicide which had no effect on this fungus but which was toxic to *T. viride.*

Benomyl at 3 to 30 kg ha^{-1} had no effect on the decomposition of organic matter by surviving species of fungi and other microorganisms.[51] Hofer et al.[51] also demonstrated that benomyl in humus sand did not affect dehydrogenase, amylase, or catalase activity. Confirmation of the lack of effect of benomyl on amylase activity in sandy soils was shown by von Faassen.[53] Benomyl at 5 or 25 kg ha^{-1} in grass soil caused qualitative and quantitative changes in the free amino acid content of the soil. High rates generally reduced the amount of amino acid nitrogen, but low rates caused increases.

Field application of benomyl (20 kg ha^{-1}), captan (9 kg ha^{-1}), dicloran (2 kg ha^{-1}), formalin (500 1 ha^{-1}), quintozene (5.6 kg ha^{-1}), and thiram (6.7 kg ha^{-1}) applied at twice the normal rates resulted in an increase in bacterial and fungal numbers after 28 days.[47] The large amount of bacteria which follow the application of fungicides is probably responsible for many of the biochemical changes that occur, particularly in relation to nitrogen mineralization, whereas the changes in fungal number, on the other hand, may affect the mineralization of carbon.[47] The growth of *T. harzianum* in the presence of other soil fungi on PDA was highly depressed by benomyl, while thiram enhanced its growth even at 300 ppm.[62] It was observed that *Gliocladium, Penicillium,* and *Trichoderma,* which are typical fungal recolonizers of partially sterilized soils, predominated in the treated soil. *Trichoderma* spp. have been consistently encountered as major recolonizers of fungicide-treated soil.[31,38,44,63-68] Mughogho[69] analyzed the species composition of *Trichoderma* found in treated soils and showed that several species, including *T. hamatum, T. harzianum, T. koningi,* and *T. viride,* were repeatedly recorded. Evidence was provided to prove that *T. viride* is a recolonizer rather than a survivor that invades and proliferates on the removal of a toxicant. However, Kuthubutheen and Pugh,[13] while acknowledging the problems of extrapolating from the results of in vitro/in vivo conditions, maintained that the growth of the fungus in soils following application of thiram and Verdasan was most probably due to recolonization from outside the treated soil. When *T. hamatum* was similarly treated, it was found to be more tolerant to the four fungicides (captan, thiram, dicloran, and Verdasan) used. The growth of this fungus after treatment of soil with mercurial fungicides, therefore, could be due to surviving inocula and to recolonization. There is therefore a need to bear in mind the different behavioral pattern of growth of the species of *Trichoderma.*

Numerous theories have been proposed to explain the increase in *Trichoderma* which are based on the initial tolerance to the fungicide and a fast growth rate.[64,65] However, the increased incidence of this fungus may also be influenced by changes in the nitrogen status of treated soils. Danielson and Davey[67] have demonstrated that *Trichoderma* spp. preferentially utilize NH_4^+-N. The presence of large amounts of nitrogen in this form in treated soils may indicate that *Trichoderma* has marked advantages over other fungi less well equipped to utilize NH_4^+-N, which enables *Trichoderma* to recolonize rapidly in partially sterilized soils.[13]

Triarimol application in the field at the rate of 2.0 kg ha[-1] under grass reduced fungi and recovery was observed after 28 days.[34,70] It was observed that at times the fungal population increased beyond the normal level.

Treatment of soil with triazbutyl at 0.5 and 10 ppm a.i. had a disruptive effect on soil microbes, particularly in fungi,[71] for 57 days. While using several fungicides for the control of *Rhizoctonia solani* in rice[72] at 0.025% after 20, 25, and 30 days of sowing, it was found that Kitazin, edifenfos, benomyl, carboxin, and carbendazim were most effective in reducing the rhizospheric fungal population and stimulated bacteria and actinomycetes.

Application of fungicides results in changes in fungal population. Thus, fungi could be divided into two distinct groups: those which disappeared from treated soils and those which were still present after treatment.[73] Those which disappeared, such as *Botryotrichum pilluliferum, Chaetomium funicolum,* and *Gliomastix murorum,* were obviously intolerant of the fungicide being applied.[14] However, those which could be isolated from treated soils may have been intolerant, but able to colonize the soil rapidly once the toxic effects were removed (e.g., *T. viride).*[13]

More recently, work on fungicides has been concerned with plant pathogenic species of *Fusarium,* to see what levels of fungicidal application are necessary to kill the pathogens and also how these levels may affect the nonpathogenic species, which may be producing beneficial effect in the soil by taking part in the decomposition of plant debris. The results obtained using different isolates of *F. oxysporum* and *F. solani* showed significant differences between the strains. There were also marked differences in the reaction of the species to different fungicides.[74] With captan, *F. nivale* was the most tolerant and *F. oxysporum* var. *dianthi* was the most susceptible. However, *F. oxysporum* var. *dianthi* was most tolerant to Verdasan, while *F. solani* was the most sensitive. Thus, useful species could be eliminated from soil by using dose rates which do not necessarily get rid of the pathogenic species. In general, the application of fungicides has been shown to have marked effects on soil microbial populations, and in some instances, the changes produced can be significant. There is always the potential for adverse effects. This potential needs to be constantly monitored as new fungicides are introduced.[73]

B. Bacteria

The bacterial population of soil outnumbers the population of all other groups of microorganisms in both number and variety. Many kinds of bacteria — autotrophs, heterotrophs, mesophiles, thermophiles, psychrophiles, aerobes, nitrogen fixers, and protein digesters — are found in soil.[6] The role of bacteria in the maintenance of soil fertility is well documented. Several species of *Rhizobium* symbiotically fix atmospheric nitrogen in the nodules of the leguminous plants, while *Azotobacter* spp., a free-living nitrogen fixer, fixes atmospheric nitrogen. Many of the bacteria play a significant role in the maintenance of soil fertility by decomposing organic matter into simple forms that are utilized by plants and also in the retention of nutrients such as sulfur and phosphorus. Toxicants can affect either their population or their related functions. Probably of more significance than the quantitative effects of fungicides on soil population are the qualitative effects. The nitrifying bacteria, which oxidizes ammonia to nitrate and nitrite to nitrate, are relatively sensitive and may be inactivated in the soil for several weeks to several months following treatment.[9,75-77] Wensley[78] noted that nitrifiers and certain cellulose decomposers were quite susceptible to methyl bromide fumigation, whereas spore-forming bacteria are quite resistant to ethylene oxide. This does not imply, however, that spore-forming bacteria account for the large numbers of bacteria which quickly develop in fumigated soil. Of the bacteria that increased following Trichoderma, *Pseudomonas* spp. often constitute the greatest number of colonies developing on the plates.[79]

1. Nonsystemic Fungicides

The application of fungicides in soil results in the increase or decrease of bacterial population. Out of the eight fungicides bioassayed on three bacterial species, only methylmercurydicyandiamide (MMDD) inhibited bacteria at 1 ppm.[80] The toxicity of other fungicides was found in the following descending order: MMDD > pyrazophos > thiram > maneb > captan. However, the inhibition occurred only at concentrations above 100 ppm. MMDD was the only fungicide that inhibited *Sarcina hitea,* and stimulated the growth of *Bacillus subtilis.* The bacterial population decreased to 60% as compared to the control after the application of captan (20 kg ha^{-1}) during the first week, then increased over the next 3 weeks to 180%. Houseworth and Tweedy[48] observed fluctuations in bacterial population in silt loam soil at 0.1 or 1 kg ha^{-1}, which were inversely related to fungal populations.

Isobac (0.1 to 1.6 kg ha^{-1}) altered bacterial populations and favored the growth of a bacterium which was antagonistic to fungal pathogens.[79]

Quintozene (20 kg ha^{-1}) had no effect on soil microorganisms or their metabolic activities in the absence of an added energy substrate.[22,23] However, when an energy source in the form of glucose was supplied externally, the fungicide increased bacterial population.[4] With quintozene (1.0 kg ha^{-1}) in loam soil of pH 6.7 and 3.8% organic matter, Katan and Lockwood[26] observed that the rate of colonization of alfalfa was adversely affected and the decrease in fungi and actinomycetes by fungicidal application was both qualitative and quantitative. Working with fenaminosulf, Agnihotri[38] found that an application at 1.1 to 4.4 kg ha^{-1} in soil of pH 6.4 had no effect on the number of bacteria; however, at 8.8 to 35 kg ha^{-1} there was significant effect. Wainwright and Pugh[47] reported that captan, thiram, and quintozene at 9, 6.7, or 13.4 and 5.6 or 11.2 kg ha^{-1} in soil of pH 5.3 and 3.1% organic matter increased the number of bacteria. In later studies, captan at the rate of 0.5 to 10 kg ha^{-1} incorporated to soil of 6.7 and 2.9% organic matter also increased bacterial populations. Agnihotri[17] also reported similar results with captan. However, Mahmoud et al.[81] reported that captan adversely affected bacterial number at the recommended rate.

Dithiocarbamates were intensively evaluated for their effects on microbial populations, but diverse reports are available on their effect on microbes and their activities. The first report on the effect of maneb, nabam, and zineb (up to 100 kg ha^{-1}) indicates that the fungicides had no effect on fungal populations of a fine sandy soil.[82] Chandra and Bollen[21] reported a decrease in the number of bacteria and *Streptomyces* by application of nabam at 225 kg ha^{-1} for the first 30 days and in fungi for 45 days in four different types of soils; however, the bacterial and fungal populations increased significantly at 60 days. Aeratan at 30 kg ha^{-1} stimulated the total population of bacteria for 15 days and thereafter decreased.[83] Similarly, the total bacterial population of a sandy soil increased due to an application of 100 kg ha^{-1}, while other microbes were reduced during the first month of application.[20] An increase in the bacterial population of an acid lateritic clay and an alluvial loam by the application of maneb (1.5 to 96 kg ha^{-1}) was observed, but the effect disappeared after 10 months of application at higher rates of maneb.[84] The same compound at 10 ppm but not at 5 ppm decreased bacterial and other microbes of a sandy loam,[85] an organic soil,[86] and a clay soil[87] after 24 hr application, but they recovered after 150 hr to their original levels in the sandy loam and organic soil. However, they increased in the clay soil. Zineb alone and in various pesticide combinations at operational rates was ineffective on bacteria and other microbes for a period of 4 months.[88] However, Dithane Z-78, a formulated product of zineb, applied along with malathione to a loam soil at 12,500 kg ha^{-1} decreased the bacteria.[89] Mancozeb, another dithiocarbamate fungicide widely used for the control of foliar diseases, at 10 ppm decreased the bacterial and microbial populations of soils for 9 weeks.[90,91]

In a study using thiram at 20 kg ha^{-1} in loam soil,[92] the bacterial populations decreased to 60% of the control during the first week after treatment, but after 3 weeks the population was restored to the control level. The increase was 180% of untreated soil populations. The bacterial populations increased in contrast to a decrease of actinomycetes in sandy, but not in compost, soil.[40] Thiram at 0.1 to 1.0 kg ha^{-1} in a silt loam soil resulted in erratic behavior in bacterial populations; therefore, its effects could not be determined.[48] The fungicide at 6.7 or 13.4 kg ha^{-1} significantly increased the populations of heterotrophic soil bacteria[47] after 28 days of treatment. Thiram was ineffective in an organic soil[86] but was inhibitory in the sandy loam soil.[85] At 94, 188, and 375 ppm it decreased bacterial population of a soil, but after 1 week the population increased.[31]

Naumann[92] reported that vapam at 1500 ppm increased bacterial population for a considerable period of time, while Teuber and Poschenrieder[93] found an increased number of bacteria in soil treated with methyl isothiocyanate. Sinha et al.[94] reported that vapam, only in the early part of the experiment and at higher concentrations (500 ppm and above), appreciably reduced the population of bacteria, although later their population gradually increased.

Anilazine and maneb (1.5 to 96 kg ha^{-1}) in acid laterite clay with pH 5.0 and alluvial loam with pH 7.7 increased bacterial population; however, the effect disappeared after 300 days of treatment, except with 24 and 96 kg ha^{-1} of maneb.[95] The inhibition was less in the rapidly nitrifying loam than in the slow nitrifying clay, and in both cases the fungicides inhibited *Nitrosomonas* spp. but not *Nitrobacter* spp. They found that maneb was more toxic.

2. Systemic Fungicides

Benomyl is being used for the control of soil-borne diseases caused by fungal pathogens. Peeples[44] found that the fungicide at 2.2 to 89.6 kg ha^{-1} in silt loam fine sand and sandy loam soils had no effect on bacterial population. Benomyl application (3.0 to 30 kg ha^{-1}) in soil with different humus contents behaved differently and in humus sand it had little effect on bacterial population of soil.[51] However, these applications decreased nitrification after 4 weeks of incubation. In sandy soil not previously treated with benomyl, bacterial population increased with increasing concentration (2.0 to 10 kg ha^{-1}). An application of benomyl at the same rate in vitro decreased the total bacterial count 200 times.[53] Benomyl inhibited mycelial growth of *Agaricus bisporus*. Foster and McQueen[96] reported no alteration in the bacterial population after benomyl application. Smiley and Craven[97] reported that after repeated application for 3 years at the recommended rates used in practice, ethazole, thiophanate-methyl + maneb, and thiophanate-ethyl + thiram consistently caused two- to fivefold increases in the total bacterial number. Other fungicides did not greatly or consistently alter the numbers of the general bacterial population. It was observed that between 0.4 and 2% of the bacteria were sporing *Bacillus* spp. and their numbers remained remarkably uniform, regardless of the treatment. *Pseudomonas* spp. comprised 0.2 to 0.6% of the bacteria and were always more numerous in treatment with benzimidazoles.

Sinha and Singh[71] reported a significant reduction in bacterial population up to 37 days after application of triazbutyl at the rate of 0.5 and 1 ppm (a.i.). This fungicide is reported to be highly specific to wheat rust *(Puccinia recondita)*.[98]

C. Actinomycetes

The actinomycetes or ray fungi are usually abundant in well-fertilized cultivated soils. The most prominent genera of this group are *Nocardia*, *Streptomyces*, and *Micromonospora*.[7] They are capable of degrading many complex organic substances such as lignin and cellulose and consequently play an important role in maintaining soil fertility. The actinomycetes are also known to synthesize and excrete antibiotics, both antifungal and antibacterial. Thus, fungicides that disrupt their populations in soil may also adversely affect their functions.

1. Nonsystemic Fungicides

Actinomycetes appeared to be more sensitive to fenaminosulf than bacteria and fungi.[38]

Vapam and aeratan reduced the actinomycete population initially, but later their population was gradually increased. Stimulation was also observed.[45,83] Application of captan (62.5, 125, and 250 mg acre^{-1} a.i.) in soil initially decreased the population of actinomycetes and then gradually increased; the general decrease of pathogenic fungi, accompanied by an increase in the population of bacteria and actinomycetes, supports the hypothesis of direct biological control.[99] Therefore it may be concluded that the extraordinary success of captan in controlling damping-off is not solely due to its fungicidal properties.

The fungicides are in general toxic toward actinomycetes, even at normal rates, as indicated by the work of Cordon and Young[20] with metasodium, nabam, and dazomet and that of Agnihotri[31] with captan. A temporary effect was observed by Agnihotri[31] with thiram.[34] However, a stimulating effect with a low dose of captan was observed.[34]

2. Systemic Fungicides

Benomyl treatment (up to 100 ppm) of soil did not influence the number of actinomycetes that were counted after 6, 21, 50, and 80 days of incubation and it was found that the number was increased with increasing benomyl concentration.[53] Similar findings were observed by Hofer et al.[51] Such findings suggest that this can be explained by the fact that microbes utilize compounds added with the benomyl formulation or that benomyl has a vitamin-like action. Unfortunately the composition of the inert materials of benomyl formulation usually is not given. Weeks and Hendricks[100] have suggested that in soils where benomyl has been applied some detoxification will occur due to bacterial utilization of the active ingredient as a source of carbon. The increased number may also result from a change in the population, with the number of actinomycetes increased and fungi reduced. Benomyl at high rates and repeated application results in no change in soil microflora.[58]

A temporary inhibition of actinomycetes was observed by Siegel[52] with benomyl.

D. Miscellaneous (*Azotobacter*)

Species of *Azotobacter* are abundantly present in cultivated fertile soils where they play a prominent role in the nitrogen economy of soil. They fix atmospheric nitrogen nonsymbiotically. The bacterium does not survive well in soils having pH below 6.0.[7] The number of *Azotobacter* in soil amended with vapam[45] and aeratan[83] apparently decreased in the first 15 days, but later was stimulated to a number greater than in the control treatment. Carbendazim markedly reduced the population of *Azotobacter* at 10 and 40 ppm for 45 days and toxic effects were clearly discernible at 20 and 40 ppm.[50] Benomyl had no effect on *Azotobacter* spp. in sandy soil.[53]

Gil Diaz-Ordonez et al.[101] studied the effect of several fungicides at 4, 40, and 400 ppm, but found no effect on *A. chroococcum* culture. Copper oxychloride completely inhibited growth; *A. chroococcum* was more sensitive to thiram at pH 7 and other rhizosphere isolates varied in sensitivity to thiram.[102] *A. chroococcum* was inhibited by zineb, captan, ferbam, thiram, folpet, and maneb at comparatively high doses not used in plant disease control,[103] but captan applied at operational rates to cotton cultivars inhibited *Azotobacter* spp. in the rhizosphere soil.[81]

III. SOIL MICROBIAL ACTIVITY

Large quantities of fungicides ultimately reach the soil either by direct application or as run-off from the treatment of foliage, fruit, and seed. Once the fungicides are in the soil they may affect the activity of soil microorganisms, which are largely responsible for the maintenance of soil fertility.[104] With the increased use of fungicides, there is a growing danger that soil fertility and elemental cycling processes may be adversely affected; thus, much attention must be focused on the possible effects of these fungicides on soil microorganisms and their functions. The metabolic activity of a population depends on the number of individuals as well as on the activity of each individual; therefore, the effect of a fungicide on a given activity need not be correlated with a change in the number of individuals that are responsible for the activity.[24] The majority of biochemical transformations in soil result from microbial activity.[1] These include ammonification or liberation of ammonia from proteinaceous matter, nitrification or oxidation of ammonia to nitrate, sulfonification or oxidation of sulfur to sulfate, and decomposition of organic matter and the concomitant formation of humus. Microorganisms mineralize, oxidize, reduce, and immobilize elements in soil and also indirectly influence their solubilities. Among these factors, ammonification and nitrification play an important role because they convert organic nitrogen into soluble nitrogen which can be readily assimilated by the green plants. Any compound which alters the activity of microorganisms could therefore affect soil biochemical processes and ultimately influence soil fertility and plant growth. Fungicides seem to be more detrimental to biological transformations in soil because of their selective toxicity to microorganisms.[1] Moreover, in effective control of pathogens, recommended concentrations for fungicides are generally high. The influence of fungicides on soil microbial processes has been reviewed.[10,105]

A. Nitrification

Nitrification is the mechanism of converting ammonium ions via nitrite into nitrate ions, which are readily consumed by green plants. The oxidation of ammonium ions is apparently catalyzed by physiologically similar groups of bacteria belonging to the family Nitrobacteraceae. *Nitrosomonas* spp. oxidize ammonium ions to nitrite ions and *Nitrobacter* spp. oxidize nitrite to nitrate. The process seems to be very sensitive to microbial activity. The soil bacteria which oxidize ammonia to nitrite and nitrite to nitrate are relatively sensitive to soil fungicides, whereas organisms which release ammonia from organic nitrogenous complexes quickly return to the soil following treatment. The activity of nitrifiers may be checked for several weeks to several months, depending upon the chemicals used, dosages, and soil properties. Ammonia accumulates from the decomposing organic fraction of the soil and from the dead bodies or the organisms killed by the chemical. Also, ammonia-nitrogen added as an inorganic fertilizer will remain in the reduced form.

1. Nonsystemic Fungicides

The effect of several fungicides on the nitrification of ammonium sulfate and ammonium hydroxide was studied by several workers with a variety of soil types and environments. Inhibition of nitrification in soils was observed with mylone and nabam;[21] nabam, vapam, mylone, and allyl alchol;[109-112] vapam;[113,114] maneb;[87,95,106] zineb;[77,87,107,108] and mancozeb[91] at operational concentrations. Nonvolatile fungicides such as zineb, maneb, and nabam[87] also inhibited nitrification. The particular effect of maneb, as with Dyrene, is to suppress NH_4-oxidizing *Nitrosomonas*, but not the NO_2-oxidizing *Nitrobacter*.[95] Audus[2] reported that nabam, vapam, thiram, maneb, and zineb initially have a strong depressive effect on the nitrification process and the adverse effect increased as its concentration increased. Wilson[77] and Audus[2] observed that most of the dithiocarbamates are inhibitory to denitrification: nabam and maneb are more toxic than ferbam, thiram, and ziram. The dithiocarbamate radical governs the toxicity; thus, ferbam is more toxic than thiram and the latter more so than ziram.[2] The toxic effects of vapam on nitrifiers were perhaps due to its decomposition into a highly toxic compound, methyl-isocyanate.[115] Ferbam, ziram, and zineb were shown to have a toxic effect on nitrifying bacteria in soil for a period greater than 40 days when applied at 50 ppm, whereas at 500 ppm the inhibition lasted even after 180 days.[77] Inhibition of nitrification may be considerably less than had been expected on the basis of in vitro experiments; the nitrogen in soil, which was treated with zineb and maneb, was found to be correlated with the dissipation of carbon disulfide, which is one of the breakdown products in soil treated with the fungicides.[116] Maneb at 15 ppm inhibited NH_4^+ oxidation for up to 8 weeks.[95] Zineb (65%) when applied at the rate of 1.88 ml ℓ^{-1} to the root system reduced the relative number of rhizosphere bacteria.[117] Low dosages of fenaminosulf (10.9 to 43.7 ppm) seem to have no pronounced effect on nitrate formation, but as the concentration of fungicide increased, nitrification was inhibited. This resulted in an accumulation of ammonium nitrogen. The length of the inhibition period was dependent on the concentration of the fungicide.[38] This inhibition of nitrification was temporary and later a significant increase was observed.[21,38]

Fungicides such as anilazine, benomyl, maneb, quintozene, and thiram have inhibited nitrification for several weeks to several months.[1,47,95,96] In contrast, 14 weekly surface applications of anilazine, benomyl, and maneb to turfgrass had no effect on nitrification rates;[116] Smiley and Craven[60] also observed noninhibitory effects of several fungicides, especially benzimidazoles (benomyl, thiophanate-methyl, and thiophanate-ethyl). The lack of inhibition from these surface-applied chemicals, which are inhibitory if incorporated into soil, presumably related to their low solubility in water, low volatility, and sorption by organic compounds and clay minerals.[5] Smiley and Craven[54] demonstrated that fungicides caused differences in pH to occur among the treated soils. The concentration of NH_4^+-N was significantly related to the pH of the fertilizers on turfgrass soils, which suggests that nitrification may be more affected by fungicidal activity.

Wainwright and Pugh[33] reported that captan, thiram, and Verdasan at lower rates (0.5 to 10, 0.5 to 5, and 0.1 to 0.5 kg ha^{-1}, respectively) either stimulated or had no effect on nitrification, but at higher rates captan, thiram, and Verdasan (viz., 25, 10, and 25, and 1.0 to 5.0 kg ha^{-1}, respectively) inhibited nitrification. In soil of pH 5.3 and 3.1% organic matter, captan, thiram, and quintozene at 9, 6.7, and 5.6 kg ha^{-1}, respectively, significantly increased the number of heterotrophic bacteria. Captan and thiram at the same rates, dicloran at 2.0 kg ha^{-1}, and formalin inhibited nitrification markedly, whereas quintozene (5.6 kg ha^{-1}) had only a slight effect.[47] It was found that captan at 0.5 to 10 kg ha^{-1} in soil of pH 6.7 and 2.9% organic matter increased bacterial numbers.

Dicloran at 1.0 to 10 kg ha^{-1} inhibited nitrification in a fine sandy loam soil; fentin acetate (1 kg ha^{-1}) added to a loam soil initially decreased and then increased nitrification. However, consistent results could not be obtained in further experiments.[118]

Few reports are available on the effect of fungicides on the nonsymbiotic nitrogen-fixing bacteria. Captan (20 kg ha^{-1}) and thiram (40 kg ha^{-1}) supressed the number of *A. chroococcum* at least 5 weeks after application. On the contrary, Hoflich[119] reported that application of an even higher dose of thiram to straw-amended soil resulted in an initial decrease in numbers of *A. chroococcum,* which was followed by an increase lasting up to 13 weeks. While studying the effect of nine fungicides at 4, 40, and 400 ppm, Gil Diaz-Ordonez et al.[101] found no effect on the *A. chroococcum* free-living nonsymbiotic bacterium, but copper oxychloride at 40 and 400 ppm completely inhibited growth. Sud and Gupta[102] studied the sensitivity of thiram and its degradation products at different pH levels of 5 to 9 and found that *A. chroococcum* was more sensitive than other isolates, especially at pH 7. The isolates from roots of different plants varied in their susceptibility to these chemicals. Langkramer[103] reported that pure cultures of *A. chroococcum* were inhibited by zineb (5000 and 50,000 ppm), captan and ferbam (3000 and 30,000 ppm), thiram (5000 ppm), and folpet and maneb (3000 ppm). An inhibition of nitrogen fixation was showed by thiram in *A. vinelandii.* A mixture of zinc, maneb, and ferbam (Tricarbamix) at 130 ppm increased the cell yield of *A. vinelandii* and inhibited nitrogenase activity by 50 ppm;[120] however, at 320 ppm no increase in the cell yield occurred and nitrogenase activity further decreased.[120] In clay soil treated with thiram (5 and 10 ppm), the population of nonsymbiotic N_2-fixers decreased initially and then increased significantly, but nitrogenase activity was not affected.[85] In sandy loam and organic soils, thiram was ineffective at both concentrations.[85] Dinocarp stimulated growth but reduced nitrogen fixation.[24]

Munnecke and Ferguson[121] studied the influence of vapam on nitrification and found that it exerted no influence at 125 kg ha^{-1}; at 225 kg ha^{-1} partial inhibition occurred and at 450 and 900 kg ha^{-1} inhibition was complete. Similar results were also observed by Wilson[77] for other fungicides.

Nitrification was depressed by high and medium concentrations of Bordeaux mixture (BM), zerlate, and Peronex (50 to 500 ppm) on Tarneb soil in vitro for periods of 14 to 84 days; BM at 50 ppm stimulated decomposition of organic matter considerably. Peronex and zerlate depressed the process at higher concentrations but stimulated it at lower ones.[122]

Agnihotri[17] reported that almost no nitrification occurred after the application of captan (62.5, 125, and 250 ppm acre^{-1} a.i.), but when ammonium sulfate was added it was restored. On the other hand, in soil containing ammonium sulfate, NO_3-N accumulated to a maximum of 100 ppm in 21 days and then sharply declined. However, a probable explanation for this sharp decline could not be given. This may be due either to denitrification or to immobilization of NO_3-N. All the concentrations tested impaired nitrification for varying periods.

2. Systemic Fungicides

Among systemic fungicides, benomyl has been extensively evaluated for its effect on nitrification. The potential toxicity of benomyl to nitrification was also demonstrated in a pure culture of nitrifying bacteria. Addition of benomyl at 20 μg g^{-1} to liquid media with a mixed culture of the autotrophic nitrifiers, *Nitrosomonos* sp. and *Nitrobacter* sp. isolated from soil caused a delayed oxidation of NO_2 to NO_3; with 20 mg ℓ^{-1} benomyl in the medium only the oxidation of NO_2 was delayed.[53] At 200 mg ℓ^{-1} oxidation of NH_4 to NO_2^- as well as the oxidation of NO_2^- to NO_3^- was delayed.[53] In pure culture, benomyl inhibited the oxidation of ammonium ions by *Nitrosomonos* sp. and nitrite by *Nitrobacter agilis*, even at the concentration of 10 μg g^{-1}.[123] Among the hydrolysis products of benomyl, amino-butane (AB) showed toxicity to both groups of nitrifiers, but only at the higher concentration of 100 μg g^{-1}, whereas carbendazim was virtually innocuous, as in soils, to both groups of nitrifiers, irrespective of its concentration. Undoubtedly, benomyl was far more toxic than its degradation products, carbendazim and AB, at least with respect to autotrophic nitrification.[123] Thus, the oxidation of nitrite to nitrate and other nitrification processes in vitro was found to be more sensitive to benomyl than in vivo.

Benomyl application at 3.0 to 30.0 kg ha^{-1} in humus sand decreased nitrification after 4 weeks of incubation.[51] After 12 weeks of incubation significantly more (NO_2 + NO_3)-N was found in the benomyl-treated sandy soils than in the control; nitrite nitrogen after 12 weeks was less than 2.5 mg kg^{-1} and no ammonium nitrogen was found at any time during the experiment. Thus, the ammonification was immediately followed by nitrification.[53] In contrast to this, Hofer et al.[51] found a somewhat considerable inhibition of nitrification after 4 weeks incubation with 15 to 150 ppm concentration of benomyl in humidified sandy soil. There was no effect of the mineralization of benomyl on nitrite, as the higher amounts of nitrite found in the treated samples were independent of the benomyl concentration of (NO_2 + NO_3)-N found in the treated soil and may have resulted from an enhanced ammonification of organic nitrogen in soil.

Benomyl stimulated the rate of NO_3^- accumulation, 3.5 μg day^{-1} in control and 4.5 μg day^{-1} after treatment with carbendazim.[124] However, Torstensson and Wessen[125] observed no signs of the stimulation of inhibition by benomyl at 10 μg g^{-1} soil. This could be explained in different ways. As benomyl seems to inhibit *Nitrosomonas* sp. and *Nitrobacter* sp.,[53] it would appear more probable that the process of nitrification is inhibited rather than stimulated. However, Torstensson and Wessen[125] showed that when an increase in NO_3^- accumulation is registered, it may instead be the result of an increase in ammonification. This could be the result of a generally stimulated nitrogen mineralization or could depend on the mineralization of fungi killed by the fungicide, although the latter seems less probable.

The increase in NO_3^--N after benomyl treatment as compared to the control in a laboratory, if the value of a dose of 2 kg ha^{-1} could be transferred to field conditions and calculated on a scale that takes into account the extra NO_3^--N released in the uppermost 5 cm of the soil, corresponds to something in the range of 5 to 10 kg ha^{-1}. About the same amount was calculated when the dose was 150 g carbendazim per hectare.[124]

Benomyl application in the soil at the recommended rates sometimes stimulates ammonification and nitrification. Processes in the soil after the application of benomyl at 1000 ppm[126] and the effect of benomyl on soil microflora and certain biochemical transformations, such as ammonification and nitrification in nonflooded soil systems, were studied.[127] A stimulation of *A. carneus* was observed in a red sandy loam soil amended with 5000 ppm benomyl; NO_2^- accumulated in the soil which received peptone or $(NH_4)_2SO_4$ in addition to benomyl. Whether NO_2^- accumulation was due to heterotrophic nitrification by the predominant fungus *A. carneus* or to the inhibition of the NO_2^--oxidizing autotroph *Nitrobacter* was not clear. However, Gowda et al.[128] found that this was due to the proliferation of the fungus *A. carneus*. The data presented by these workers provide evidence for heterotrophic nitrification of NH_4^+-NO_2^- by this benomyl-tolerant predominant fungus isolated from benomyl-amended soil both in vitro and in soil incubation. However, the oxidation of the end product was not evident. Gowda[127] observed no accumulation of NO_2^- at lower concentrations of benomyl. In view of the known toxicity of benomyl to autotrophs even at 200 ppm, NO_2^- accumulation with the abnormally high concentration of 5000 ppm benomyl is assumed to be due to heterotrophic nitrification. The ability of the predominant fungus *A. carneus* from benomyl-amended soil to oxidize NH_4^+ both in pure cultures and soils tends to support this. Moreover, application of benomyl at 5000 ppm to a stimulated oxidized zone of a flooded soil together with $(NH_4)_2SO_4$ and a nitrification inhibitor, N-Serve, caused NO_2^- accumulation.[129] Since N-Serve is specific in inhibiting nitrification by *Nitrosomonas* and not heterotrophic nitrification, accumulation of NO_2^- in benomyl-amended soils apparently results from heterotrophic nitrification. The possible mechanism of the selective stimulation of heterotrophic nitrification in benomyl-treated soil is not known.

The ability of the fungus to form the biologically and chemically unstable compound NO_2^- from NH_4^+ as well as from NO_3^- indicated that the striking increase in its number in benomyl-amended soil[127] could eventually result in N loss from the soil. Moreover, NO_2^- accumulation in soil, though temporary, is undesirable, because of its known toxicity to plants and microorganisms. On the other hand, in addition to its direct toxic action, the fungicide may control plant parasites through NO_2^- accumulation, as indicated by the inhibition of the soil-borne pathogen *Fusarium oxysporum* f. *cubense* by NO_2^- formed in urea-amended soils.[130]

Benomyl, carbendazim, and AB exert both inhibitory and stimulatory effects on nitrification in soils. These observations could at best be ascribed to different soil properties and different test conditions in view of their importance in determining the overall persistence of benomyl in the soil environment. The effects of benomyl on nitrification may be related to the different rates of benomyl conversion to carbendazim, since nitrification was considerably inhibited by benomyl and not at all by carbendazim, both in soils and pure cultures of autotrophic bacteria.[123] It is generally found that large losses of applied ammonium nitrogen can occur in flooded soils because of its high rate of diffusion from the reduced soil layer to the oxidized surface. This leads to its nitrification in the oxidized surface and to the eventual loss in volatile forms by denitrification in the reduced layer. The results of inhibitory action of benomyl on nitrification may have applied significance in preventing nitrogen loss from flooded soils.

Ramakrishna et al.[123] studied the effect of benomyl, carbendazim, and technical grade AB (a hydrolysis product of benomyl) on nitrification in a simulated oxidized surface of flooded soils. They observed that nitrification proceeded rapidly in unamended soils and in soils amended with 10 and 100 $\mu g\ g^{-1}$ benomyl, which was evident by the disappearance of ammonia with the concomitant formation of nitrate during the 30-day incubation period. However, benomyl at 1000 $\mu g\ g^{-1}$ greatly inhibited the oxidation of NH_4 to NO_3 and this inhibitory action was well pronounced even after 30 days. AB-inhibited action was quite noticeable even after 30 days. AB also inhibited nitrification in soils at concentrations of 100 and 100 $\mu g\ g^{-1}$. In terms of nitrate formed, AB was relatively less toxic than benomyl.[123] Carbendazim was relatively nontoxic to nitrification in soils as compared to benomyl; no appreciable inhibition of nitrification occurred in soils amended with 10 and 100 $\mu g\ g^{-1}$ carbendazim, whereas 1000 $\mu g\ g^{-1}$ nitrification was slightly reduced by carbendazim and greatly by benomyl.[123] In contrast to this, Sinha et al.[50] observed that carbendazim at 5 to 40 $\mu g\ g^{-1}$ reduced the nitrification process for 45 days, although significant differences occurred only at 20 and 40 ppm. These authors found large differences among the benomyl-treated soils. This may be explained by the differences in the soil humus content that are expected to influence strongly the availability of benomyl in soil.

The effect of benomyl at 5 ppm (a rate equivalent to recommended field levels) on heterotrophic N_2 fixation in five air-dried, cellulose-amended, submerged, tropical soils using the ^{15}N tracer technique in vitro was studied.[131] Benomyl at 10, 20, and 100 ppm exhibited a marked stimulation on nitrogen fixation in four out of five soils by *Azospirillum* and *Azotobacter* spp. in flooded paddy fields. Another marked stimulation of population and nitrogen fixation by a N_2-fixing *Azospirillum* sp. was observed in a benomyl-amended paddy soil.[132] Likewise, 2-AB at 10, 20, and 100 ppm stimulated the population of *Azospirillum* sp. in a submerged soil, but inhibited the population of symbiotrophic nitrogen fixers.[122] However, these bacteria from 2-AB-amended soil exhibited more significant nitrogenase activity than the population from the control soil. Carbendazim inhibited the growth of *Westiellopsis* sp. at 0.1% but *Aulosira fertilissima* and *Tolypothrix tenuis* tolerated this concentration. However, stimulatory effects were observed in *Nostoc* and *Tolypothrix* spp. Since benomyl maintains the submerged soil at relatively high potentials for prolonged periods,[133] the provision of favorable redox potential* by benomyl application might have favored the heterotrophic nitrogen fixation in these cellulose-amended fields of paddy. Thus, *Azospirillum* was stimulated by benomyl in alluvial, laterite, and karapadam soils; however, with respect to anaerobic N_2 fixers, benomyl was stimulatory only in alluvial soil. *Azotobacter* was not markedly influenced by benomyl, irrespective of the soil used. These results suggest that the different effects of fungicides on specific groups of N_2 fixers account for differences in N_2 fixation in flooded soils.[131]

The possibility exists that the decrease in nitrification which results from fungicidal treatment may be due to changes in denitrification. Saive,[134] however, found that while the fungicides benomyl and fentin hydroxide led to similar changes in nitrate concentration in soils, they affected different processes: benomyl enhanced the oxidation of nitrite to nitrate, while fentin hydroxide appeared to decrease the rate of denitrification. Kouba[135] demonstrated a more rapid and complete conversion of fertilizer ammonium nitrogen into NO_3 when benomyl was added to soils.

Treatment of soil with triazbutyl at 0.5 and 1.0 ppm a.i. affected the physiological process of soil microbes. The fungicides at those rates impaired the process of nitrification for 67 days, whereas lower concentrations did not produce any detrimental effect on nitrifiers. This indicates that soil bacteria that oxidizes NH_3 to NO_2 and NO_3 are highly sensitive to triazbutyl.[71] This is because they are usually nonsporulating.[76]

The inhibition of nitrification need not be adverse to soil fertility; it can result in preservation of N as plant nutrient. This is because of the accumulation of ammonia which is less susceptible to leaching than nitrate. Ethazole, a nitrification inhibitor, offers potential

---

* Redox potential (oxidation + reduction potential) is a chemical reaction in which one or more electrons are transferred from one atom or molecule to another.

for reducing N losses in soils from denitrification and leaching, thereby improving N utilization by crops in wheat.[136]

Fungicides generally have a greater initial and longer lasting effect on nitrification than other pesticides. This is mostly true of the older broad-spectrum compounds such as maneb, nabam, vapam and Verdasan which are applied at high rates. The new systemic fungicides, however, appear to have only a marginal effect on nitrification. Low concentrations of fungicides, on the other hand, may stimulate nitrate formation. This is presumably because these concentrations stimulate ammonification and thereby provide increased substrate for nitrification. Nitrite + N rarely accumulates in soils treated with fungicides, which would suggest that fungicides, unlike other pesticides, mainly inhibit ammonium-oxidizing microflora.

B. Denitrification

The increased use of fungicides with nitrogenous fertilizers has emphasized the need for evaluating their side effects on transformations of fertilizer nitrogen in soil. The influence of fungicides on denitrification in soil has been largely undetermined, probably because conditions most often met in soil do not favor these anaerobic organisms. Denitrifying bacteria are more affected by fungicides. Dithiocarbamates such as nabam and maneb are more toxic than ferbam, thiram, or ziram;[2] monoalkyldithiocarbamates such as maneb and nabam are more toxic than their dialkyldithiocarbamate counterparts, ferbam and ziram. Their toxicity is proportional to the number of dithiocarbamate radicals. The descending order of toxicity of the fungicides is ferbam > thiram > ziram.[2] It was reported that, in contrast to the above, chloropicrin in paddy soils and ethylene bromide increase the denitrifying microflora.

1. Nonsystemic Fungicides

Folpet and captafol applied at the rate of 10 μg g^{-1} soil have been reported to have no effect on denitrification in soil.[37] Similarly, nabam applied at the rate of 100 μg g^{-1} also had no inhibitory effect.[107] However, ziram, dithane, thiram, ferbam, maneb, and captan have been reported to inhibit denitrification in soil when applied at the rate of 50 or 100 μg g^{-1} soil.[108] Zineb at 20 μg g^{-1} had no effect on denitrification in paddy soil.[138] The descending order of toxicity is in the following order: nabam > maneb > ferbam > thiram > zineb > ziram.[108] Later, Audus[2] confirmed these findings. Sodium diethyldithiocarbamate applied at the rate of 20 μg mℓ^{-1} was also inhibitory.[139] Inhibitory effects of captan, nabam, and maneb at 100 μg g^{-1} on denitrification in a silt loam soil with transitory accumulation of NO_2 and N_2O have been reported.[140] It was further observed that bacterial denitrification was inhibited even at 5 μg g^{-1} and the end product was N_2O. The accumulation of NO_2 by bacterial culture from NO_3 in a medium containing captan, nabam, or maneb at concentrations of 10 μg g^{-1} and above has been observed.[140] The maximum accumulation of NO_2 from NO_3 was recorded when maneb was dispensed in the medium, especially at 25 μg g^{-1}. An inhibitory effect of phenylmercuryacetate on denitrification has been reported.[141] Contrary to these findings, the fungicides mancozeb, maneb, thiram, and captan at 10 μg g^{-1} soil had no appreciable effect on denitrification.[142] However, thiram increased the ratio of N_2 to N_2O in the gaseous products of denitrification; captan inhibited denitrification in some soils when applied at the rate of 50 μg g^{-1} soil.

2. Systemic Fungicides

Mills and McElhannon[143] reported that terrazole inhibited denitrification in soil when applied at rates as low as 0.5 μg g^{-1} soil and inhibited denitrification in a liquid medium inoculated with soil bacteria when applied at rates as low as 0.02 μg mℓ^{-1} medium. Contrary to these findings, terrazole and benomyl had no effect on denitrification when applied at the rate of 10 μg g^{-1} and either had no effect on or enhanced denitrification when applied at the rate of 50 μg g^{-1}.[142] Benomyl decreased the ratio of N_2 to N_2O in the gaseous products of denitrification. Since denitrification essentially causes a loss of nitrates as plant nutrients, its inhibition is regarded as a favorable effect with respect to the preservation of nitrogen.

C. Ammonification

Ammonification in soil is brought about mostly by spore-forming bacteria, actinomycetes, and fungi. These can degrade proteinaceous matter with the concomitant production of ammonia. Bacteria and fungi break down protein molecules rapidly, whereas actinomycetes do so slowly.[6,8] Any fungicide that has the potential to induce physical, chemical, and/or microbiological changes can affect the population of ammonifiers in soil and also the final organic compounds produced during the degradation of proteinaceous material. The rate of the ammonification process depends largely on the nature and concentration of the fungicides and also the moisture content, pH and other characteristics of the soil.

1. Nonsystemic Fungicides

As with other microorganisms, various studies show a diversity in the effect of fungicides. Actually, the same product may have different effects, depending on experimental conditions. Lower concentrations of vapam (125 to 500 ppm) stimulated the production of ammonical nitrogen, whereas a considerable reduction occurred at higher concentrations. After 30 days, more ammonical nitrogen was detected in vapam-treated soil as compared to the control soil.[83] This may have resulted from the reestablishment of bacteria that regulate the process. Temporary inhibition of ammonification in vapam-amended soils has also been reported.[121] Aeratan effected an initial depression of ammonification but later stimulated it.[83] Fenaminosulf was not inhibitory to ammonification.[38] The buildup of ammonia in soil following fenaminosulf treatment was advantageous to crops which transform ammonium nitrogen to nitrate nitrogen. Ammonification was not affected by captan at 40 and 60 ppm because ammonical nitrogen was accumulated in soils. This may have been due to the liberation of ammonia from organic residues of the affected microflora.[17] Application of thiram increases ammonification for 6 weeks and its intensity depends on the concentration of thiram used.[31]

Captan, thiram, and an organomercuric fungicide at low doses in in vitro experiments stimulate ammonification.[33] At higher concentrations of thiram and the organomercurials, the stimulation becomes inhibition; in the field, captan, thiram, dicloran, and quintozene stimulated ammonification at twice the rate.[33] Nitrogen mineralization of field soil was decreased if the dithiocarbamate fungicides maneb and zineb were applied repeatedly. Ammonification was not affected in an acid lateritic clay or an alluvial soil by maneb and anilazine at 1.5 to 24 kg ha^{-1}, but was affected by 96 kg ha^{-1}.[95] No synergistic effect was discernible when both fungicides were applied simultaneously. Maneb (50 to 200 ppm) increased the ammonical nitrogen content of a soil.[106] Doneche et al.[91] reported that mancozeb (10 ppm) inhibited amylolytic microorganisms and nitrogen mineralization for 1 month in vineyard soil.

Ammonification was inhibited in humus soil by TCMTB (5 to 30 kg ha^{-1}) at pH 5.7 with increasing concentrations up to 30 kg ha^{-1},[43] by maneb and anilazine used at higher rates in vitro, or by metam-sodium, which suppresses peptone ammonification for a certain time.[95]

2. Systemic Fungicides

Carbendazim accelerated the process of ammonification and the degree of stimulation increased as the concentrations of toxicant increased in soil. Application of benomyl *in situ* at twice the rate stimulated ammonification.[47,116] Benomyl at 50 and 250 μg g^{-1} in virgin soil resulted in qualitative and quantitative changes in the free amino acid content of the soil and the amount of amino acid nitrogen in the soil increased at 50 μg g^{-1} and decreased at 500 μg g^{-1}.[34]

Benomyl at 10, 100, or 1000 ppm was toxic to the ammonifying bacteria in soil.[126] However, the amylolytic bacteria were not affected in the humus-rich sandy soil.[53] Gowda[127] observed the inhibition of ammonification at the same concentrations in three soils in the presence of peptone. Treatment of soil with triazbutyl at 0.5 and 1.0 ppm a.i. significantly stimulated ammonification for up to 67 days.[71]

That the ammonifying microorganisms are relatively tolerant to fungicidal application at operational rates is probably due to their diversity and heterogeneity. They degrade protein molecules under both aerobic and anaerobic conditions with the concomitant formation of ammonia. Further, many ammonifying bacterial species are also heterogeneous and spore formers and are thus difficult to kill.[24] Fungicide treatment led to both quantitative and qualitative changes in the free amino acid composition of soil. Larger amounts of free amino acid were found in soils treated with low concentrations of fungicides, whereas high concentrations resulted in smaller amounts of free amino acids being extracted.

D. Soil Respiration

Since soil may be considered an organism, it is appropriate to study its reactions to fungicides with physiological techniques. An indicator of overall soil activity is soil respiration, which is manifested by oxygen consumption or carbon dioxide evolution. Several investigators have reported that the evolution of carbon dioxide is directly related to the nitrifying capacity of the soil. With soil fungicides, soil respiration is inhibited for a short time after application, but surviving microflora recover soon and the respiration rate rises to high levels, exceeding even those of untreated soil.[10] The same is true even for nonspecific fungicides. This is not due to the return of the original microflora population, however, but to the increased activity of a few resistant species.[25] Respiration increases can also result from microbial utilization of fungicides rather than stimulation of the population.

1. Nonsystemic Fungicides

Broad-spectrum fungicides undoubtedly have the most marked inhibitory effect on soil respiration.[144,145] After the application of captan at the recommended rates caused a depression lasting approximately 24 hr, soil respiration increased; the lag phase of the respiration rate depends upon the concentration of the fungicide applied.[10] Captan initially decreased soil respiration and the depression in carbon dioxide production was proportional to its concentration. The increase in carbon dioxide production in the later part of the experiment was probably due to the use of decomposition products of captan by microorganisms. The production of carbon dioxide in the control soil decreased after 28 days.[17] This suggested a reduction in the amount of food materials necessary for microbial growth and/or activity. Vapam (125 to 1000 ppm) slightly stimulated carbon dioxide production for 32 days, indicating that either certain groups of microorganisms were not inhibited or they utilized the chemical for their growth and reproduction.[45] Aeratan adversely affected soil respiration, since the initial depression in carbon dioxide production was directly proportional to the concentration of the fungicide applied.[83] PCNB, effective against *Rhizoctonia* and *Sclerotium* spp., has no effect on soil respiration.[10] However, PCNB affected oxygen uptake at 100 kg ha^{-1}.[118]

Dithiocarbamates generally retarded carbon dioxide production after their application in the soil for the control of plant diseases. Maneb, nabam, and zineb at the rate of 100 kg ha^{-1} reduced carbon dioxide production in sandy soil;[82] a similar effect for maneb and mancozeb has been reported.[106] Soil respiration was reduced by nabam for 1 month; the toxic effect decreased after 42 days and at 56 days the amended soil evolved more carbon dioxide than unamended soils.[21] Carbon dioxide production in soil, inhibited by zineb, was reinitiated by incorporating inorganic nitrogen, zinc, and sulfur.[146]

Fenaminosulf is a potent fungicide used for the control of *Pythium* spp., *Phytophthora* spp., and *Aphanomyces* spp. The fungicide is applied directly to the soil and is supposed to be fairly specific against the above pathogens. The effect of fenaminosulf on carbon dioxide production in soil, amended or not with organic matter, was studied for 60 days. It was found that it retarded the breakdown of glucose and paddy straw added to soil; the inhibition of carbon dioxide production occurred during the early stages of glucose decomposition and in the case of straw throughout the period of study.[147] The study indicated that fenaminosulf treatments affected the fate of glucose and straw in soil in different ways. It is also apparent that the organisms or the biochemical mechanisms involved in the degradation of straw are different and more susceptible to the fungicide than those capable of activity utilizing glucose in an easily metabolized substrate. However, Domsch[10] did not observe any effect on soil respiration.

Fungicidal effects on the decomposition of crop residues have been mainly investigated under in vitro conditions; application of captan, fenaminosulf, thiram, and zineb caused a delay.[10,145] The process depends on the composition of the fungicide-resistant population, the kind of the substrate to be decomposed, and the persistence of the fungicide. Resistant mycoflora with prominent antagonistic properties may ward off more effective decomposers from the substrate. Such interactions may play a role where selective fungicides with low persistence cause a delay in decomposition; such a situation has been reported in paddy straw for fenaminosulf.[147] Plant residues accumulate in fields where broad-spectrum fungicides have been used frequently. The accumulation of grass litter on a golf course was due to the frequent use of mercurial fungicides; however, the accumulation was not observed in adjacent untreated areas.[148] In the fungicide-amended field, the numbers of cellulolytic fungi were greatly reduced.[24]

Dlvang[149] and Wessen et al.[124] found no influence of benomyl and carbendazim on straw decomposition in soils. Similar results were also observed for benomyl (0 to 10 kg ha^{-1}) in clay soils.[125] However, in sandy soils benomyl doses of 2 kg ha^{-1} affected straw decomposition. The inhibiting effect appeared in the early phase of decomposition. Berg and Staaf[150] suggested a model for litter decomposition in which the early phase was regulated by nutrient concentration and lignin decomposition was the regulating factor. One explanation[125] could be that benomyl acts primarily on deuteromycetes, which are fast-growing fungi and are thereby nutrient-dependent. The fungi which are least affected by benomyl are basidiomycetes, which are slow-growing and not as nutrient dependent. Many have the ability to decompose lignin.[125]

Broad-spectrum fungicides have also been known to cause pronounced changes in soil structure. A surface mat of grass appeared in an orchard which had been treated continuously with BM.[151] The soils were rich in copper, which presumably inhibited the cellulose-degrading ability of microorganisms and also reduce activity.[105] A similar increase in the litter layer on golf greens was observed with organomercurials.[148] This may be due to reduction in the fungal pathogens on treated grass, but reduction in the inhibition of cellulolytic fungi by organomercurials must be a major contributing factor in its appearance.[46]

Studies have been carried out to determine the effects of fungicides on the utilization of specific substrates in soils. Captan at the operational rates completely inhibited cellulose degradation for 6 weeks following treatment. Even more severe inhibition occurred in the breakdown of chitin and cutin.[25] These processes are especially sensitive to fungicides because they involve a limited number of species of microorganisms. Elimination of one substrate group is, however, often compensated for by an increase in another.[25,105]

2. Systemic Fungicides

Benomyl had no effect on the decomposition or organic matter by surviving (resistant) species of fungi and other organisms[51] at the rate of 3 to 30 kg ha^{-1}. In autoclaved soil seeded with *Achromobacter* sp. and treated with benomyl, oxygen uptake was greater than in soil with the active ingredient only. In contrast, another *Achromobacter* sp. isolated from an orchard soil had a higher oxygen uptake in soil + benomyl, inorganic salts, and glucose than in a soil containing benomyl alone.[100] This indicates the potential differences of isolates and also the effects of the soil nutrient status and the presence of formulation chemicals on the activity of microorganisms. Neither benomyl[44,51,53,126] nor carbendazim[126] was inhibitory to soil respiration. Benomyl reduced the respiration rates of *Fusarium oxysporum* f. sp. *melonis* and *Saccharomyces cerevisiae* by 50% at 3.5×10^{-6} *M*.[152] It was also reported that benomyl and carbendazim at 250×10^{-6} *M* inhibited mitochondrial respiration and oxidative phosphorylation. Benomyl at 3 to 30 kg a.i. ha^{-1} was ineffective on the decomposition of organic matter in humus.[51] Similarly, at 2 kg a.i. ha^{-1} it was nontoxic to the cellulolytic activity of microorganisms in sandy soils.[53] Benomyl at 100 μg g^{-1} completely stopped the decomposition of labeled cellulose in acid soils but had no effect in neutral soil.[51]

Caseley and Broadbent[118] reported that a fine sandy soil treated with Lanstan (a formulated product of carbendazim) at 1.0 and 10 kg ha^{-1} exhibited a depressed respiratory activity. Carbendazim (20 to 40 ppm) adversely affected the soil respiration process for varying periods of time. Higher concentrations were found to be more inhibitory than lower concentrations.[50]

E. Enzymatic Activity

Experiments on the effect of fungicides on enzyme activity are very limited, but earlier results show considerable diversity. Most studies are concerned with the dehydrogenase activity associated with total respiratory activity.

1. Nonsystemic Fungicides

An assay of dehydrogenase activity provides a good index of overall microbial activity in soil. Fenaminosulf decreased dehydrogenase activity of soil at the rate recommended for plant disease control.[147] The inhibition was more pronounced in soils where initial activity itself was higher, as in garden soil, than virgin soil. Further, it was demonstrated that the inhibitory activity of fenaminosulf depends on its concentration and on the incubation time. Thus, invertase, urease, and tryptophanase are stimulated by 20 ppm of the fungicide, but are inhibited at 100 and 200 ppm.[147] TCMBT (5 to 30 kg ha^{-1}) in arable soil inhibits saccharase, urease, phosphatase, and β-glucosidase activity in proportion to the concentration of the compound.[53] Degradation of cutin, pectin, and glucose was severely delayed in soil treated with captan at 200 ppm.[25] Dehydrogenase activity was significantly lower in captafol- and folpet-treated soil during the 21-day incubation following the fungicide application at 1 to 10 ppm.[49] The depression in the dehydrogenase activity may be a direct consequence of the adverse effect on the fungal population. The degree of inhibition was greater at 10 ppm than at 1 ppm. No adverse effects on soil phosphatase activity were observed.

Mineralization of chitin and cellulose was depressed following the application of captafol and folpet, respectively. This depression was probably the result of the inhibition of cellulolytic or chitinolytic fungi. Some caution may be warranted concerning the application of fungicide during periods when high mineralization rates of plant residue are desirable.[49]

Not much work has been carried out with regard to the effects of dithiocarbamate fungicides on the soil enzymatic studies. Maneb was ineffective upon phosphatase and urease activities of clay soil although it stimulated dehydrogenase activity.[153] In other studies, similar results were obtained in organic soils.[87] Mancozeb at 10 ppm showed inhibitory effects on phosphatase and urease activity.[91] Thiram (5 ppm) was stimulatory to dehydrogenase, urease, and phosphatase activities in clay soils, but at 10 ppm it was innocuous.[153] However, in an organic soil, urease and dehydrogenase activities were little affected and phosphatase activity was retarded by thiram.[87] Urease activity was not affected after the application of thiram at 100 ppm.[119]

2. Systemic Fungicides

Hydrogenase activity of soil was not affected significantly by 20 and 200 ppm benomyl in soil; similar results were also obtained without further addition of any substrate and with the simultaneous addition of 200 ppm glucose.[53] Benomyl at 3.0 to 30 kg ha^{-1} in humus sand did not affect dehydrogenase, amylase, or catalase activity.[51] Gowda[127] noted an inhibition of dehydrogenase activity of peptone-amended soils by benomyl at 100 to 10,000 μg g^{-1} a.i. It seems from the result that systemic fungicides are not toxic; however, nonsystemic fungicides are harmful to soil enzyme activity. Therefore, they should be used judiciously in the control of plant diseases, especially soil-borne.

F. Sulfur Transformation

Since crop plants generally use sulfate (SO_4^{2-}-S) as the principal source of the element available within the plant, it is desirable that elemental S or $S_2O_3^{2-}$-S which is added as fertilizer be rapidly oxidized to sulfate. The reduction of sulfate is an important transformation; this process in soil is mediated by chemotrophic bacteria of the genus Thiobacillus and a range of heterotrophic microorganisms. The oxidation principally involves the activity of bacteria. The effect of fungicides on sulfur oxidation is particularly important because of sulfur deficiencies in various parts of the world. Such deficiencies could be overcome by adding sulfur to the soil, which must be oxidized before it can be used by crops.[12]

1. Nonsystemic Fungicides

The fungicides thiram, dicloran, and Verdasan at 50 ppm added to the soil at commercial rates inhibited S oxidation; the inhibition presumably resulted from the effects of these compounds on *Thiobacillus* or S-oxidizing heterotrophic microorganisms. Thiram blocked oxidation for 14 days with the corresponding accumulation of intermediates such as thiosulfate and tetrathionate, coupled with the decrease in the formation of sulfate.[12] The fungicide dicloran was highly toxic; the slow decline in pH due to this treatment is the major reason for its toxicity. No significant effects were observed on microorganisms responsible for the nitrogen and sulfur cycle due to the application of folpet and captafol at 10 ppm.[49]

2. Systemic Fungicides

The addition of the commercial formulation of benomyl inhibited sulfur oxidation in both alluvial and laterite soils, although the inhibition was more pronounced in alluvial soil. However, benomyl was stimulatory in alluvial soil only with concentrations of 5 and 10 μg g^{-1}, whereas higher concentrations were inhibitory.[154] In a later study, benomyl at 50 and 100 μg g^{-1} but not at 10 μg g^{-1} inhibited sulfur reduction in flooded soil.[155] The inhibition of sulfur reduction was related to high redox potentials and a decrease in the population of sulfate reducers. However, Wainwright[12] observed no effect of benomyl on sulfur oxidation.

IV. SOIL CHEMICAL PROPERTIES

Treatment of soil with fungicides may have a beneficial effect on plant growth.[9,156] The increased growth response (IGR) which follows fungicidal application is generally thought to be due to several factors. Numerous explanations have been proposed,[79] including correction of wide soil-borne phytostasis,[157] increased mobilization and availability of NH_4^+-N,[158] and decreased parasitic activity.[79] However, the most common view is that the effect is due largely to the increased release of plant nutrient elements, especially nitrogen, from organic and inorganic soil constituents.[39,75,159] Microorganisms play a major role in the mineralization of such nutrients, so that any change in the microbial population brought about by fungicidal treatment will be reflected in the nutrient status of the soil.[1] Changes in the physicochemical properties of the soil due to the use of fungicides are difficult to separate from indirect biotic influences. The possibility that the change in the solubility or availability of inorganic nutrient elements may derive from the decomposition of the fungicides has been noted.[33] Similarly, changes in physical factors such as pH and cation exchange capacity may have a direct effect on soil fertility.

A. Nitrogen and Nitrogenous Compound Transformation

The availability of plant nutrients in soil depends mainly on the activity of microorganisms and the extent of their activity depends on the balance between microbial mobilization and immobilization.[1] Although the increase in cations following fungicide treatments is significant, it is relatively small and is likely to affect the growth of higher plants. However, extended application of fungicides, particularly systemic fungicides, may raise cation content sufficiently to increase plant growth. This would be particularly true in otherwise nutrient-starved conditions or if the increase in cation content occurred in close proximity to roots.

Plant species vary widely in their ability to utilize NH_4^+ and NO_3^- as a source of nitrogen for growth. Cultural conditions have also been shown to have a pronounced effect on NH_4^+ and NO_3^- uptake and utilization. The extensive and repeated use of fungicides on turfgrass has created an interest in their possible effects on soil microorganisms. Controlled-release fungicides are used on turfgrass and, in general, these and fertilizers initially release nitrogen as NH_4^+ rather than as NO_3^-. Thus, any fungicide that had an effect on soil microorganisms involved in decomposition or organic matter, ammonification, or, most importantly, nitrification could have a corresponding effect on soil fertility and plant nutrition.[34,47,116] Naftel[160] observed that NO_3^- was readily absorbed by plants under acidic conditions, whereas NH_4^- uptake was greater under neutral conditions; the absorption of ammonia and nitrate was shown to be affected by the proportion of other monovalent cations present. Some young plants absorb ammonium more readily than nitrate, but the reverse occurs in later growth.[160] In laboratory studies, the fungicides benomyl, Dyrene, and maneb were incorporated at the rates of 2, 25, 75, and 150 ppm and the soil incubated at 21°C and moisture content of 80%; in an analysis for NH_4^+-N and NO_2^--NO_3^--N at various intervals during a 16-week incubation period, almost no effect was detected for benomyl, but a complete blockage occurred at 150 ppm of maneb. Dyrene had an intermediate effect based on the concentration of NH_4^+-NO_2^--NO_3^- in untreated soil as compared to treated soil.[116] It was observed that benomyl was stimulatory to N mineralization by Dyrene and maneb were inhibitory for 4 weeks and the effects disappeared after 16 weeks. In field studies, benomyl (90 kg ha^{-1} a.i.), Dyrene (90 kg ha^{-1} a.i.), and maneb (135 kg ha^{-1} a.i.) were applied to pennocrose creeping bent grass *(Agrostis palustris)* and soil samples were assayed for ammonical and nitrite-nitrate nitrogen concentrations at weekly intervals during a month incubation. No effect was observed on the nitrification, but enhanced nitrogen metabolism occurred with the application of all the fungicides.[116] The differences in the effects of these fungicides on nitrification and nitrogen mineralization in the laboratory as compared to field application were ascribed to more rapid rates of degradation under field conditions as contrasted with the high rate of a single application under laboratory conditions.[116]

Wainwright and Pugh[47] reported changes in the exchangeable cation status of soils following application of the fungicides benomyl, captan, dicloran, quintozene, and thiram at field rates. Significant increases in NH_4^+-N, K, and Mn occurred. They postulated that some of these changes may have resulted from the presence of the cation in the fungicide formulations. The high levels of sodium in the Verdasan-treated soil, for example, probably resulted from the large amounts of sodium in the fungicide.[47] The different effects of Mn-oxidizing and Mn-reducing bacteria may account for the increase of manganese in treated soils.[34] Cations will increase the soil following lysis of the microorganisms killed by fungicidal treatment.[33,34] The increase of bacteria will produce copious amounts of organic acids (2-ketogluconic acid), which in turn causes the release of cations to stimulate plant vigor.[161,162]

The addition of a fungicide to a soil has marked qualitative and quantitative effects on the free amino acids present. The changes which occur are dependent on the concentration and activity of the fungicide and also on the size and nature of the microbial increase which follows its application. The increases in free amino acid-N in soil following treatment with low dosages may contribute directly to increased plant growth and vigor; at high rates of application, however, a smaller amount of free amino acid-N becomes available. If plant vigor is increased after the heavy application of fungicides, this would more likely be caused by increases in NH_4^+-N rather than by direct uptake of free amino acids.[34]

Free amino acids are present in the soil in very small amounts, rarely exceeding 2.0 μg g^{-1} soil, yet they are a potentially important source of nutrients for both plants and microbes.[163] The uptake of amino acids by plants from soil is well documented.[164] Similarly, amino acid-dependent segments of the soil microflora have been recognized.[165] While some free soil amino acids are stable, most are extremely labile; they are in a dynamic state which probably results mainly from excretory or autolysis products of microorganisms and plant root.

Qualitative and quantitative changes have been observed in the free amino acid content of the soil following application with the fungicides benomyl, thiram, and Verdasan.[34] The amount of amino acid-N extracted following the addition of high concentrations of fungicides was low; however, low rates resulted in a marked increase. A total of nine amino acids was extracted. Glycine and threonine were markedly favored by all the treatments, whereas the frequency of extraction of the other seven amino acids was dependent upon the fungicide used and its concentration. Wainwright and Pugh[34] discussed the quantitative changes brought about by fungicidal application in the microflora increase after microflora were initially killed by the application. First, after treatment with the fungicide the amount of free amino acid-N_2 decreased, a finding contrary to the theory that large amounts of amino acid are freed following partial sterilization. However, van Driel[166] reported that the amount of mineralization of added amino acid-N_2 was dependent on the concentration. The mineralization increased with the increasing concentration; any large amounts of amino acid released into the soil due to the lysis of microorganisms following fungicidal treatment would then be quickly mineralized. Second, low concentrations of fungicides tended to result in increased levels of extracted amino acid-N_2, although the converse was true of high concentrations. These findings indicate that large concentrations of fungicide added to the soil favor amino acid mineralization. This agrees with the result of Wainwright and Pugh[33] who found that large amounts of NH_4^+-N accumulated in soils treated with high concentrations of fungicides, whereas only small amounts of NH_4^+-N were found following low rates of application.[166]

B. Iron and Manganese Transformation

Within a few days after flooding a soil, trapped oxygen is exhausted, causing some soil components to be reduced.[167] One of the obvious and sometimes rapid consequences of flooding the soil is the reduction of Fe^{3+} to Fe^{2+} and Mn^{4+} to Mn^{2+}, which is attributed to chemical and/or biological processes. Microorganisms have been implicated directly in Fe^{3+} reduction through anaerobic respiration,[168] while chemical reduction is catalyzed by anaerobic decomposition products of organic matter.[169,170] On application of benomyl (100 ppm) to alluvial soil under flooded conditions, the treated soil largely retained its original reddish-brown color, even after 24 days of flooding; the soils with no fungicide started to turn gray within 5 to 7 days after flooding, as is characteristic of a reduced soil.[167] This indicated that benomyl retarded the reduction of the soil following flooding.[133] Later workers further observed that the redox potential of soil treated with 5 and 50 ppm benomyl was higher as compared to the control throughout the 24-day incubation period. The effect of benomyl in maintaining the flooded soil at higher potentials was evident even at concentrations as low as 5 and 10 ppm. These concentrations are comparable to field application rates. However, no significant difference in pH occurred between benomyl-treated and control soils; pH values of benomyl-treated and control soils ranged between 6.9 and 7.2 at the end of 24 days.[133] The mechanism of benomyl in maintaining the flooded soil at the higher redox potential is not clear; apparently, oxygen depletion by aerobic respiration is delayed by the inhibitory action of benomyl on the fungal population of the soil.[171] Thus, the reduction of the flooded soil can be hindered if benomyl functions as a respiratory inhibitor of both aerobes and anaerobes.

Benomyl retarded the decrease of redox potential and prevented the accumulation of both Mn and Fe in solution after soil submergence, despite the presence or organic matter. The reduction of minerals is one of the most predominent reactions of flooded soils. Benomyl restricted the formation of Fe^{2+} from Fe^{4+}, even at 5 ppm, and this effect persisted for up to 24 days of floodings.[133] It was observed that benomyl inhibited iron reduction at 20 and 100 ppm and virtually no accumulation of Fe^{2+} was observed at these concentrations. Likewise, benomyl inhibited soluble Mn formation in flooded soils, but this effect was less pronounced than its action on Fe^{3+} reduction. The addition of decomposable organic matter to flooded soil further accentuates the reduction processes, including Fe^{3+} reduction.[167] The authors studied the influence of benomyl in the presence of rice straw: Fe^{2+} accumulated in larger quantities in soil amended with rice straw, but benomyl retarded the Fe^{2+} accumulation, particularly at concentrations of 20, 50, and 100 ppm. Benomyl also reduced the buildup of soluble manganese in flooded soils with rice straw. The redox potential decreased rapidly in rice straw-amended soils, but benomyl prevented this sharp drop, especially at concentrations of 50 and 100 ppm. Despite the presence of rice straw, soils treated with benomyl at a concentration as low as 5 ppm showed higher potentials without the fungicide, even after 24 days of flooding. The pH values with and without straw ranged from 7.2 to 7.4 after 24 days.[133] The reduction of Mn and Fe is both chemical and biological;[168-170] however, it is not clear whether benomyl depresses the activity of anaerobic microorganisms involved in Mn^{4+} and Fe^{3+} reduction in flooded soil.

Reduction of Fe^{3+} and Mn^{4+} after the flooding of soil has both beneficial and harmful consequences. Flooding increases the availability of Fe^{2+} to the rice plant, but excess Fe^{2+} causes iron toxicity in rice, especially in acid tropical soils.[172,173] Also, highly reduced conditions aggravate zinc deficiency in wetland rice. The effect of benomyl in retarding the reduction of a flooded soil may, therefore, have some applied significance in minimizing the problems of iron toxicity and zinc deficiency in rice.[133]

C. Phosphate Transformation

Microorganisms responsible for phosphate solubilization belong to a Gram-negative group, *Bacillus* sp., and a *Penicillium* sp. Benomyl at 100 μg g^{-1} increased the number of phosphate-solubilizing bacteria and fungi 30 and 4 times, respectively, as compared to the control.[174] Benomyl increased the amount of calcium chloride-extractable phosphorus after a 2-week application; of note was the fact that the fungicide enhanced the solubilization of added insoluble calcium-hydrogen phosphate.

V. FUTURE OUTLOOK

The influence of fungicides on soil microorganisms and microbial transformations in soil is dependent on physical, chemical, and biochemical conditions in the soil, in addition to the nature and concentration of the fungicide in the ecosystem.

The relationship between the microflora of soil and fungicides is, therefore, complex and often indirect. The complete effects of these interactions must be more fully explored if greater efficiency is to be achieved in overall management of plant diseases. Changes in the plant ecosystem would be expected to affect costly production management practices. Avoidance of certain deleterious or nontarget effects of fungicides and exploitation of desirable effects could lead to the development of an integrated plant disease management program which offers a performance superior to that of chemical disease control programs that are simply oriented toward the target pathogen.

REFERENCES

1. **Alexander, M.,** *Introduction to Soil Microbiology,* 2nd ed., John Wiley & Sons, New York, 1977.
2. **Audus, L. J.,** The action of herbicides and pesticides on the microflora, *Meded. Fac. Landbouwwet. Rijksuniv. Gent,* 35, 465, 1970.
3. **McCallan, S. E. A. and Miller, L. P.,** Innate toxicity of fungicides, in *Advances in Pest Control Research,* Vol. 2, Metcalf, R. L., Ed., Interscience, New York, 1958, 107.
4. **Farley, J. D. and Lockwood, J. L.,** Reduced nutrient competition by soil microorganisms as a possible mechanism for pentachloronitrobenzene induced disease accentuation, *Phytopathology,* 59, 718, 1969.
5. **Munnecke, D. E.,** Factors affecting the efficacy of fungicides in soil, *Annu. Rev. Phytopathol.,* 10, 375, 1972.
6. **Waksman, S. A.,** *Soil Microbiology,* John Wiley & Sons, New York, 1952.
7. **Rangaswami, G.,** *Agricultural Microbiology,* Asia Publication House, Bombay, 1966.
8. **Agnihotri, V. P., Sinha, A. P., and Singh, K.,** Influence of insecticides on soil microorganisms and their biochemical activity, *Pesticides,* 15(9), 16, 1981.
9. **Warcup, J. H.,** Chemical and biological aspects of soil sterilization, *Soils Fert.,* 20, 1, 1957.
10. **Domsch, K. H.,** Soil fungicides, *Annu. Rev. Phytopathol.,* 2, 293, 1964.
11. **Kreutzer, W. A.,** The reinfestation of treated soil, in *Ecology of Soil Borne Plant Pathogens,* Baker, K. F. and Synder, W. C., Eds., University of California Press, Berkeley, 1965.
12. **Wainwright, M.,** Effects of fungicides on the microbiology and biochemistry of soils — a review, *Z. Pflanzenernaehr. Bodenkd.,* 140, 587, 1977.
13. **Kuthubutheen, A. J. and Pugh, G. J. F.,** The effects of fungicides on soil fungal population, *Soil Biol. Biochem.,* 11, 297, 1979.
14. **Pugh, G. J. F. and Williams, J. I.,** Effects of an organo-mercury fungicide on saprophytic fungi and on litter decomposition, *Trans. Br. Mycol. Soc.,* 57, 164, 1971.
15. **Domsch, K. H.,** Die Wirkung von Bodenfungiziden. III. Quantitative Veranderungen der Bodenflora, *Z. Pflanzenkr. Pflanzenschutz,* 66, 17, 1959.
16. **Vaartaja, O. and Agnihotri, V. P.,** A comparison of rhizosphere flora in conifer beds treated with fungitoxicants, *Zentralbl. Bakteriol. Parasitenkd. Infektionskr. Hyg. Bd.,* 124(2), 157, 1969.
17. **Agnihotri, V. P.,** Persistence of captan and its effects on microflora, respiration and nitrification of a forest nursery soil, *Can. J. Microbiol.,* 17, 377, 1971.
18. **Domsch, K. H.,** Die Wirkung von Bodenfungiziden. V. Empfindlichk eit von Bodenorganismen in vitro, *Z. Pflanzenkr. Pflanzenpathol. Pflanzenschutz,* 67, 213, 1960.
19. **Cordon, M. E. and Young, R. A.,** The fungicidal activity and sorption of nabam in soil, *Phytopathology,* 50, 83, 1960.
20. **Cordon, M. E. and Young, R. A.,** Changes in the soil microflora following fungicide treatment, *Soil Sci.,* 99, 272, 1965.
21. **Chandra, P. and Bollen, W.,** Effects of nabam and mylone on nitrification, soil respiration and microbial numbers in four Oregon soils, *Soil Sci.,* 92, 387, 1961.
22. **Farley, J. D. and Lockwood, J. L.,** The suppression of actinomycetes by PCNB in culture media used for enumerating soil bacteria, *Phytopathology,* 58, 714, 1968.
23. **Ko, W. H. and Farley, J. D.,** Conversion of pentachloronitrobenzene to pentachloroaniline in soil and the effect of these compounds on soil microorganisms, *Phytopathology,* 59, 64, 1969.
24. **Bollen, G. J.,** Side-effects of pesticides on microbial interactions, in *Soil-Borne Plant Pathogens,* Schippers, B. and Gams, W., Eds., Academic Press, New York, 1979.
25. **Domsch, K. H.,** Effects of fungicides on microbial populations in soil, in *Pesticides in the Soil: Ecology, Degradation and Movement,* Guyer, G. E., Ed., Michigan State University, East Lansing, 1970, 42.
26. **Katan, J. and Lockwood, J. L.,** Effect of pentachloronitrobenzene on colonization of alfalfa residues by fungi and streptomyces in soil, *Phytopathology,* 60, 1578, 1970.
27. **Popov, V. I. and Zdrozhevskaya, S. D.,** Characteristics of the action of fungicides in soil, *Tr. Vses. Inst. Zashch. Rast.,* 35, 270, 1972.

28. **Lockwood, J. L.,** Ecological effects of PCNB, in *Pesticides in the Soil: Ecology, Degradation and Movement,* Guyer, G. E., Ed., Michigan State University, East Lansing, 1970, 47.

29. **Kesavan, R.,** Toxicological effects of pentachloronitrobenzene, its degradation products and thiabendazole on soil bio-activity, *Fitopatologia,* 8(1), 13, 1983.

30. **Kimura, Y. and Miller, V. L.,** The degradation of organomercury fungicides in soil, *J. Agric. Food Chem.,* 12, 253, 1964.

31. **Agnihotri, V. P.,** Thiram induced changes in soil microflora, their physiological activity and control of damping-off in chilies *(Capsicum annum), Indian J. Exp. Biol.,* 12, 85, 1974.

32. **Anderson, J. P. E., Armstrong, R. A., and Smith, S. N.,** Methods to evaluate pesticide damage to the biomass of the soil microflora, *Soil Biol. Biochem.,* 13, 149, 1981.

33. **Wainwright, M. and Pugh, G. J. F.,** The effects of three fungicides on nitrification and ammonification in soil, *Soil Biol. Biochem.,* 5, 577, 1973.

34. **Wainwright, M. and Pugh, G. J. F.,** Effect of fungicides on numbers of microorganisms and frequency of cellulolytic fungi in soils, *Plant Soil,* 43, 561, 1975.

35. **Hills, F. J. and Leach, L. D.,** Photochemical decomposition and biological activity of p-dimethylaminobenzenediazo sodium sulfate (Dexon), *Phytopathology,* 52, 51, 1962.

36. **Tammen, J., Muse, D. P., and Haas, J. H.,** Control of *Pythium* root diseases with soil fungicides, *Plant Dis. Rep.,* 45, 858, 1961.

37. **Alconero, R. and Hagedorn, D. J.,** The persistence of Dexon in soil and its effects on soil mycoflora, *Phytopathology,* 58, 34, 1968.

38. **Agnihotri, V. P.,** Effect of dexon on soil microflora and their ammonification and nitrification activities, *Indian J. Exp. Biol.,* 11, 213, 1973.

39. **Waksman, S. A. and Starkey, R. L.,** Partial sterilization of soil, microbiological activities and soil fertility, *Soil Sci.,* 16, 247, 1923.

40. **Richardson, L. T.,** The persistence of thiram in soil and its relationship to the microbiological balance and damping off control, *Can. J. Bot.,* 32, 335, 1954.

41. **Martin, J. P., Baines, W. C., and Ervin, J. O.,** Influence of soil fumigation for citrus replant on the fungus population of the soil, *Proc. Soil Sci. Soc. Am.,* 21, 163, 1957.

42. **Martin, J. P.,** Use of acid, rose bengal and streptomycin in the plate method for estimating soil fungi, *Soil Sci.,* 69, 215, 1950.

43. **Voets, J. P. and Vandamme, E.,** Effect of 2-(thiocyanomethylthio) benzothiazole on the microflora and enzymes of the soil, *Meded. Fac. Landbouwwet. Rijksuniv. Gent,* 35, 563, 1970.

44. **Peeples, J. L.,** Microbial activity in benomyl-treated soil, *Phytopathology,* 64, 857, 1974.

45. **Sinha, A. P., Agnihotri, V. P., and Singh, K.,** Persistence of aeratan in soil and its effect on soil bioecosystems and their related biochemical activity, *Proc. Indian Natl. Acad. Sci. B,* 45(3), 261, 1979.

46. **Pugh, G. J. F., Williams, J. I., and Wainwright, M.,** The effect of fungicides on microbial activities in the soil, Proc. 5th Int. Congr. Soil Zoology, Prague, 1973, 17.

47. **Wainwright, M. and Pugh, G. J. F.,** The effects of fungicides on certain chemical and microbial properties, *Soil Biol. Biochem.,* 6, 263, 1974.

48. **Houseworth, L. D. and Tweedy, B. G.,** Effect of atrazine in combination with captan and thiram upon fungal and bacterial population in the soil, *Plant Soil,* 38, 439, 1975.

49. **Atlas, R. M., Pramer, D., and Barcha, R.,** Assessment of pesticide effects on non-target soil microorganisms, *Soil Biol. Biochem.,* 10, 231, 1978.

50. **Sinha, A. P., Agnihotri, V. P., and Singh, K.,** Effect of Bavistin on soil microflora, their related physiological activity and growth of sugarbeet seedlings, *Indian J. Exp. Biol.,* 18, 489, 1980.

51. **Hofer, I., Beck, T., and Wallnofer, P.,** Effect of the fungicide benomyl on the microflora of soil, *Z. Pflanzenkr. Pflanzenschutz,* 78, 399, 1971.

52. **Siegel, M. R.,** Benomyl soil microbial interaction, *Phytopathology,* 65, 219, 1975.

53. **von Faassen, H. G.,** Effect of the fungicide benomyl on some metabolic processes and on numbers of bacteria and actinomycetes, *Soil Biol. Biochem.,* 6, 131, 1974.

54. **Smiley, F. W. and Craven, M. M.,** *Fusarium* species in soil, thatch and crowns of *Poa pratensis* turfgrass treated with fungicides, *Soil Biol. Biochem.,* 11, 355, 1979.

55. **Edgington, L. V., Khew, E. L., and Barron, G. L.,** Fungitoxic spectrum of benzimidazole compounds, *Phytopathology,* 61, 42, 1971.

56. **Foster, M. G. and McQueen, D. J.,** Multiple application of benomyl and effects on nontarget soil fungi, *Bull. Environ. Contam. Toxicol.,* 17, 468, 1977.

57. **Winley, C. L. and San Clemente, C. L.,** Effects of pesticides on nitrite oxidation by *Nitrobacter agilis, Appl. Microbiol.,* 19, 214, 1970.

58. **Ponchet, J. and Tramier, R.,** Effects du benomyl sur la croissance de l' oeiller et la microflora des soils traits, *Ann. Phytopathol.,* 3, 401, 1971.

59. **Raynal, G. and Ferrari, F.,** Persistence of soil-incorporated benomyl and its effects on soil fungi, *Phytiatr. Phytopharm.,* 22, 259, 1973.

60. **Smiley, R. W. and Craven, M. M.,** Microflora of turfgrass treated with fungicides, *Soil Biol. Biochem.,* 11, 349, 1979.

61. **Backman, P. A., Rodriguez-Kabana, R., and Williams, J. C.,** The effect of peanut leaf spot fungicides on the nontarget pathogen, *Sclerotium rolfsii, Phytopathology,* 65, 773, 1975.

62. **Davet, P.,** Effets de quelques pesticides sur la colonisation d' un substrat par le *Trichoderma harzianum* Rifai en presence des autres champignons du sol, *Soil Biol. Biochem.,* 13, 513, 1981.

63. **Bliss, E. D.,** The destruction of *Armillaria mellea* in citrus soils, *Phytopathology,* 41, 665, 1951.

64. **Evans, E.,** Survival and recolonization by fungi in soil treated with formalin or carbon dioxide, *Trans. Br. Mycol. Soc.,* 38, 335, 1955.

65. **Sakesena, S. B.,** Effect of CS_2 fumigation *on Trichoderma viride* and other fungi, *Trans. Br. Mycol. Soc.,* 43, 111, 1960.

66. **Moubasher, A. H.,** Selective effects of fumigation with carbon disulphide on soil fungus flora, *Trans. Br. Mycol. Soc.,* 46, 338, 1963.

67. **Danielson, R. M. and Davey, C. B.,** Carbon and nitrogen nutrition of *Trichoderma, Soil Biol. Biochem.,* 5, 505, 1973.

68. **Catska, V. and Vrany, J.,** Rhizosphere mycoflora of wheat after foliar application of chlorocholin chloride, urea and 4-chloro-2-methyl-phenoxyacetic acid, *Folia Microbiol.,* 21, 268, 1976.

69. **Mughogho, L. K.,** The fungus flora of fumigated soils, *Trans. Br. Mycol. Soc.,* 51, 441, 1968.

70. **Wainwright, M. and Pugh, G. J. F.,** Changes in the free amino acid content of soil following treatment with fungicides, *Soil Biol. Biochem.,* 7, 1, 1975b.

71. **Sinha, A. P. and Singh, A.,** Effects of Indar on soil microflora and their nitrification and ammonification activities, *Proc. Indian Natl. Acad. Sci. B,* 45(3), 255, 1979.

72. **Kannaiyan, S. and Prasad, N. N.,** Effect of fungicidal spray on population of *Rhizoctonia solani* and on rhizospheric microflora of rice, *Int. Rice Res. Inst. Newsl.,* 4(1), 33, 1979.

73. **Pugh, G. J. F.,** Strategies in fungal ecology, *Trans. Br. Mycol. Soc.,* 75, 1, 1980.

74. **Pugh, G. J. F.,** The effect of agrochemicals on microbial processes in the soil, The Malaysian Microbiology Symp., Kebangsaan, Malaysia, August 17 to 19, 1981, 1.

75. **Aldrich, D. G. and Martin, J. P.,** Effect of fumigation on some chemical properties of soils, *Soil Sci.,* 73, 149, 1952.

76. **Waksman, S. A. and Starkey, R. L.,** Partial sterilization of soil, microbiological activities and soil fertility, *Soil Sci.,* 16, 247, 1923.

77. **Wilson, H. A.,** The effect of certain pesticides on nitrification in the soil, *W. Va. Agric. Exp. Stn. Bull.,* No. 566T, 1954.

78. **Wensley, R. N.,** Microbiological studies of the action of some selected soil fumigants, *Can. J. Bot.,* 31, 277, 1953.

79. **Martin, J. P.,** Influence of pesticides on soil residues on soil microbiological and chemical properties, *Residue Rev.,* 4, 96, 1963.

80. **Chinn, S. H. F.,** Effect of eight fungicides on microbial activities in soil as measured by a bioassay method, *Can. J. Microbiol.,* 19, 771, 1973.

81. **Mahmoud, S. A. Z., Selim, K. G., and El-Mokadem, T.,** Effect of some pesticides on rhizosphere microflora of cotton plants. I. Insecticides and fungicides, *Egypt. J. Microbiol.,* 7, 39, 1972.

82. **Eno, C. F.,** Effect of soil applications of carbamate fungicides on the soil microflora, *Fla. Agric. Exp. Stn. Bull.,* No. 142, 1957.

83. **Sinha, A. P., Agnihotri, V. P., and Singh, K.,** Persistence of aeratan in soil and its effect on soil bioecosystems and their related biochemical activity, *Proc. Indian Natl. Acad. Sci. B,* 45(3), 261, 1979.

84. **Dubey, H. D. and Rodriguez-Kabana, R.,** Changes in soil microflora following application of fungicides Dyrene and maneb to tropical soils, *J. Agric. Univ. P.R.,* 58, 78, 1974.

85. **Tu, C. M.,** Effect of pesticides on acetylene reduction and growth of microorganisms in a sandy loam, *Soil Biol. Biochem.,* 10, 451, 1978.

86. **Tu, C. M.,** Influence of pesticides on acetylene reduction and growth of microorganisms in an organic soil, *J. Environ. Sci. Health B,* 14, 617, 1979.

87. **Tu, C. M.,** Effects of some pesticides on enzyme activities in an organic soil, *Bull. Environ. Contam. Toxicol.,* 27, 109, 1981.

88. **Hubbel, D. H., Rothwell, D. F., Wheeler, W. B., Tappan, W. B., and Rhoads, F. M.,** Microbial effects and persistence of some pesticides combination in soil, *J. Environ. Qual.,* 2, 96, 1973.

89. **Stojanovic, B. J., Kennedy, M. V., and Shuman, F. L.,** Edaphic aspects of the disposal of unused pesticides, pesticide wastes and pesticide containers, *J. Environ. Qual.,* 1, 54, 1972.

90. **Doneche, B.,** Effects du "Mancozebe" sur la microflora des sols du vignoble bordelais, premiers resultats, *C. R. Acad. Sci. Ser. D,* 278, 3011, 1974.

91. **Doneche, B., Seguin, G., and Ribereau-Gayon, P.,** Mancozeb effect on soil microorganisms and its degradation in soil, *Soil Sci.,* 135, 361, 1983.

92. **Naumann, K.,** The effect of some environmental factors on the reaction of the soil microflora to plant protection agents, *Zentralbl. Bakteriol. Parasitenkd. Infektionskr. Hyg. Abt. 2,* 127, 379, 1972.

93. **Teuber, M. and Poschenrieder, H.,** Investigations on the effect on the microflora of cultivated fen soil of nematodes that contain mustard oil, *Bayer. Landwirtsch. Jährb.,* 41, 350, 1964.

94. **Sinha, A. P., Agnihotri, V. P., and Singh, K.,** Effect of soil fumigation with vapan on the dynamics of soil microflora and their related biochemical activity, *Plant Soil,* 53, 89, 1979.

95. **Dubey, H. D. and Rodriguez-Kabana, R.,** Effect of Dyrene and maneb on nitrification and ammonification and their degradation in soils, *Proc. Soil Sci. Soc. Am.,* 34, 435, 1970.

96. **Foster, M. G. and McQueen, D. J.,** The effects of single and multiple applications of benomyl on nontarget soil bacteria, *Bull. Environ. Contam. Toxicol.,* 17, 477, 1977.

97. **Smiley, R. W. and Craven, M. M.,** Fungicides in Kentucky bluegrass turf: effects on thatch and pH, *Agron. J.,* 70, 1013, 1978.

98. **von Mayer, W. C., Greenfield, S. A., and Siedel, M. C.,** Wheat leaf rust control by 4-n-butyl-1,2,4-triazole, a systemic fungicide, *Science,* 169, 997, 1970.

99. **Vaartaja, O.,** Pythium and Mortierella in soils of Ontario forest nurseries, *Can. J. Microbiol.,* 14, 265, 1968.

100. **Weeks, R. W. and Hendricks, H. G.,** Screening microorganisms from two soils for utilization of the systemic fungicide benlate, *Bacteriol. Proc.,* (Abstr.) p. 79, 1971.

101. **Gil Diaz-Ordonez, I., Morales, J., and Martin Gonzales, A.,** Effect of copper oxychloride on the respiration of *Azotobacter, Meded. Fac. Landbouwwet. Rijksuniv. Gent,* 35, 493, 1970.

102. **Sud, R. K. and Gupta, K. G.,** On the sensitivity of isolates of *Rhizobium* spp. and *Azotobacter chroococcum* to TMTD and its degradation products, *Arch. Mikrobiol.,* 85, 19, 1972.

103. **Langkramer, O.,** Determination of the effect of pesticides on soil microorganisms in pure culture by means of a laboratory technique, *Zentralbl. Bakteriol. Parasitenkd. Infektionskr. Hyg. Abt. 2,* 125, 713, 1970.

104. **Russell, E. W.,** *Soil Conditions and Plant Growth,* Longman, London, 1973.

105. **Ingham, E. R.,** Review of the effects of 12 selected biocides on target and nontarget soil microorganisms, *Crop. Prot.,* 4(1), 3, 1985.

106. **Saive, R., Eteve, G., and Brakel, J.,** The effect of four commercial fungicides on the soil microflora, *Rev. Ecol. Biol. Sol.,* 12, 557, 1975.

107. **Mitsui, S., Watanabe, I., and Honda, S.,** Action of pesticides on denitrification in paddy soil. II, *J. Soil Sci.,* (Tokyo), 34, 407, 1964.

108. **Mitsui, S., Watanabe, I., and Honda, S.,** The effect of pesticides on denitrification in paddy soil, *Soil Sci. Plant Nutr. (Tokyo),* 10, 15, 1964.

109. **Goring, C. A. I.,** Control of nitrification by 2-chloro-6-(trichloromethyl) pyridine, *Soil Sci.,* 93, 211, 1962.

110. **Nishihara, T.,** The search for chemical agents which efficiently inhibit nitrification in soil and studies on its utilization in agricultural practices, *Bull. Fac. Agric. Kagoshima Univ.,* 12, 107, 1962.

111. **Zebtmyer, G. A. and Kendrich, J. B.,** Fungicidal action of volatile soil fumigants, *Phytopathology,* 39, 864, 1949.

112. **Koike, H.,** The effect of fumigants on nitrate production in soil, *Proc. Am. Soil Sci. Soc.,* 25, 204, 1961.

113. **Gasser, J. K. R. and Peachey, J. E.,** A note on the effects of soil sterilants on the mineralisation and nitrification of soil nitrogen, *J. Sci. Food Agric.,* 15, 142, 1964.

114. **Gregory, K. F., Allen, O. N., Riker, A. J., and Peterson, W. H.,** Antibiotics and antagonistic microorganisms as control agents against damping off of alfalfa, *Phytopathology,* 42, 613, 1952.

115. **Turner, N. J.,** Decomposition of Sodium N-Methyl Dithiocarbamate, Vapam, in Soil, Ph.D. thesis, Oregon State University, Corvallis, 1962.

116. **Mazur, A. R. and Hughes, T. D.,** Nitrogen transformations in soil as affected by the fungicides benomyl, Dyrene and maneb, *Agron. J.,* 67, 755, 1975.

117. **Halleck, F. E. and Cochrane, V. W.,** The effect of fungistatic agents on the bacterial flora of the rhizosphere, *Phytopathology,* 40, 715, 1950.

118. **Caseley, J. C. and Broadbent, F. E.,** The effect of five fungicides on soil respiration and some nitrogen transformations in fine sandy loam, *Bull. Environ. Contam. Toxicol.,* 3, 58, 1968.

119. **Hoflich, G.,** Einstaz von Bioziden zur Beeinfludsung der Bodenmikroflora und deren Umsetsungen. IV. Einfluß von strohabbauhemmenden auf den Stickstoffumsatz, *Zentralbl. Bakteriol. Parasitenkd. Infektionskr. Hyg. Abt. 2,* 132, 155, 1977.

120. **Peters, J. L., von Rossen, A. R., Heremans, K. A., and Delcambe, L.,** Influence of pesticides on the presence and activity of nitrogenase in *Azotobacter vinelandii, J. Agric. Food Chem.,* 23, 404, 1975.

121. **Munnecke, D. E. and Ferguson, J.,** Effect of soil fungicides upon soil-borne plant pathogenic bacteria and soil nitrogen, *Plant Dis. Rep.,* 44, 552, 1960.

122. **Azad, M. I., Khan, A. A., and Saleem, M.,** Effect of copper and zinc fungicides on soil microbiological processes, *J. Agric. Res. (Lahore),* 9, 218, 1971.

123. **Ramakrishna, C., Gowda, T. K. S., and Sethunathan, N.,** Effect of benomyl and its hydrolysis products, MBC and AB, on nitrification in flooded soil, *Bull. Environ. Contam. Toxicol.,* 21, 328, 1979.

124. **Wessen, B., Helweg, A., and Torstenssen, L.,** Sidoeffckter av carbendazim, Plant Protection Conf., Uppsala, *Vaextskyddsrapporter (Jordbruk),* 4, 193, 1978.

125. **Torstenssen, L. and Wessen, B.,** Interactions between the fungicide benomyl and soil microorganisms, *Soil Biol. Biochem.,* 16, 445, 1984.

126. **Helweg, A.,** Influence of the fungicide benomyl on microorganisms in soil, *Tidsskr. Planteavl,* 77, 375, 1973.

127. **Gowda, T. K. S.,** Studies on the Effect of Benlate on Microflora and Other Biological Properties of Soils of Mysore State, M.Sc. thesis, University of Agricultural Sciences, Bangalore, India, 1973.

128. **Gowda, T. K. S., Siddaramappa, R., and Sethunathan, N.,** Heterotrophic nitrification and nitrite tolerance by *Aspergillus carneus* (Van Tiegh Blochwitz), a predominant fungus isolated from benomyl-amended soil, *Soil Biol. Biochem.,* 8, 435, 1976.

129. **Gowda, T. K. S. and Sethunathan, N.,** Effect of Benomyl Residues on Heterotrophic Nitrification, Working Paper No. 22, 1st FAO/IAEA/GSF Research Coordination Meeting, Vienna, 1975.

130. **Sequeira, L.,** Effect of urea application on survival of *Fusarium oxysporum* f. *cubense* in soil, *Phytopathology,* 53, 332, 1963.

131. **Nayak, D. N. and Rao, V. R.,** Pesticides and heterotrophic nitrogen fixation in paddy soils, *Soil Biol. Biochem.,* 12, 1, 1980.

132. **Charyulu, P. B. B. N. and Rao, V. R.,** Nitrogen fixation by *Azospirillum* sp. isolated from benomyl amended rice soil, *Curr. Sci.,* 47, 822, 1978.

133. **Pal, S. S., Suchaker-Barik, K., and Sethunathan, N.,** Effect of benomyl on iron and manganese reduction and redox potential in flooded soil, *J. Soil Sci.,* 30, 155, 1979.

134. **Saive, R.,** Action of pesticides on soil microorganisms, *Ann. Gembloux,* 80, 55, 1974.

135. **Kouba, N. R.,** Method of increasing the beneficial oxidation of a biological substrate with 2-aminobenzimidazole derivatives, U.S. Patent 3,649,530, 1972.

136. **Liu, S. L., Varsa, E. C., Kapistn, G., and Mtruweu, D. N.,** Effect of elridiazol and nitrapsum treated N fertilizers on soil mineral N status and wheat yields, *Agron. J.,* 76, 265, 1984.

137. **Atlas, R. M., Pramer, D., and Bartha, R.,** Assessment of pesticide effects on non-target soil microorganisms, *Soil Biol. Biochem.,* 10, 231, 1978.

138. **Mitsui, S., Watanabe, I., and Honda, S.,** The action of pesticides on denitrification in paddy soil. I, *J. Soil Sci.,* (Tokyo), 33, 469, 1963.

139. **Matsubara, T. and Mori, T.,** Studies on denitrification. IX. Nitrogen oxide, its production and reduction to nitrogen, *J. Biochem. (Tokyo),* 64, 863, 1968.

140. **Bollag, J. M. and Henninger, N. M.,** Influence of pesticides on denitrification in soil and with an isolated bacterium, *J. Environ. Qual.,* 15, 1976.

141. **Sandhu, M. S. and Morgham, J. T.,** Influence of three chemicals on soil biological activity, *Commun. Soil Sci. Plant Anal.,* 3, 439, 1972.

142. **Yeomans, J. C. and Bremner, J. M.,** Denitrification in soil: effects of insecticides and fungicides, *Soil Biol. Biochem.,* 17, 453, 1985.

143. **Mills, H. A. and McElhannon, W. S.,** Terrazole suppression of denitrification, *Hortic. Sci.,* 19, 54, 1984.

144. **Parr, J. F.,** Effects of pesticides on microorganisms in soil and water, in *Pesticides in Soil and Water,* Guenzi, W. D., Ed., American Society of Soil Science, Madison, Wisc., 1974, 315.

145. **Wainwright, M.,** A review of the effects of pesticides on microbial activity in soil, *J. Soil Sci.,* 29, 287, 1977.

146. **Iley, J. R.,** Studies on the effects of zinc ethylenebisdithiocarbamate (zineb) on citrus seedlings grown in solution cultures and soil and on its degradation by sunlight and microbial action, *Diss. Abstr.,* 24, 1779, 1963.

147. **Karanth, N. G. K. and Vasantharajan, V. N.,** Persistence and effect of Dexon on soil respiration, *Soil Biol. Biochem.,* 5, 679, 1973.

148. **Woodcock, D.,** Microbial degradation of fungicides, fumigants and nematicides, in *Pesticides Microbiology,* Wright, S. J. L., Ed., Academic Press, New York, 1978, 731.

149. **Dlvang, H.,** Benomyl-restmangder i halm samt inverkan pa deggmask och halmeedbrytnig, *Vaextskyddsnotiser,* 41, 115, 1977.

150. **Berg, B. and Staaf, H.,** Decomposition rate and chemical changes of Scots pine needle litter. II. Influence of chemical composition, *Ecol. Bull.,* 32, 373, 1980.

151. **Hirt, J. M., Le Riche, H. H., and Bascomb, C. L.,** Copper accumulation in the soils of apple orchards near Wisbech, *Plant Pathol.,* 10, 105, 1961.

152. **Decallone, J. R., Genot, M., and Meyer, J. A.,** Effects of benomyl, carbendazim, thiophanates on respiration and oxidative phosphorylation of *Fusarium oxysporum* and *Saccharomyces cerevisiae, Pestic. Sci.,* 6, 113, 1975.

153. **Tu, C. M.,** Effects of some pesticides on enzyme activities in an organic soil, *Bull. Environ. Contam. Toxicol.,* 27, 109, 1981.

154. **Ray, R. C. and Sethunathan, N.,** Effect of commercial formulations of hexachlorocyclohexene and benomyl on the oxidation of elemental sulphur in soil, *Soil Biol. Biochem.,* 12, 451, 1980.

155. **Ray, R. C. and Sethunathan, N.,** Effect of hexachlorocyclohexene and benomyl on sulphate reduction in flooded acid sulphate soil, *Environ. Pollut. Ser. B,* 5, 91, 1983.

156. **Cooke, D. A. and Hull, R.,** The effect of soil fumigation with D-D on the yield of sugar beet and other crops, *Ann. Appl. Biol.,* 71, 59, 1972.

157. **Rawlinson, C. J. and Calhoun, J.,** Chemical treatment of cereal seed in relation to plant vigour and control of soil fungi, *Ann. Appl. Biol.,* 65, 459, 1970.

158. **Jenkinson, D. A., Nowakowski, T. Z., and Mitchell, J. D. D.,** Growth and uptake of nitrogen by wheat and rye-grass in fumigated and irrigated soil, *Soil Sci.,* 36, 149, 1972.

159. **Smith, H. D.,** Effect of fumigants on the soil status and plant uptake of certain elements, *Proc. Soil Sci. Soc. Am.,* 27, 538, 1963.

160. **Naftel, J. A.,** The absorption of ammonium and nitrate N_2 by various plants at different stages of growth, *J. Am. Soc. Agron.,* 23, 142, 1931.

161. **Webley, C. M. and Duff, R. B.,** The incidence in soils and other habitats of microorganisms producing 2-ketogluconic acid, *Plant Soil,* 22, 307, 1965.

162. **Barber, D. A.,** Microorganisms and the inorganic nutrition of higher plants, *Annu. Rev. Plant Physiol.,* 19, 71, 1968.

163. **Putnam, H. D. and Schmidt, E. L.,** Studies on the free amino acids from soils, *Soil Sci.,* 87, 22, 1959.

164. **van Andel, O. M.,** Amino acids and plant diseases, *Annu. Rev. Phytopathol.,* 5, 349, 1966.

165. **Wallace, R. H. and Lochhead, A. G.,** Qualitative studies of soil microorganisms. IX. Amino acid requirement of the rhizosphere bacteria, *Can. J. Res.,* 28, 1, 1950.

166. **van Driel, W.,** Studies on the conversions of amino acids in soil, *Acta Bot.,* 10, 209, 1961.

167. **Ponnamperuma, F. N.,** The chemistry of submerged soils, *Adv. Agron.,* 24, 29, 1972.

168. **Kimura, T., Takai, Y., and Ishikawa, K.,** Microbial reduction mechanism of ferric iron reduction in paddy soils, I, *Soil Sci. Plant Nutr. (Tokyo),* 9, 171, 1963.

169. **Bloomfield, C.,** Experiments on the mechanisms of grey formation, *J. Soil Sci.,* 2, 196, 1951.

170. **Takai, Y. and Kimura, T.,** The mechanism of reduction in waterlogged paddy soil, *Folia Microbiol. (Prague),* 11, 304, 1966.

171. **Gowda, T. K. S., Rao, V. R., and Sethunathan, N.,** Heterotrophic nitrification in the simulated oxidized surface of a flooded soil amended with benomyl, *Soil Sci.,* 123, 171, 1977.

172. **Tanaka, A. and Yoshida, S.,** Nutritional disorders of rice plant, *Int. Rice Res. Inst. Tech. Bull.,* 10, 57, 1970.

173. **van Breemen, N. and Moormann, F. R.,** Iron toxic soils, in *Soils and Rice,* International Rice Research Institute, Philippines, 1977.

174. **Wainwright, M. and Sowden, F. J.,** Influence of fungicide treatment on $CaCl_2$ extractable phosphorus and phosphate-solubilizing microorganisms in soil, *Plant Soil,* 48, 335, 1977.

175. **Wainwright, M. and Sowden, F. J.,** Influence of fungicide treatment on $CaCl_2$ extractable phosphorus and phosphate-solubilizing microorganisms in soil, *Plant Soil,* 48, 335, 1977.

Chapter 11

RHIZOSPHERE MICROFLORA

I. INTRODUCTION

The rhizosphere is the area of soil that surrounds and is influenced by the roots of a plant. The rhizoplane is the external surface of the root, together with closely adhering soil particles and debris. The term "rhizosphere soil" refers to the thin layer adhering to the root after the loose soil and clumps have been removed by shaking. The soil coating varies in thickness, which may influence this magnitude of the rhizosphere effect but will neither obscure it nor alter its characteristic features. The rhizoplane-root surface is a more sensitive index of the specific qualitative effects of roots on soil microorganisms. The most important aspect of the rhizosphere, or zone of root influence, is the greater number and activity of soil microorganisms in this region than in root-free soil. Between the rhizoplane and the rhizosphere is an area of transition in which the root influence diminishes with distance. The soil-root interface is a complicated environment for microbial growth, both physically and chemically.

The actively growing plant roots exert a distinct selective action on soil microorganisms, resulting in the stimulation of certain groups and the suppression of others. Significant qualitative and quantitative effects in microorganisms have been demonstrated in the rhizosphere and root-free soil. There is mounting interest in pesticide application because this practice may be a more direct means of influencing the rhizosphere population. The increasing use of fungicides and other related chemicals, their possible effects on physiological processes in the plant, and their direct effects in the root zone after excretion by the roots suggest the importance of fungicides in the biological control of plant disease. That the application of fungicides on foliage results in the release of these substances in root exudates has been established. This effect occurs in the early stages of plant growth, even from the beginnings of seed germination. The microbial equilibrium in the rhizosphere as usually observed is due to microbial as well as plant activity, both of which are influenced by pesticides. The microorganisms that comprise the rhizosphere and rhizoplane are bacteria, fungi, actinomycetes, algae, protozoa, nematodes, and other microfauna. There is no evidence that plant species support specific microbial populations, aside from the well-established symbiotic relations between rhizobia and legumes and in mycorrhizal associations. Mosse,[1] Foster and Rovira,[2] Rovira,[3] Greaves and Darbyshire,[4] Rodriguez-Kabana and Curl,[5] and Gray et al.[6] reviewed associated microorganisms and effects of pesticides on them. This review is limited to the nontarget effects of agricultural fungicides on the saprophytic rhizosphere populations of intact roots. The rhizobia or plant pathogenic organisms are discussed in Chapter 14.

II. NONSYSTEMIC FUNGICIDES

Application of Bordeaux mixture (BM), malachite green, and zineb at rates recommended for disease control reduced the number of bacteria in the rhizosphere of bean plants, whereas Spergon and related compounds increased counts.[7]

Seed treatment with fungicides is likely to affect the nature of the rhizosphere, but little is known about the role of these treatments in promoting or reducing the populations of free-living nitrogen-fixing bacteria. In the rhizosphere of barley plants derived from thiram-treated (0.3%) seeds, bacteria and actinomycetes were initially stimulated, then 2 weeks later declined to the original level.[8] However, rhizosphere fungi were little affected by seed treatment.

Pseudomonas sp. isolated from corn rhizosphere developed resistance to 175 ppm mancozeb after a gradual series of transfers in a mancozeb-amended medium. The number of propagules decreased significantly when it was incorporated into unamended soil after 6 days, but in soils treated with 175 ppm mancozeb, its number exceeded 10^5 g^{-1} soil for 12 days.[9] Corn seed treated at the rate of 2 mg/10^5 seed, inoculated with the resistant strain of this bacterium, and planted increased the number of propagules in the rhizosphere two- to fourfold. This technique involving a fungicide-resistant population may assist in the successful establishment of associated microorganisms, specifically N_2-fixers, on the roots of crop plants.[9]

Plant treatment with captan strongly reduced both the total number of fungi and number of species present in the rhizosphere of *Allium cepa*.[10] Application at 1.5 mg captan per milliliter reduced the total number of fungi to 90,000 and the difference from the control was +76%. Although captan is more inhibitory, its effect seems to be limited in time, since it is greatly reduced within 2 months after the final treatment. The strong inhibitory effect observed during captan treatment dissipated, probably due to the lack of persistence of this fungicide in soil. The nontarget effect of captan on rhizosphere fungi seemed to be different from other fungicides. It greatly reduced the number of species and did not seem to have a selective effect. Ascomycetes were strongly inhibited by captan; in hyphomycetes, a 50% reduction in species isolation was observed.[10] Evidence is widely reported that root mycoflora exert a beneficial effect on plant growth.[11] De Bertoldi observed that the application of captan largely reduced the rhizosphere microflora. The most important consequence of this treatment was the reduction of the growth of onion plants. Differences from the control in this experiment were significant: the diameters of bulbs and dry matter were reduced by 22.3 and 31.1%, respectively. This decreased growth of the onion seedling is associated with the captan-induced effect on endotrophic mycorrhiza of *Allium cepa*.[12]

Foliar application of fungicides is a modern practice in agriculture for controlling diseases caused by microorganisms. This practice apparently exerts a direct influence on the quality and quantity of root exudates[3] and consequently on the rhizosphere microflora of plant subjects.[13] Indeed, Kutznelson[11] observed that foliar application of chemicals was a more direct method of influencing the rhizosphere microflora than soil treatment. Thus, the use of fungicides as foliar sprays undoubtedly has far-reaching nontarget effects, since the rhizosphere microflora have several beneficial effects on plant physiology.

Studies were conducted on the effect of foliar application of BM, zineb, mancozeb, copper oxychloride, dinocap, carboxin, and aureofungin at rates recommended for plant disease control and found that the fungal populations in the rhizosphere and rhizoplane of okra *(Hibiscus esculantus)* were reduced qualitatively. Maximum reduction was observed in the case of aureofungin followed by carboxin. *Gliocladium roseum, Penicillium granulatum,* and *Trichoderma harzianum* were isolated from the rhizosphere of sprayed plants but not from controls. This is highly desirable; since *T. harzianum* is antagonistic to several foliar and soil-borne pathogens, its survival on the sprayed plant is important in the control of plant diseases. *Thielavia terricola* was isolated from the rhizoplane of zineb-sprayed plants but not from controls. *Aspergillus sydowi, Macrophomima phaseolina,* and white sterile mycelial forms were isolated from the control plant.[14]

A study was undertaken on the effects of captan and zineb foliar application at the rate of 0.2% on root exudates and the rhizosphere microflora of chick pea *(Cicer arietinum L.)*, cv. L-550.[15] The seeds were inoculated with the *Rhizobium* culture. The fungal numbers were reduced significantly in the rhizospheres of treated plants, thus establishing the fungicidal properties. However, an increase was later recorded. A similar trend was recorded in bacteria, although Gram-negative bacteria showed an increase over the controls during the 13th to the 40th day in zineb-treated plants, while they decreased significantly in the rhizospheres of captan-treated plants. Actinomycetes increased in plants treated with zineb over controls and were variable in plants treated with captan.[15]

The effect of phenylmercury acetate and captan at 1000 ppm on soil and cotton rhizospheres and rhizoplanes was studied.[16] These two fungicides were highly toxic to almost all soil and rhizosphere fungi 3 days after application at 400 ppm, but were almost completely nontoxic in the soil. However, their toxicity persisted in the rhizosphere. The total count of rhizoplane fungi was not affected, although some individual species were either promoted or inhibited. Foliar application of benomyl (0.12%) and captan and zineb at the recommended rates (0.25%) resulted in a change in the rhizosphere microflora in cauliflower; the fungal counts were significantly reduced whereas bacteria and actinomycetes increased. The corresponding change in the microflora also resulted in a reduction in enzymatic activities.[17]

Despite this positive effect of reduction and/or alteration in the rhizosphere population, which is responsible for a great many activities in the plant-root interface, little attention has been paid to these nontarget interactions of agricultural fungicides in the soil. It is suggested that the utmost care be undertaken in the selection of fungicides for the control of diseases.

III. SYSTEMIC FUNGICIDES

Benomyl, a systemic fungicide, is most commonly used in agriculture and is widely applied as a spray to control spoilage of fruits and vegetables both before and after harvest. The fungicide is highly specific and strongly persistent in plant tissues as well as in soil.[18] The effects of soil treatment with benomyl on the rhizosphere fungi of onion have been investigated.[10] A reduced population of fungi was observed in plants treated with benomyl. The total number of fungi was 1.6×10^5 cells per gram of soil in the plants in which benomyl was applied at the rate of 0.4 mg active ingredient (a.i.) mℓ^{-1}, whereas in controls it was 3.8×10^5 cells per gram of dry soil. A difference of -58% was recorded. The influence of benomyl on rhizosphere fungi seems to have been more persistent, since the number of rhizosphere fungi was still reduced even 2 months after the final treatment, as compared to captan application whose effect dissipated after 2 months.[10] Foliar application of benomyl (0.6%) to wheat plants increased the rhizosphere bacterial count (43%) and decreased the occurrence of *Pseudomonas fluorescens* (16%), *Agrobacterium* (50%), and fungi (67%).[19] The persistence of benomyl in soil and plant tissues has been reported.[12,18] In plants treated with fungicide the number of genera isolated from the rhizosphere was largely reduced to 43 in the control and 27 in benomyl-treated plants. A similar reduction was observed in the number of species:[10] 91 different species have been isolated from the rhizospheres of treated and untreated plants, 37 of which (40.7%) were isolated from the plant rhizosphere alone, confirming that these species were sensitive to benomyl; 18 species (19.0%), including phycomycetes, ascomycetes, and hyphomycetes, were isolated from rhizosphere soil taken from benomyl-treated plants, which was the only observation supporting the suggestion that benomyl affects soil fungi differently.[20]

The effects of benomyl on rhizosphere and root-colonizing fungi under wheat were studied by Weber-Czerwinska.[21,22] She reported that treatments at dose rates up to 0.6 kg ha^{-1} only slightly affected the composition of the population. An analysis of the root-colonizing microflora of rye grown in the greenhouse was made.[23] The results were similar to those for wheat found by Weber-Czerwinska. In both experiments, few fungi were inhibited or stimulated for a long period of time after the treatment. Plantenkamp and Bollen[23] observed that 3 weeks after application of the fungicide the population was markedly affected, but 6 weeks later this effect had almost disappeared. However, at that time, a significant inhibition was only recorded for *Microdochium bolleyi* and *Papulaspora* sp.

Bollen et al.[24] studied the effect of benomyl on soil fungi associated with rye roots immediately after the harvest date. The crop was sprayed three times with benomyl at the rate of 1.6 kg ha^{-1} and the last spraying was given 65 days before sampling. It was observed that the colonization ratio of roots was unaffected, but the species composition was still markedly influenced by benomyl. The fungi which were isolated from the treated roots were *Mucor* spp. and *Mortierella elongata.* In addition, a tendency toward a higher frequency of resistant *Alternaria* spp. was observed in the plots treated with 1.20 kg ha^{-1} or more but not in those that received 0.24 kg ha^{-1}, which is the recommended dose for the control of eyespot. None of the treatments resulted in a decreased fungal recolonization of roots, which was reported for benomyl-treated onions.[10] This probably occurs only when extremely high dose rates are used. The colonization ratio of roots in treated plots equaled that in untreated plots, but the species composition was different.[24]

The responses of the root mycoflora to treatment of the crop with benomyl are inconsistent. Malone et al.[25] reported that the mycoflora of grass roots were unaffected even after the application of 25 kg ha^{-1}, which is more than five times the highest dose rate used for the control of plant diseases. The increased yield of grass swards was therefore attributed to factors other than the control of root-colonizing fungi. The effects of benomyl on the rhizosphere and root-colonizing fungi under wheat were studied by Weber-Czersinska.[21,22] She reported that treatments at dose rates of 0.15 to 0.60 kg ha^{-1} left most rhizosphere and root-colonizing fungi unaffected, but that the *Penicillium* population was markedly increased in treated plots. The increase in these sensitive fungi might have been due to the development of resistance.[26,27] Unlike this frequent occurrence on wheat roots in Weber-Czerwinska's experiments, penicillia were rare on roots of rye.[24] On the other hand, *Mortierella* spp. were not observed in her study, but were consistently observed by Bollen et al.,[24] especially in benomyl-treated plots. The prevalent species, *M. elongata,* was reported to be one of the most common rhizosphere fungi of wheat, or of potatoes grown after wheat, in the soils studied.[28]

Fusarium spp. were unaffected or showed a tendency to increase after benomyl application.[24] Earlier, van der Hoeven and Bollen[29] reported that *F. culmorum* was reduced by high dose rates of benomyl. Application of a low dose rate of carbendazim to wheat (0.18 kg ha^{-1}) resulted in an increased incidence of Fusarium foot rot.[30] Similar results were observed by Weber-Czerwinska.[22] She found that fusaria were stimulated on roots of wheat by a treatment with benomyl at 0.15 kg ha^{-1} but were inhibited at 0.60 kg ha^{-1}. This can be explained by the moderate sensitivity of predominant *F. avenaceum* and *F. culmorum;* when low dose rates are applied, they may benefit from the inhibition of their more sensitive competitor *M. bolleyi.*[22,24] Resistance towards fungicides has been reported in the antagonists.[31]

The rhizosphere was affected when roots of yellow poplar seedlings were dipped in a 150-μg m$\ell$$^{-1}$-solution of benomyl and transplanted into pots of nursery soil for 10 days.[32] The effect was an increase in microbial populations, particularly species of *Trichoderma* and *Streptomyces.* It is not clear whether the effect was of a direct fungicidal nature favorable to these fungi or due to increased exudations from roots.[5] According to the researchers, the latter seems probable, since subsequent nonrhizosphere studies revealed either little direct effect on microbial populations in the field[33] or a reduction in numbers of fungi and bacteria.[34]

In soybean, the number of *Rhizobium japonicum* in the rhizosphere increased rapidly during 2 days after sowing seed inoculated with a strain resistant to benomyl.[35] However, the increase was followed by a rapid decrease which was less rapid for seeds coated with benomyl, which decreased the number of bacteria and protozoa in the rhizosphere. An increase in dry matter yield and nodulation in sterile soil inoculated with *R. japonicum* and a mixture of microorganisms was observed.[35]

The effect of foliar application of carbendazim (0.1%) on the mycoflora of nonrhizosphere and rhizosphere soils of groundnut was studied.[36] The fungicide markedly increased the growth of shoot and root systems of groundnut and the fungal number in the rhizosphere. However, the differences in the nonrhizosphere soil of treated and control plants were not significant. Carbendazim reduced the propagules of pathogenic fungi such as *Sclerotium oryzae* and *F. oxysporum* in the rhizospheres of treated plants.[36]

As far as distinct taxonomic groups are concerned, the isolation of Phycomycetes appeared to have been positively affected by benomyl treatment, confirming what had been previously observed by other workers.[12,37,38] Ascomycetes are strongly inhibited by benomyl. In fact, out of nine species isolated, eight were from control plants and one from benomyl-treated plants. Species of *Aspergillus* and *Penicillium* were differently inhibited by benomyl, apparently without a clear specificity within the two groups. This can be explained by the selective effect of this compound, which is active only against certain groups of fungi.[10] Root microorganisms have a beneficial effect on plant growth. Any disturbance in their activity and number adversely affects the normal functioning of the plant. De Bertoldi et al.[10,12] observed that the reduction in the rhizosphere microflora due to benomyl resulted in a reduction in the growth of onion plants. The diameter of bulbs as compared to controls was reduced by 27%; similarly, the dry matter was reduced by 36%.[10] De Bertoldi et al.[12] reported that this reduction in the growth of onion seedlings treated with benomyl is also associated with its effect on the endophyte mycorrhiza of onions.

Wheat (cv. Sonalika) plants from seeds treated or untreated with systemic fungicides at the rate of 2.5 g kg^{-1} were raised in pots.[39] In the rhizosphere of plants from untreated seeds, the populations were higher in the second week than at other intervals. The persistent increase in the activity of nonsymbiotic N-fixing bacteria in the rhizosphere of wheat raised from seeds treated with carbendazim and benomyl reveals that the living rhizosphere has accentuated their favorable effect on dead wheat seeds. The significant increase recorded in populations of free-living N-fixing bacteria in soil adhering to autoclaved seeds treated with carboxin was altered when recorded in soil adhering to active seeds. Chaube[39] observed that carboxin did not cause any appreciable change in the rhizosphere population of nonsymbiotic N-fixing bacteria. Chloroneb, which was stimulatory to dead seeds from the second to the fourth week, failed to show any favorable effect in the rhizosphere. This suggests that there is an interaction among roots, microorganisms, and the fungicide.[39]

Carboxin and oxycarboxin sprayed at the rate of 0.15% on tea *(Thia chinansis)* significantly reduced the fungi population counts in the root rhizosphere.[40] Kitazin, a fungicide used specifically for the rice blast, applied at the rate of 0.01 to 0.02 kg a.i. ha^{-1} significantly reduced the number of fungi in the rhizosphere and the rhizoplane of rice.[41]

IV. MECHANISM OF ACTION

Several reasons have been suggested for the reduction and alteration of rhizosphere and rhizoplane microflora induced by the application of fungicides. Halleck and Cochrane[7] reported that fungicides after foliar application are absorbed, translocate downward, and ultimately are exuded by the roots; however, this could not later be proved. Recent evidence with the labeled compound did support the finding. The application of fungicide modifies root exudates which support the rhizosphere microflora.[42] Wallen and Hoffman[43] reported that degradation products after foliar application of captan translocate downward and Annapurna and Rao[15] ascribed the observed reduction of mycroflora in the Gram-negative rhizosphere to this. With the introduction of systemic fungicides, the translocation theory[7,43] has been widely supported. Systemic fungicides travel upward via the xylem to the leaves and downward via the phloem to the roots of the plant. The potential for movement into the soil via root exudation raises the possibility of nontarget effects on mycorrhizal symbionts as well as rhizosphere microbes. Smith and Heeple[44] reported that although the amount of carbendazim was small, it was consistent. If it is assumed that a positive detection was equivalent to approximately 1 µg and that this was obtained from roots weighing 174 mg more than the maximum, release under these conditions may have been 5 µg mg^{-1} dry root. The specific chemistry of carbendazim release is not known. In aqueous solutions, carbendazim may exist in the anion, cation, or unionized form. The preferential uptake of the molecular form of carbendazim has been observed in excised corn roots.[45] If the mechanism of root exudate loss is passive, the form of exudate may well be molecular. Most of the effects of systemic fungicides that have been discussed were caused by chemicals sprayed directly into the plant parts. The importance of systemic movement, however, was not considered.

The colonization of the rhizosphere by *Rhizobium* spp. is probably important in determining the extent of the nodulation of legumes. The need for rhizobia to proliferate as a necessary step for successful nodulation was stressed[46] and a correlation was found between the failure to nodulate and low rhizobial numbers in the rhizosphere. Microbial antagonism may also contribute to the extent of colonization of the rhizosphere and nodulation on the host. The application of fungicide also enhanced the nodulation in several crops by suppressing microbial activity. Lennox and Alexander[46] reported that application of thiram to bean seeds suppressed protozoa and allowed for more extensive development of thiram-tolerant *R. phaseoli*. Treatment of seeds with mancozeb allowed for the establishment of mancozeb-resistant bacterium on corn roots.[9] In order to increase the colonization of the rhizosphere of soybean by *R. japonicum,* a strain of this bacterium resistant to benomyl was used.[35] The number of rhizobia rose quickly in the first 2 days after soybean seeds were planted in soil and then rapidly fell. The decline was slower if the seeds were coated with benomyl. This fungicide reduced the number of bacteria and protozoa in the rhizosphere, but the effect became less or disappeared as the plants grew. Nodulation and plant yield were increased by the addition of benomyl to soybean seeds sown in sterile soil inoculated with *R. japonicum* and a mixture of microorganisms.[35] Inoculation of thiram-treated seed with a thiram-resistant strain of *R. phaseoli* resulted in higher numbers of *R. phaseoli* in the spermosphere and rhizosphere of kidney beans as compared with seedlings that grew from untreated seeds.[9,47] The favorable effect of benomyl on colonization by *R. japonicum* probably results from its effects on microorganisms antagonistic to the rhizobium.

V. FUTURE OUTLOOK

The nature and importance of the rhizosphere and nontarget effects of pesticides on the rhizosphere microflora have been reviewed.[11,48] Various arguments have been advanced contending that the root microflora exert beneficial effects on plant growth: increased amount of microbial carbon dioxide in the root zone; greater number, activity, and turnover of microorganisms therein; increased solubilization of mineral nutrients; and greater synthesis of vitamins, amino acids, auxins, and gibberellins, which may stimulate the growth of plants and antibiotics which will protect plants from the attack of plant pathogens in the root zone.[11] Pesticides can cause physiological and biochemical alterations to plant defense mechanisms. Both decreases and increases in the effectiveness of host defense mechanisms have been observed.[49,50] The application of pesticides results in drastic changes in root exudates, which often play an important role in the nonspecific resistance of the plant to several plant pathogens.[48] Enhanced exudation increases the nutritional level in the rhizosphere and, consequently, can influence the infection process. Enhanced exudation has been associated with (1) a stimulation of spore germination and mycelial growth of the pathogen, (2) an alteration in the rhizosphere microflora with a higher incidence of fungi and actinomycetes at the cost of bacteria, and (3) a reduction of root colonization by mycorrhizae. The latter fungi often contribute to the resistance of plant roots to pathogens.[51] These observations suggest that the rhizosphere microflora greatly influence plant growth and disease reaction and any change in their behavior caused by fungicides may alter crop growth and thus yield.[11,48]

REFERENCES

1. **Mosse, B.,** A microbiologist's view of root anatomy, in *Soil Microbiology,* Walker, N., Ed., Butterworths, London, 1975, 39.
2. **Foster, R. C. and Rovira, A. D.,** The rhizosphere of wheat roots studied by electron microscopy of ultra thin section, *Bull. Ecol. Res. Commun. (Stockholm),* 17, 92, 1973.
3. **Rovira, A. D.,** Plant root exudates, *Bot. Rev.,* 35, 35, 1969.
4. **Greaves, M. P. and Darbyshire, J. F.,** The ultrastructure of the mucilaginous layer on plant roots, *Soil Biol. Biochem.,* 4, 443, 1972.
5. **Rodriguez-Kabana, R. and Curl, E. A.,** Nontarget effects of pesticides on soilborne pathogens and disease, *Annu. Rev. Phytopathol.,* 18, 311, 1980.
6. **Gray, T. R. G., Gwynefryn-Jones, J., and Wright, S. J. L.,** Microbiological aspects of the soil, plant, aquatic, air and animal environments, in *Pesticides Microbiology,* Hill, I. R. and Wright, S. J. L., Eds., Academic Press, New York, 1978, 17.
7. **Halleck, F. E. and Cochrane, V. W.,** The effect of fungistatic agents on the bacterial flora of the rhizosphere, *Phytopathology,* 40, 715, 1950.
8. **Murthy, N. B. K. and Raghu, K.,** Effect of thiram on plant growth and rhizosphere microflora of barley and nodulation in cowpea, *Plant Soil,* 44, 491, 1976.
9. **Mendoz-Castro, F. A. and Alexander, M.,** Method for establishing a bacterial inoculum on corn roots, *Appl. Environ. Microbiol.,* 45, 248, 1983.
10. **de Bertoldi, M., Rambelli, A., Giovannetti, M., and Griselli, M.,** Effects of benomyl and captan on rhizosphere fungi and the growth of *Allium cepa, Soil Biol. Biochem.,* 10, 265, 1978.
11. **Katznelson, H.,** Nature and importance of the rhizosphere, in *Ecology of Soilborne Plant Pathogens,* Baker, K. F. and Synder, W. C., Eds., University of California Press, Berkeley, 1965, 187.
12. **de Bertoldi, M., Giovannetti, M., Griselli, M., and Rambelli, A.,** Effects of soil application of benomyl and captan on the growth of onions and the occurrence of endophytic mycorrhizae and rhizosphere microbes, *Ann. Appl. Biol.,* 86, 111, 1977.
13. **Bloom, J. K. and Walker, J. C.,** Effect of nutrient sprays on Fusarium wilt of tomato, *Phytopathology,* 45, 443, 1955.
14. **Shrivastav, L. S. and Dayal, R.,** Studies on rhizosphere micoflora of *Abelmoschus esculentus.* X. Effect of fungicidal spray on rhizosphere and rhizoplane mycoflora, *Indian Phytopathol.,* 34, 426, 1981.
15. **Annapurna, Y. and Rao, P. R.,** Influence of foliar sprays on the rhizosphere microflora of *Cicer arietinum* L., *Biol. Bull. India,* 4(2), 113, 1982.
16. **Abdel-Kader, M. I. A., Moubasher, A. H., and Abdel-Hafez, S. I.,** Selective effects of five pesticides on soil and cotton rhizosphere and rhizoplane fungus flora, *Mycopathologia,* 66(1/2), 117, 1979.
17. **Rao, A. V. and Sharma, R. L.,** Influence of chemicals on the microflora and enzyme activities in the cauliflower rhizosphere, *Acta Bot. Ind.,* 6(1), 71, 1979.
18. **Frahm, J.,** Verhalten und Nebenwirkkungen vor Benomyl, *Z. Pflanzenkr. (Pflanzenpathol.) Pflanzenschutz,* 80, 431, 1973.
19. **Vrany, J., Stanek, M., and Vancura, V.,** Rhizosphere microflora and colonization of wheat roots by *Gaeumannomyces graminis* var. *tritici* after foliar application of urea and benomyl, *Folia Microbiol.,* 25, 476, 1980.
20. **Edgington, L. V., Khew, K. L., and Barron, G. L.,** Fungitoxic spectrum of benzimidazole compounds, *Phytopathology,* 61, 42, 1971.
21. **Weber-Czerwinska, E.,** Grazyby wystepujace na korzeniack pszenicy pdmiany Grana traktowanej fungicydem benomyl, *Zesz. Nauk. Uniw. Mikolaja Kopernika Toruniu Nauki Mat. Przyr. Biol.,* 20, 41, 1977.
22. **Weber-Czerwinska, E.,** Dalsza badania nad oddzisttwaniem benomylu na mikroflora korzeniowa pszenicy ozimej odminay Grana, *Zesz. Probl. Postepow Nauk Roln.,* 230, 157, 1979.
23. **Plantenkamp, R. G. H. and Bollen, G. J.,** The effect of benomyl sprays on the root mycoflora of rye, *Acta Bot. Neerl.,* 22, 168, 1973.
24. **Bollen, G. J., van der Hoeven, E. P., Lamers, J. G., and Schoonnen, M. P. M.,** Effect of benomyl on soil fungi associated with rye. II. Effect on fungi of culm bases and roots, *Neth. J. Plant Pathol.,* 89, 55, 1983.
25. **Malone, J. P., McGimpsey, H. C., and McLiwaine, R. S.,** Effect of some fungicides on yield of grass swards, *Ann. Appl. Biol.,* 90, 65, 1978.
26. **Bollen, G. J.,** Resistance to benomyl and some chemically related compounds in strains of *Penicillium* species, *Neth. J. Plant Pathol.,* 77, 187, 1971.
27. **Kuramoto, T.,** Resistance to benomyl and thiophanate methyl in strains of *Penicillium digitatum* and *P. italicum,* in Japan, *Plant Dis. Rep.,* 60, 168, 1976.
28. **van Emden, J. H.,** Soil microflora in relation to some crop plants, *EPPO Bull.,* 7, 17, 1972.
29. **van der Hoeven, E. P. and Bollen, G. J.,** Effect of benomyl on soil fungi associated with rye. I. Effect on the incidence of sharp eyespot caused by *Rhizoctonia cerealis, Neth. J. Plant Pathol.,* 86, 163, 1980.

30. **Hanuss, K. and Oesau, A.,** Einfluß von Fungiziden auf Erreger parasitarer Halmbasiserkrankungen und auf den Ertrag, *Mitt. Biol. Bundesanst. Land Forstwirtsch.*, 178, 118, 1977.

31. **Bollen, G. J.,** Fungicide resistance and microbial balance, in *Fungicide Resistance in Crop Protection*, Dekker, J. and Georgopoulos, S. G., Eds., Pudoc, Wageningen, 1982, 161.

32. **Patterson, M. E.,** Effect of Benomyl on the Rhizosphere Microflora of *Liriodendron tulipifera* and Spore Germination of *Cylindrocladium floridanum*, M.S. thesis, Auburn University, Auburn, Ala., 1972.

33. **Peeples, J. L.,** Microbial activity in benomyl treated soils, *Phytopathology*, 64, 857, 1974.

34. **Siegel, M. R.,** Benomyl-soil microbiology interactions, *Phytopathology*, 65, 219, 1975.

35. **Hossain, A. K. M. and Alexander, M.,** Enhancing soybean rhizosphere colonization by *Rhizobium japonicum*, *Appl. Environ. Microbiol.*, 48, 468, 1984.

36. **Khavi, M. A. and Ramarao, P.,** Effect of foliar application of Bavistin on the mycoflora of phylloplane and rhizosphere and nonrhizosphere soils of groundnut, *Proc. Natl. Acad. Sci. India Sect. B*, 51, II, 129, 1981.

37. **Delp, C. J. and Klopping, H. L.,** Performance attributes of a new fungicide and mite ovicide candidate, *Plant Dis. Rep.*, 52, 95, 1968.

38. **Bollen, G. J. and Fuchs, A.,** On the specificity of the in vitro and in vivo antifungal activity on benomyl, *Neth. J. Plant Pathol.*, 76, 299, 1970.

39. **Chaube, H. S.,** Effect of fungicidal seed treatment on rhizosphere populations of free living N-fixing bacteria, *Indian Phytopathol.*, 38, 346, 1985.

40. **Ram, C. S. V.,** Microorganisms in the root region of tea plants and their importance, *Indian Phytopathol.*, 20, 715, 1968.

41. **Sullia, S. B.,** The fungicide Kitazin and the mycoflora of rice, *Proc. Indian Acad. Sci.*, 69, 295, 1969.

42. **Smirnoha, V. I.,** The effect of herbicides on the development of microflora of the corn rhizosphere, *Agrobiologia*, 1, 88, 1963.

43. **Wallen, V. R. and Hoffman, J.,** Fungistatic activity of captan in pea seedlings after treatment of the seeds or roots of seedlings, *Phytopathology*, 49, 680, 1959.

44. **Smith, W. H. and Heeple, D. M.,** Carbendazim exudation from roots of American elm, *Soil Biol. Biochem.*, 11, 687, 1979.

45. **Leroux, P. and Gredt, M.,** Absorption of carbendazim by corn roots, *Pestic. Biochem. Physiol.*, 5, 507, 1975.

46. **Lennox, L. B. and Alexander, M.,** Fungicide enhancement of nitrogen fixation and colonization of *Phaseolus vulgaris* by *Rhizobium phaseoli*, *Appl. Environ. Microbiol.*, 41, 404, 1981.

47. **Ramirez, C. and Alexander, M.,** Evidence suggesting protozoan predation on Rhizobium associated with germinating seeds and in the rhizosphere of beans *(Phaseolus vulgaris* L.), *Appl. Environ. Microbiol.*, 40, 492, 1980.

48. **Bollen, G. J.,** Non-target effects of pesticides on soil-borne pathogens, Proc. FAO Eur. Coop. Network Pesticides with Special Reference to Their Impact on the Environment, Versailles, France, June 4 to 8, 1984, 11.

49. **Heitefuss, R.,** Der Einfluß von Herbiziden auf bodenburtigen Pflanzenkrankheiten, Proc. Eur. Weed Res. Coun. Symp. Herbicides and Soil, 1973, 97.

50. **Altman, J. and Campbell, C. L.,** Effect of herbicides on plant diseases, *Annu. Rev. Phytopathol.*, 15, 361, 1977.

51. **Schonbeck, F.,** Endomycorrhiza in relation to plant diseases, in *Soil Borne Plant Pathogens*, Shcippers, B. and Gams, W., Eds., Academic Press, New York, 1979, 271.

Chapter 12

MYCORRHIZAE

I. INTRODUCTION

Mycorrhiza, the symbiotic association of the mycelium of a fungus with the roots of most plant species, including angiosperm, gymnosperm, pteridophyte, and thallophyte, occurs not only among forest crops but also among shrubs and herbs. In 1885, Fraser discovered such mutual associations in temperate forest trees and named them mycorrhizae (fungus-root), believing that the fungi performed the function of root hairs, which were lacking in these much modified dual structures.[1] During the prolonged period of physiological interaction, the mycorrhizal association is characterized by a high degree of specificity and specialization of the higher and lower symbionts. Physiologically, mycorrhizae represent a case of symbiosis where:

1. There is an increased uptake of nutrients and water transport from soil, particularly in infertile soil, and thus an increase in plant growth.
2. Mycorrhizal roots may afford protection to trees from infection by feeder root pathogens which either reduce or kill growth in plants (biological control).
3. Mycorrhizae make plants drought and frost resistant.

Studies conducted by many workers on the mechanism leading to mycorrhizal development indicate that both the higher and lower symbionts produce a complex of substances: an accumulation of soluble carbohydrates in roots of plants and production of one or more growth-promoting as well as growth-inhibiting metabolites. In addition to thiamine, one or more B-vitamins attract fungal symbionts from the soil to grow on the surface of roots and cause infection. The fungi in turn produce substances related to auxins that result in characteristic root morphogenesis in short roots in which the symbiotic relationship is established.[2-4]

Mycorrhizal association is of two types: ectomycorrhiza and endomycorrhiza. The ectomycorrhiza is characteristic of all genera belonging to conifers and hardwood; they can be distinguished from the heterozoic root system, comprising long roots of unlimited length which remain uninfected, and mycorrhizal (short) roots of limited growth which are ephemeral. The infected roots are restricted in growth; they are branched and colored and the root hairs are absent. The mycorrhiza is enveloped completely by a fungal mat or mantle. Endomycorrhizae are characterized by intracellular infection within the root, absence of any organized fungus growth on the root surface, and little, if any, change in morphology. They may be formed by septate or nonseptate fungi; those formed by nonseptate fungi, referred to as phycomycetous or vesicular-arbuscular mycorrhizae (VAM), are most widely prevalent not only among conifers in all tree species but also in most agricultural crops throughout the world. The fungi are less abundant in forests though by no means less prevalent.[5] VAM have been shown to increase plant growth in soil and are increasing utilization of less available forms of phosphorus. They are abundant in forests, and widely distributed in soils in the top 15 cm, and are more numerous and varied in cultivated than in noncultivated soils.[6] The vesicles are intracellular, subglobose, and usually single, although sometimes in groups; they are terminal with open connections, and generally in the outer cortex, rarely both outer and inner. The arbuscles are compound and are usually found in the inner cortex.[4]

While considering the nontarget effects of agricultural fungicides, several interactions between mycorrhizal fungi, their hosts, and environment must be identified. Certainly, cause and effect can be difficult to determine, because any nontarget effect on mycorrhizae may indirectly influence the host and vice versa. A fungicide that affects the host will almost certainly affect the mycorrhizae and consequently the mycorrhizal fungi. If the mycorrhizal fungus is eradicated, on the other hand, the host can suffer nutrient deficiency quite unrelated to the direct effects of the fungicide on the host.

The persistence of fungicides is yet another variable that influences the ultimate side effects on either mycorrhizal fungi, mycorrhiza formation, or host growth. Persistence is a complex phenomenon in itself and its effect on mycorrhizal fungi is difficult to separate from totally extraneous factors, such as those that affect the timing and amount of propagule reintroduction into soil from which mycorrhizal fungi have been affected. Soil fungicides, which may not persist for a long period of time, may result in the prompt reintroduction of fungi; however, soil fumigants, which also may not persist for long periods, may delay the reintroduction of mycorrhizal fungi which produces stunting of host plants for several years.

II. ECTOMYCORRHIZAE

The importance of ectomycorrhizae for the survival and growth of tree seedlings is well documented.[7] In many parts of the world, special procedures are used in tree nurseries to encourage ectomycorrhizal development.[8,9] Such procedures include the addition of soil containing these fungi, maintenance of high organic matter and low to moderate fertility, and growing host trees near the nursery to encourage production of fruiting bodies for natural colonization. The ectomycorrhizae fungi include *Pisolitus tinctorius* and *Thelephora terrestris*. Studies indicate that introducing pure cultures of these fungi into nursery soil is more beneficial to tree seedlings after outplanting than are naturally occurring nursery fungi.[10]

A. In Vitro

General biocides (broad spectrum and nonspecific) will eliminate or kill an organism exposed to them. Hence, it is not surprising that mycorrhizae fungi exposed to fungicides at various concentrations in vitro in different media will show variable growth patterns. The available literature is briefly reviewed in the following paragraphs.

1. Nonsystemic Fungicides

Nonsystemic fungicides are broad spectrum and nonspecific toward mycorrhizae fungi. Captafol (100 to 200 μg g^{-1}),[11] captan (1 to 15,000 μg g^{-1}),[7,12,13] copper oxychloride (500 to 3000 μg g^{-1}),[13,14] and chlorothalonil (100 to 200 μg g^{-1})[11] reduced the growth of fungi in medium. Sobotka[15] evaluated different dithiocarbamates — thiram, ferbam, zineb, and folpet — using drenched paper strips in medium and found that ferbam and folpet reduced growth in the medium. However, thiram (100 to 200 μg g^{-1}),[11] zineb (500 to 3000 μg g^{-1}),[13,14] ziram (500 to 3000 μg g^{-1}),[14] mancozeb (1 to 500 μg g^{-1}),[16] and maneb (100 to 200 μg g^{-1})[11,15] were toxic to fungi in different degrees.

2. Systemic Fungicides

Systemic fungicides were evaluated for their effect on mycorrhizae fungi in vitro. The benzimidazoles benomyl, fuberidazole, and thiabendazole at 1 to 200 μg g^{-1} [7,11,17] had no effect on growth in vitro. However, benodanil, another systemic fungicide, was toxic to fungi at 1 to 17 μg g^{-1}.[7,18] Hymexazol, still another systemic fungicide, was also toxic at 200 μg g^{-1}.[11]

The inhibitory or other effects of fungicides on mycorrhizae in vitro may not tell us exactly what side effects may occur in nature. Since mycorrhizal associations in forests and fields are the most important, future studies should focus on the beneficial or deleterious effects of fungicides in nature. Nonetheless, the in vitro studies will provide background information for field experiments on fungicide mycorrhizae interactions/associations.

B. In Vivo

1. Nonsystemic Fungicides

Many different fungicides are used to control diseases in the production of nursery seedlings[19] and many of them effect ectomycorrhizal development. Bakshi and Dobriyal[20] found that captan and pentachloronitrobenzene (PCNB) used to control damping-off in nurseries in India delayed the development of naturally occurring ectomycorrhizae on seedlings of *Pinus patula* until the fungicides had lost their effectiveness. In Malaysia, Hong[21] observed that ectomycorrhizal development on *P. caribaea* seedlings was suppressed by chlorothalonil and thiram and stimulated by captafol and captan. PCNB inhibited the growth of several ectomycorrhizal fungi in pure cultures but did not alter the development of naturally occurring ectomycorrhizae of pine and spruce seedlings when they were applied to soils in nurseries. In nursery and greenhouse tests in Australia, Theodorou and Skinner[22] found that seedcoat dressing of captan, zineb, and thiram inhibited ectomycorrhizal development of *P. radiata* seedlings grown from seed inoculated with basidiospores of different fungi, but had no effect on the subsequent development of ectomycorrhizae formed by naturally occurring fungi. Pawuk et al.[23] found that the development of *Pisolithus* ectomycorrhizae on *Pinus palustris* seedlings grown for 14 weeks in a pine bark medium in containers was completely inhibited by PCNB and reduced by captan and fenaminosulf. Because of these known effects, fungicides that are currently used or are being considered for use to control diseases in forest nurseries were evaluated.[7] A comparison of the results of the laboratory and nursery tests shows that mycelium of *Pisolithus* and *Thelephora* is affected more by fungicides in agar medium than soil.[7] The limited persistence of these fungicides in soil probably accounts for some of this difference in sensitivity. The lack of leaching, microbial activity by associated microflora, and barriers to fungicide diffusion in agar permits the hyphae to endure long exposure at higher concentrations in agar culture than in soil. Due to the higher nutrient concentrations in agar medium than in soil, the mycelium grows at a faster rate which would increase the metabolic uptake of the fungicides. Also, the hyphae of ectomycorrhizal fungi grown in vermiculite particles as the initial inoculum may be protected to a certain extent from high concentrations of fungicides in soil. These differences between nursery and laboratory studies on fungicides indicate the limited value of studies on agar medium that are not supported by studies carried out in soil.

The fungitoxicants arasan, botron, lanstan, and ethazole mixed at 25, 50, 100, and 250 ppm, respectively, captan at 50 and 100 ppm, and preplanting compounds mylone, vapam, and vorlex at 40 ppm restricted mycorrhizal development of *Endogone fasciculata* on corn to some extent, even at the lowest concentration.[24] Of the five comparable toxicants, captan was least injurious while vorlex was somewhat less so than mylone and vapam. It was also observed that root volume and branch root development were severely limited by arasan, botran, and ethazole at 100 ppm and moderately so at 50 ppm. There was a slight suppression of both root development and volume by ethazole at 25 ppm. Botran at all concentrations caused some distortion of root hairs. Arasan reduced the number of root hairs, but only at 50 and 100 ppm. The treatment of soil with fungitoxicants is frequently beneficial to crop plants. However, these toxicants may sometimes kill both detrimental and favorable soil microorganisms indiscriminately.

The extreme effects of eight fungitoxicants found by Nesheim and Linn[24] in reducing mycorrhizal infections may be attributed to their use of a sterilized soil and mixture. In natural soils, fungal toxicants can be depleted by microbial action, hydrolyses, and chemical reactions, but the time of their disappearance depends on the compound and the physical condition of the soil. This suggests that root hairs and their epidermal cells, which are the potential initials, should be protected from the toxic effects of fungicides and fungitoxicants.

The target organisms of captan are primarily hyphomycetes, ascomycetes,[25] and certain phycomycetes.[26] A lower population of these fungi and other microorganisms may increase the effectiveness of inocula of ectomycorrhizal fungi. A reduction in populations of hyphomycetes and other fungi in the rhizosphere of onion plants by captan has been demonstrated.[27] Captan temporarily decreased soil populations of *Pythium* of pecan trees and stimulated the development of ectomycorrhizae formed by naturally occurring *Scleroderma bovista*. The significance of changes in rhizosphere microbe populations, such as an ectomycorrhizal development by *Pisolithus tinctorius* and *T. terrestris* caused by the indirect action of fungicides, needs further investigation.

In nursery soil infested with mycorrhizal fungi, ectomycorrhizal development on seedlings of *Pinus taeda* and basidiocarp production by *Pisolithus tinctorius* were greater than in fungicide-free controls when the plots were treated with benomyl and captan and less than in those treated with benodanil. PCNB had no effect on the ectomycorrhizal development of *T. terrestris*. However, the development was greater in plots with double doses of benomyl and captan. PCNB and benodanil decreased basidiocarp production. The mycelium of both fungi was affected by the fungicides in agar medium more than in soil.[28]

Where maneb, quintozene, captan, thiram, zineb, and ziram were used for the control of damping-off at the rate of 0.7 to 60 kg ha^{-1}, they showed toxicity against mycorrhizal development in *Pinus insularis, P. patula,* and *P. radiata.*[20] Captafol, chlorothalonil, copper oxychloride, folpet, and triphenyltin acetate at 4.5 to 24.5 kg ha^{-1} showed slight to no toxicity to *P. caribaea* and *P. oocarpa.*[29] Captafol, chlorothalonil, and captan had no effect on mycorrhizal development when applied as foliar fungicides at the operational rate for disease control.[21]

2. Systemic Fungicides

A range of fungicides are used in the control of diseases in forest nurseries and may affect the development of ectomycorrhizae. Nesheim and Linn[24] reported that the systemic fungicide ethazole applied at 25, 50, and 100 ppm in soil has a deleterious effect on the ectotrophic fungus *Glomus fasciculatus* on corn but enhances the growth of forest trees. There was a slight suppression of both root development and volume by ethazole at 25 ppm. Benomyl had no effect on the development of ectomycorrhizae on *Pinus caribaea.*[21] Benodanil, a systemic fungicide undergoing testing for the control of fusiform rust on pines in the southern U.S., prevented pure culture growth of several ectomycorrhizal fungi. When applied as a drench to nursery soil it delayed the development of naturally occurring ectomycorrhizae on *Pinus taeda.*[30] The development of *Pisolithus* ectomycorrhizae on *Pinus palustris* seedlings was stimulated by ethazole and thiabendazole.[23]

Many factors that influenced the formation of ectomycorrhizae affect either the susceptibility of the host or the survival and infective potential of the fungal symbionts.[9] Although certain fungicides used in this test may have affected the susceptibility of pine seedlings to root infection by fungi, the results of other research suggest that the action of fungicides is more likely to have a direct or indirect effect on the fungi. Benodanil directly inhibits the vegetative growth of symbiotic fungi and causes a decrease in ectomycorrhizal development.[7,30] As discussed by Kelley,[31] this type of action by benodanil is an example of its detrimental effect on beneficial, nontarget organisms. However, Marx and Rowan[7] reported that benomyl indirectly stimulated ectomycorrhizal development by specific nontarget fungi. The target organisms of benomyl are primarily hyphomycetes and ascomycetes.[25]

The influence of nursery practices and soil amendments with benomyl, ethazole, metalaxyl, and propamocarb for the control of nursery diseases was assessed on rhododendron mycorrhizae.[32] The foliar application of fungicide had no adverse effect on mycorrhizal formation. Mycorrhizal studies were conducted on the effect of benomyl, chloroneb, and thiophanate-methyl on exotic conifers in west Malaysia.[29] Except for chloroneb, the other fungicides had no effect on the mycorrhizae. A weekly spray of triadimefon at 0.28 kg ha^{-1} for the control of rust on pine *(Pinus radiata)* seedlings was toxic to the mycorrhizae.[33] This indicates the downward translocation of the fungicide.

When onion seedlings were grown for 6 months in the absence of pathogens, repeated soil applications of benomyl or captan significantly and simultaneously decreased the diameter and the dry weight by 22 to 25% and 31 to 34%, respectively. The losses were associated with differing effects on soil organisms. Captan had no effect on the occurrence of VAM but appreciably decreased the population of rhizosphere fungi. In contrast, benomyl inhibited mycorrhiza formation but had relatively little effect on rhizosphere fungi.[27]

These studies indicate that the utmost care should be taken in the selection of fungicides for control of diseases in tree nurseries. The qualitative and quantitative effects of fungicides on the development of ectomycorrhizae should be considered when any fungicide is selected to control specific diseases. Disease-free seedlings lacking adequate ectomycorrhizae may be grown to plantable size in a nursery following fungicide application, but the lack of adequate ectomycorrhizae on such seedlings will often result in substandard survival and growth performance following field outplanting. The stimulating effect of certain fungicides on ectomycorrhizal development, especially those formed by artificially introduced fungi, should be considered an added benefit of these fungicides.[7,10]

III. ENDOMYCORRHIZAE

VAM are formed by *Glomus* spp. with the roots of most wild and important agronomic crop plants. Economically important crops with VAM include peas, maize, wheat, soybeans, potatoes, tomatoes, strawberries, apples, citrus, grapes, tobacco, and sugar maple.

VAM, mutually beneficial symbionts, are widely distributed and physiologically unspecialized. They are so widespread in the plant kingdom that there are fewer families in which they do not occur than in which they do. In VAM, there is a nonpathogenic symbiotic association between fungi and the roots of the higher plants. The fungi are members of the endogonaceae. The VAM enhances nutrition uptake, conserve crop plant mineral nutrient resources, and are involved in the protection of the host plant from soil-borne pathogens. Mukerji and co-workers[34] reviewed taxonomy, distribution, and other recent developments on VAM. It is becoming increasingly apparent that the VAM have a probiotic influence on higher plants. The VAM association is a labile condition: environmental factors greatly influence the balanced relationship between plant roots and the endophyte.[35,36] This is of particular significance because VAM occur in a large number of commercial crops, and the wide use of toxic materials in normal agricultural practices may therefore upset this balance.

The development of a VAM relationship can be divided into the following stages:

1. Spore germination or initiation of hyphal growth from the infected root inoculum
2. Growth of hyphae through the soil to the roots
3. Penetration and successful initiation of infection in roots
4. Spread of infection, development of a mycorrhizal relationship with roots, and spore production

Burpee and Cole[37] and Smith[38] have assessed changes in the percentages of infected roots and spore production. Thus, any observed changes result from the response of each of the three preceding stages as well as the stage in which chemicals are added. With the increasing use of fungicides for the control of plant diseases, it is obligatory to review the effects of fungicides on VAM, particularly systemic fungicides which enter the plant system and affect the symbiotic relationship of the host and the VAM.

Diverse groups of fungicides are widely used in the control of plant diseases. The application of these chemicals may, however, result in indiscriminate killing of both pathogenic as well as nonpathogenic or beneficial microorganisms.[39] Several reports have demonstrated the partial or complete inhibition of vesicular-arbuscular endophytes by fungicides. The action of fungicide delays or reduces VAM infection. Nonsystemic fungicides such as PCNB, botran, and thiram are highly toxic to mycorrhizal fungi.[24,40-44]

Some fungicides may increase the infection and sporulation of the VAM fungi. Demosan, daconil, sodium azide, terrazole, captan, and copper sulfate are not very toxic to VAM fungi and may favor VAM activity under certain specific environmental conditions.[27,35,40,41,44-47] Systemic fungicides are more damaging than nonsystemic fungicides on VAM fungi. Nonsystemic fungicides can delay infection by VAM fungi but do not completely eliminate them, whereas systemic fungicides can adversely affect spore germination, infection, and growth within the root.

A. Seed Treatment

Jalali and Domsch[35] reported that seed treatment with three systemic fungicides — benomyl, thiabendazole and ethirimol — each at three different concentrations, had an adverse effect on the formation of mycorrhiza on roots of wheat by Endogone; they observed that the effect was more pronounced on 9-week-old seedlings than 6-week-old seedlings, with the exception of the thiabendazole treatment (0.6 mg/100 g seed). Ethirimol (5.0 mg a.i./ 100 g seed) was most harmful and benomyl was comparatively less toxic than the other two fungicides. They did not observe any toxic effects on the host by seed-treatment fungitoxicants.

Harvey and co-workers[48] tested the action of a number of fungicides against *Lolium* endophytes and observed that they were resistant to many of the fungicides tested. Propiconazole could reduce the level of the endophytes after foliar application. In culture, the fungus was most sensitive to prochloraz, but showed some sensitivity to propiconazole and imazalil. In seed, the fungus could be completely controlled by prochloraz.

B. Foliar Application

The response of foliar applications with systemic fungitoxicants such as triforine, chloramformethane, tridemorph, thiophanate-methyl, triadimefon, and benomyl and, for the purpose of comparison, with nonsystemic fungitoxicants such as maneb, captan, and dichlofluanid (at recommended rates) showed that most of the treatments had less of an effect on the establishment of mycorrhiza in root tissues than seed treatments. Triforine and benomyl treatments had relatively stronger inhibitory effects; triforine, tridemorph, chloramformethane, and dichlofluanid reduced the number of chlamydospores by 50% or more. Maneb and captan did not have such a pronounced effect on chlamydospore development. Foliar application of maneb at 0.8 kg a.i. ha^{-1} had no effect on mycorrhizal infection and chlamydospore formation.[35]

Only a negligible amount of the systemic fungicides used by Jalali and Domsch[35] translocated. Therefore, inhibitory effects could arise from residues of the fungicides taken up by the germinating seed and translocated passively during root growth. The formation of mycorrhizal chlamydospores was suppressed by foliar application of fungicides. This was brought about by changes in the spectrum of wheat root exudates as a result of the stress caused by the fungicides. It could be further explained by the work of Jalali and Domsch[49] in which they demonstrated that application of the systemic fungicides triforine and tridemorph on foliage results in changes in root exudations and amino acid metabolism. Alteration of host metabolism is one of the ways in which systemic fungicides may act on fungal invaders. If the concurrent changes in the metabolism of the plant persist long enough, conditions may prevent the fungus from developing normally. According to Ratanayake et al.,[50] the root exudates govern mycorrhizal symbiosis. Demosan, sodium azide, metalaxyl, and terrazole have been found to increase VAM infection and spore production and to stimulate root exudation and thus mycorrhizal infection. Atilano and van Gundy[51] observed a reduction in the number of mycorrhizal hyperparasites following treatment with fungicides and fumigants, which ultimately led to an increase in mycorrhizal infection.

Trials with foliar application of systemic fungicides (pyroxychlor, prothiocarb, ethyl aluminum phosphate, ethazole, and metalaxyl) effective against phycomycetous fungi have shown that these fungicides are almost ineffective against VAM fungi,[43,45,47,52] although these fungicides of their metabolites have been reported by several workers to translocate downward in sufficient quantity to be toxic to pythiacious fungi in vitro.[53]

C. Soil Application

Because mycorrhizal fungi are members of soil and rhizospheric microbes, the effects of fungicides on the general population may in turn affect either the mycorrhizal fungi or the host and thereby confound cause-and-effect interpretation.[11,14] The side effects of soil fungicides on soil microbes have been reviewed.[54] Trappe et al.[55] reviewed the reactions of mycorrhizal fungi and mycorrhiza formation to soil pesticides.

1. Nonsystemic Fungicides

The effects of soil fumigants and fungicides on VAM fungi *(Glomus etunicatus, G. mosseae, G. fasciculatus, G. macrocarpus,* and *G. constrictus)* have been reviewed.[56] VAM infection and chlamydospore formation and development was increased after the use of 1,3-D (Telone) in cotton and dibromochloropropane (DBCP) in soybean and sudangrass at the recommended rates of application for plant disease control. However, VAM in crops were reduced by treatment with nonsystemic fungicides such as dicloran (corn), captan (corn, citrus), captafol (citrus), dichlofluanid (wheat), korex (corn), maneb (citrus), quintozene (wheat, corn), sodium azide (peanut), and thiram (corn, bean, wheat, millet). VAM were promoted by captan (bean), and sodium azide (citrus).

Copper (up to 22.4 kg ha^{-1}), chlorothalonil (up to 22.4 kg ha^{-1}), and sodium azide (up to 31.4 kg ha^{-1}) were tested against the endomycorrhizal *G. etunicatus* and *G. mosseae* in sour orange. After 8 months, the middle and low rates of copper depressed plant growth, but not infection and sporulation by *G. etunicatus,* as compared to the control. Plant growth and sporulation after the middle and high doses of chlorothalonil were significantly less than the control, whereas plant growth was better and fungus sporulation higher in *G. mosseae* than the control after sodium azide treatment.[67] Chlorothalonil (1.2 g a.i./m^2), maneb (1.0 g a.i./m^2), and PCNB (0.9 g a.i./m^2) reduced mycorrhizal development of creeping bent grass when applied in the spring to golf course greens or 4 to 8 weeks after bent grass was seeded and inoculated with *G. fasciculatus* in the greenhouse. However, fungicides applied 16 to 20 weeks after bent grass was seeded did not affect mycorrhizal development.[58] The results suggested that spring application of several fungicides on turf may cause reduced mycorrhizal development in bent grass turfs.

Feeder root necrosis of pecans *(Carya illinoensis)* caused by *Pythium* spp. in Georgia was reduced after application of DBCP and other fungicides without the reduction of Pythium populations in soil.[59] There was no apparent overall reduction in nematode populations. Since the application of DBCP and the fungicides increased the mycorrhizal fungus *Scleroderma bovista,* it was suggested that a side effect, caused by the pesticides, on feeder root necrosis through the mycorrhiza-forming fungus, was a possibility.

Mycorrhizal endophytes are normal constituents of most field soils and the application of fungicides may affect this beneficial symbiotic association. Pitcher et al.[60] showed that six soil fumigants improved growth and increased mycorrhizal infection in field-grown apple and cherry trees. It was suggested that this was probably due to reduced microbial competition in the treated soils. Methyl bromide, a general biocide, was found to be an efficient soil fumigant in nursery beds and its application stimulated the growth of sweet gum plants.[61] It most strongly affected soil-inhabiting pathogens and also reduced soil mycorrhizal fungi, but these rapidly recolonized the host roots. In fumigated soils, citrus seedlings may show reduced rates of growth associated with the inhibition of phosphate absorption by the plant.[62] Such seedlings may be stunted, chlorotic, and nonmycorrhizal.[63]

El-Giahmi et al.[41] studied the effect of botran, arasan, and ethazole (25 and 50 ppm) and captan (50 ppm) on mycorrhizal maize. Higher levels of the fungicides depressed growth; however, preinoculation and fungicidal treatment significantly increased growth. It was observed that the toxicants reduced the growth of mycorrhizae from the indigenous source and that a fairly high level of infection can be produced in the roots if the plants are preinoculated with a selected mycorrhizal variety. Preinoculation may therefore be a way to keep the infection levels high enough to be beneficial to the host.

The influence of carbendazim and captofol in vitro on the germination of VA endophyte mycorrhizal spores of *Acaulospora laevis, G. caledonius,* and *G. monosporus* was estimated. It was observed that neither germination nor growth of the hyphae from spores was affected at soil concentrations approximately one half to ten times those likely to be found after field applications.[64] The length of germ tubes of *A. laevis, G. caledonius,* and *G. monosporus* were 8 to 13, 6 to 10, and 5 to 9 cm, respectively, for all treatments including the control.[64] Captan[24,27,44] at the recommended rates had no effect or only a slight detrimental effect on the percentage of root infection. Captan may therefore be expected to have little or no effect on stages 1, 2, or 3 of the life cycle. Spore germination and hyphal growth of the three VA endophytes tested by Tommerup and Briggs[64] were unchanged by captafol at rates ten times those of normal field application. If captafol is applied at the rates which do not affect the growth of host plants, then it probably has no adverse effect on any stage of the life cycle of VAM fungi.

Using sour orange seedlings, Nemec[45] studied the effects of copper, metalaxyl, thiabendazole, captan, captafol, chloroneb, and formaldehyde for their effects on *G. etunicatus* and chlorothalonil, sodium azide, benomyl, and maneb for their effects on *G. mosseae.* Captafol, chloroneb, metalaxyl, and captan did not reduce the growth of the mycorrhizal plant, although some adverse effect on the fungus was apparent with all the rates of captafol and the higher rate of captan. All the rates of benomyl and thiabendazole reduced plant growth, but fungal sporulation was higher than that of the mycorrhizal control. In all sodium azide treatment sporulation was higher. Copper at 112 and 224 kg ha^{-1} reduced plant growth significantly but not infection or fungus sporulation. Maneb and chlorothalonil, which had no effect on *G. mosseae* at the low rates of 5.6 kg ha^{-1}, sharply reduced fungal sporulation and mycorrhizal plant growth at 11.2 and 22.4 kg ha^{-1}. Formaldehyde was toxic to the fungus.[45] Kelley and Rodriguez-Kabana[65] reported that the application of a granular formulation of sodium azide to fine nursery beds at the rates of 0, 22.4, 67.2, and 134.5 kg a.i. ha^{-1} under water seal or plastic seal was compared over 1 year with methyl bromide applied at 650 kg a.i. ha^{-1} to determine their different effects on the development of mycorrhizal roots. An adverse effect of sodium azide was observed. The lack of activity of methyl bromide on mycorrhizal development of pine roots may be attributed to the rapid recolonization of the soil by ectomycorrhizal fungi after fumigation; basidiospores of the fungi are produced in copious quantities and are disseminated at an earlier stage by air, insects, and movement in soil.[66]

The root pathogens that compete with mycorrhizal fungi in the soil and the rhizosphere are adversely affected by fumigants and fungicides; thus, the VAM activity increases.[45,51,67,68] Some of the soil microorganisms favor VAM activity.[36]

Fungicides can have a negative effect on the development of endomycorrhizal fungi, which in turn may drastically affect the normal development of host plants under nutrient-deficient conditions. Fungicides have been examined for their effects on the inoculum and development of the VAM fungi. Menge et al.[69] reported that soaking inocula (hyphae, vesicles, arbuscles, and chlamydospores) of *G. fasciculatus* in a suspension of PCNB, ethazole, and DBCP did not impair their viability. The dosages used were 4000, 80, and 240 ppm a.i., respectively. It was further observed that when sudangrass, inoculated with the fungus, was drenched separately with PCNB at 1000 ppm at the time of inoculation, the fungicide restricted spore production by 70% after 104 days, whereas ethazole and DBCP did not significantly change spore production. When these fungicides were applied 60 days after inoculation, PCNB reduced spore production by 10%; however, ethazole and DBCP increased spore production by 76 and 63%, respectively. Menge et al.[43] reported that when inocula of *G. fasciculatus* were soaked in a PCNB solution of various concentrations they survived; however, infection and sporulation on sudangrass were considerably and consistently inhibited by soil drenches with this fungicide. Soil drenches with DBCP at 15 to 20 ppm or ethazole 10 to 40 ppm applied 30 to 60 days after inoculation increased infection and sporulation.

2. Systemic Fungicides

Several studies have indicated that benomyl and its hydrolysis products carbendazim and 2-amino-butane (AB) inhibit the formation of VAM belonging to *Endogone* and *Glomus* in various crops such as barley, clover, maize, onion, soybean, and wheat.[27,28,35,44,52,70] However, other zygomycetes fungi appear to be relatively insensitive to benomyl.[17,25] Exceptions are *Conidobolus eurymites*[71] and some species of *Mortierella,* a genus taxonomically related to the Endogonaceae.

Decreases in the growth of onions in the soil amended with benomyl at 0.6 to 6.0 kg a.i. ha^{-1} were attributed to suppression of the mycorrhizal fungus by the fungicide.[27] Similar growth decreases after captan application appeared to be associated with measured reductions in soil microbes rather than inhibition of the mycorrhizal fungus, because captan did not affect the degree of mycorrhizal colonization of the onion roots. The effects of fungicides on the mycorrhizal fungi and mycorrhiza development on hosts in different soils may partly reflect their effects on differing microbial populations.

Boatman et al.[52] studied the effect of systemic fungicides on VAM infection in clover roots and observed that soil drenches of benomyl and thiophanate-methyl prevented the formation of mycorrhiza and also the spread of established infections. The immersion of a fungal inoculum in a suspension of fungicides reduced infectivity. However, clover plants grown in benomyl-treated soil did not retain enough fungicide to affect the amount of infection after transplanting into benomyl-free soil. These experiments show that suspensions of benomyl and thiophanate-methyl are toxic to *Endogone* inocula on direct immersion and when mixed with irradiated or infested soil. Sutton and Sheppard[44] reported that benomyl was toxic to VA endophytes in a 3:1 soil-sand mixture, although other zygomycetes fungi appeared to be relatively insensitive to it.[17,25] Boatman et al.[52] observed that longer immersion of the inoculum in a suspension of the fungicides reduced its infectivity, but when different amounts of the fungicide were applied, the lower rates of application were usually as effective as the higher rates, which were occasionally phytotoxic. It was further recorded that as the solubility of both benomyl and thiophanate-methyl is low, the higher rates of application probably did not increase the concentration in the soil solution, but ensured that solid residues capable of going into solution persisted longer.[52] Benomyl was not retained in the roots in sufficient quantity to reduce infection when plants were removed from treated to untreated soil. Although the toxic principal of benomyl and carbendazim is known to be rapidly translocated from roots to leaves, it has been suggested that some may be retained on the roots which had been in contact with levels of fungicide high enough to cause considerable phytotoxicity.[72] However, even these roots rapidly become infected when transplanted into untreated soil.[52] The objective of Boatman's experiments, to find a way of estimating the efficiency of mycorrhizal endophytes *in situ,* was not achieved because the fungicide drenches, though effective in irradiated soil, were ineffective in the unsterile soils. The workers inhibited two different endophytes in two irradiated soils, but not the native endophytes in the unsterile soils. Whereas the native endophytes could have been more resistant, the more likely explanation is that the fungicides were broken down more rapidly by the general microbial population, or a particular component of it, occurring in the unsterile soil. Although the irradiated soils in open pots were by no means sterile, they may have lacked certain specifically soil-inhibiting microorganisms. The half-life of benomyl in the field is 2.7 to 3.6 months according to pH and soil organic matter content[73] or 3 to 6 months in turf and 6 to 12 months in fallow.[74] In pots kept continually moist and also probably at higher temperatures in greenhouses, breakdown may be more rapid.[52]

Benomyl applied at different growth stages has been shown to reduce the percentage of mycorrhizal infection in roots.[35,44,52,70] Suspension of benomyl was toxic to *Endogone* inoculum on direct immersion and when mixed with irradiated or infested soil, as shown by the reduced percentage of infection in roots at about 5 weeks.[52] This probably indicates that soil drenched with benomyl was toxic to established infections in the external mycelium so that new roots remained uninfected.[52] However, hyphal growth may also have been reduced because the plants did not grow well in fungicide-treated soil due to phytotoxicity caused by the fungicide and possibly aggravated by phosphorus deficiency.[52] Carbendazim at dose rates comparable to those used for benomyl did not reduce spore germination or hyphal growth of three endophytes tested by Tommerup and Briggs.[64] These two phases of growth, stages 1 and 2, are different from hyphal growth from infected roots; their response is not complicated by interactions with chemically affected roots. Benomyl and related fungicides may prevent penetration of the host roots or progression of the initial infection after penetration, and so prevent development beyond growth stage 3. They did not reduce spore germination or hyphal growth. However, Sutton and Sheppard[44] showed that the percentage of infection in roots was decreased in soil containing benomyl when spores were used as inoculum. Penetration structures, when developed from germ tubes or hyphae growing from infected roots, are probably morphogenetically different from other developmental stages and may not have the same tolerance to benomyl.[64] All rates of thiabendazole (up to 18 kg ha^{-1}) reduced plant growth and infection and sporulation decreased as rates increased.[46]

VAM fungi are major components of the soil microflora in turfgrasses. Application of recommended rates of benomyl, iprodione, triadimefon, and chloroneb reduced mycorrhizal development of creeping bent grass when applied in the growing season 4 to 8 weeks after bent grass was planted and inoculated with *G. fasciculatus*. Maximum initiation and growth of turfgrass roots usually occurs in the spring and fungicides applied then apparently prevented the establishment of mycorrhizal fungi in newly produced root tissue.[58] However, the fungicides applied 16 to 20 weeks after bent grass was seeded did not affect mycorrhizal development in the greenhouse, as measured by the length of the root colonized by fungi.[58]

According to Spokes et al.,[28] different endophyte species vary in infectivity in soils treated with different fungicides. It was reported that the application of chloroneb to inocula of *G. microcarpus* used with pot-grown lettuce stimulated mycorrhizal development; however, benomyl and triadimefon induced long-term inhibition of *G. mosseae* on onion.[28] It was further reported that the germination of spores of *G. epigaeus* was completely inhibited by terrazole and PCNB; however, mancozeb did not completely inhibit the germination. Infection by *G. fasciculatus* was not affected by terrazole, triadimefon, or chloroneb, whereas that of *G. mosseae* was reduced to 2/3 of the control by the same fungicides. It was further observed that infection by *G. microsporus* was reduced to 1/2 of the control when treated with triadimefon and was increased 3-fold by chloroneb. Terrazole was found to be ineffective to it.[28]

Many fungicides have been used to control pea root rot and several new compounds have been made available recently.[47,75] The effect of several fungicides on endemic populations of *Glomus* spp. in association with peas has been studied and the distribution of *Glomus* spp. in the soil profile was also determined. Pyroxychlor, a systemic fungicide specific for the control of *Phytophthora* and pythiaceous fungal pathogens[76] in furrow application at the rate of 1.1 kg a.i. ha^{-1} significantly reduced the numbers of chlamydospores of *G. fasci-*

culatus.[75] Captan, alone and with copper sulfate, pyroxychlor, etridiazole (as Ban-Rot 40 WP), and copper sulfate were applied to seed and furrow at several rates.[47] Afterwards, populations of *G. mosseae* and *G. fasciculatus* were studied. It was observed that the chlamydospores of *G. mosseae* were significantly more numerous (at the 5% level) in the upper 15-cm soil layer than in the lower, probably because roots of peas proliferate in the upper soil layer. Copper sulfate as a seed treatment and a furrow application had no significant effect on the reduction of chlamydospore numbers. There was a corresponding increase in chlamydospores at higher rates of fungicidal application. However, more chlamydospores of *G. mosseae* occurred in methyl bromide-treated plots than in controls, although the difference was not significant. Because this chemical is usually considered a soil sterilant, it appears that either tolerant strains were selected or that reinvasion occurred at some time after dissipation of the soil fumigant.[47] No significant effect on the number of *G. fasciculatus* chlamydospores was observed. Spores of *G. fasciculatus* were fewer than those of *G. mosseae* but were uniformly distributed throughout the soil depth. *G. fasciculatus* was less abundant, because in this combination of host and soils *G. mosseae* may have had a competitive advantage. With the increase in the concentration of fungicides, the number of *G. mosseae* chlamydospores increased correspondingly. This may be due to the suppression of competitive, nonmycorrhizal soil organisms which permitted greater sporulation of *G. mosseae*. A positive correlation between spore number and percentage of root infection has been observed after fungicidal application[77,78] but not by Stewart and Pfleger.[47]

Metalaxyl, a systemic fungicide, is highly effective for the control of oomycetes; its incorporation into maize field soil naturally infected with VAM in the greenhouse at 1.4 and 2.9 mg kg^{-1} increased VAM infection from 57% to 62 and 72%, respectively, after 30 days. When *G. fasciculatus* inoculum was added to the soil, VAM infection increased by 16% but metalaxyl had no effect. Metalaxyl at the rate of 9 kg ha^{-1} had no adverse effect on vesicle number and plant growth, but mycelial colonization and chlamydospores were higher than in the *G. etunicatus* control.[46]

Studies were conducted on the growth and development of mycorrhizae in Fraser fir after soil application of metalaxyl at 1.1 kg a.i. ha^{-1}. Seedling root and shoot dry weights were significantly greater than in controls and mycorrhizal incidence was greater and onset was earlier in the metalaxyl-treated trees. The seedlings from plots that previously had 75 to 100% mycorrhizal incidence decreased to 25%, suggesting that nonmycorrhizal roots may be more susceptible to soil-borne pathogens.[80]

Two spring applications of chloroneb (0.8 a.i./m^2) did not reduce mycorrhizal development in the field.[58] Also, chloroneb resulted in the smallest numerical reduction of mycorrhizal development as compared to benomyl and triadimefon tested by the authors. This agrees with previous findings on ethazole[81] and metalaxyl,[46] which also have activity against oomycetes but do not appear to be detrimental to mycorrhizal fungi. This indicates that it may therefore be possible to selectively control pythiaceous fungi without substantially reducing mycorrhizal development. It is further suggested that studies involving a wider range of rates and application schedules are needed before the effects of fungicides on mycorrhizal fungi can be compared adequately.

Application of systemic fungicides in soil at the operational rates reduced the VAM in soil in different crops: ethazoletridiazole (sudangrass); thiophanate-methyl (peas); benomyl (wheat, bean, onion, soybean, citrus, barley); tridemorph, triforine, ethirimol, and chloraniformethan (wheat); pyroxychlor (pea); ethazole (lettuce); thiabendazole (wheat, bean, citrus); thiophanate (wheat, onion); triadimefon (wheat, lettuce); and carboxin (bean). VAM were promoted by chloroneb (lettude), efosite-AL (lettuce), metalaxyl (citrus), ethazole (sudangrass), and thiabendazole (citrus).[56]

IV. EFFECT OF FUNGITOXICANTS ON PHOSPHATE ACCUMULATION BY MYCORRHIZAE

VAM have been demonstrated to increase plant growth in soil and utilization of less available forms of phosphorus. Gray and Gerdmann[42] observed a 16-fold reduction of ^{32}P uptake in 12-week-old mycorrhizal onions when PCNB was applied 48 hr before ^{32}P application. Similarly, Hirrel and Gerdmann[82] found that in 10-week-old onion plants PCNB (applied 2 and 5 days before ^{14}C-glucose injection into the soil) inhibited ^{14}C translocation through mycorrhizal hyphae. PCNB prevented mycorrhizal formation in maize[24] by *Endogone fasciculata* and greatly reduced phosphate uptake and accumulation by mycorrhizal roots, but did not significantly affect phosphate accumulation by mycorrhizal roots.

Benomyl and thiophanate-methyl soil drenches reduced the phosphate uptake of inoculated onion and strawberry plants grown in irradiated soil, but not in unsterile soil.[52] Although the results were variable, the possibility of measuring short-term ^{32}P uptake by suffusing parts of the soil using a muslin wick with a radioactive solution of approximately the same concentrations of phosphate as the soil solution seems viable. Its success depends on a reasonably uniform root distribution in the pot and the retention of an equilibrium P concentration little changed from that normally present in the soil solution.[52] Since it is capable only of establishing what proportion of P uptake is lost when endophyte activity is inhibited, and not obtaining an accurate measurement of total uptake, however, the technique may be useful but obviously it requires further work.

Studies on the effects of soil fungitoxicants on the development of VAM and phosphate uptake in wheat have been carried out.[83] The results clearly showed that VAM infection is considerably influenced by the soil application of fungicides. The greatest effects were observed with PCNB (100 ppm) and thiram (100 ppm) which also interfered, in varying degrees, with the growth and phosphate uptake of the plants. It was further observed that the toxic effects of fungitoxicants, particularly PCNB and thiram, persist long enough to upset the delicate symbiotic balance. Benomyl (25 and 50 ppm) and captan (50 and 100 ppm) also showed toxic effects in varying degrees over controls.

V. FUTURE OUTLOOK

Studies relating the effects of fungicides to mycorrhizal activity at the root surface are of growing interest, because it is in this zone of intense microbial activity that pathogens and/ or symbionts of either disease or symbiosis must penetrate before establishing infection. Apparently, side effects on VAM are more permanent than similar effects of ectomycorrhizal fungi, perhaps because the recolonization of amended soil by ectomycorrhizal fungi is higher than soil with VAM. Also, information is scarce on the importance of the side effects of fungicides on mycorrhizae with respect to disease development or reduction in yield.

The repeated use of soil fungicides in farming systems merits careful consideration. Some fungicides are reported to induce instability of fungal pathogens. Since VAM symbionts may become important in the biological control of plant pathogens,[84] the introduction and management of mycorrhizal endophytes could be used to the advantage of the crop. This is of practical significance, since any interference with mycorrhizal development may ultimately have a depressive effect on plant growth and development. Nutrient deficiency symptoms on plants have also been observed following soil fumigation.[82,83]

There is a wide spectrum of fungitoxicants in common use to reduce detrimental effects by soil-borne pathogens. Their use may also reduce mycorrhizal infection from the indigenous soil population. De Bertoldi et al.[27] showed that repeated applications of toxicants decreased the growth of onion seedlings and these losses were associated with differing effects on soil microbes.

Plants in fungicide-treated soil in the field would probably develop eventually a mycorrhizal association, depending to a great extent on the biological properties of the toxicant, the degree of volatility, the speed of diffusion, and the method of application. Mycorrhizal fungi might resume growth quickly following the degradation of fungicides such as botran or terraclor, but might not do so after exposure to biocides that are strongly fungicidal. Fumigants with relatively high vapor pressure, such as vorlex, vapam, and mylone, potentially are able to move over greater distances in the soil than are compounds with little if any volatility. Following band application of toxicants, the roots could grow both horizontally and vertically from the treated zone into untreated soil and mycorrhizal development might be delayed longer, since the roots could reach untreated soil only by vertical growth. Thus, the least interference with mycorrhizae theoretically would follow narrow-band application of nonvolatile fungistats that are relatively nonspecific for endotrophic fungi.

REFERENCES

1. **Mosse, B., Stribley, D. P., and Letacon, F.,** Ecology of mycorrhizae mycorrhizal fungi, *Adv. Microb. Ecol.,* 5, 137, 1981.
2. **Melin, E.,** Some effects of forest tree roots on mycorrhizal basidiomycetes, in *Symbiotic Associations,* Nutman, P. S. and Mosse, B., Eds., The University Press, Cambridge, 1963, 125.
3. **Slankis, V.,** Formation of ectomycorrhizae of forest trees in relation to light, carbohydrates and auxins, in Mycorrhizae, Hacskaylo, E., Ed., Forest Service Misc. Publ. 1189, U.S. Department of Agriculture, Washington, D.C., 1971, 151.
4. **Bakshi, B. K.,** Mycorrhiza and Its Role in Forestry, PL-480 Project Report, Indian Forest Research Institute, Dehra Dun, India, 1974.
5. **Gerdmann, J. W.,** Fungi that form vesicular-arbuscular type of mycorrhiza, in Mycorrhizae, Hacskaylo, E., Ed., Forest Service Misc. Publ., 1189, U.S. Department of Agriculture, Washington, D.C., 1971, 9.
6. **Mosse, B.,** Advances in the study of vesicular-arbuscular mycorrhiza, *Annu. Rev. Phytopathol.,* 11, 171, 1973.
7. **Marx, D. H. and Rowan, S. J.,** Fungicides influence growth and development of specific ectomycorrhizae on loblolly pine seedlings, *For. Sci.,* 27, 167, 1981.
8. **Mikola, P.,** Application of mycorrhizal symbiosis in forestry practice, in *Ectomycorrhizae — Their Ecology and Physiology,* Marks, G. C. and Kozlowski, T. T., Eds., Academic Press, New York, 1973, 383.
9. **Bowen, G. J. and Theodorou, C.,** Growth of ectomycorrhizal fungi around seeds and roots, in *Ectomycorrhizae — Their Ecology and Physiology,* Marks, D. C. and Kozlowski, T. T., Eds., Academic Press, New York, 1973, 107.
10. **Marx, D. H.,** Ectomycorrhizal fungus inoculations: a tool for improving forestation practices, in *Tropical Mycorrhizae Research,* Mikela, P., Ed., Oxford University Press, London, 1980, 13.
11. **Kawai, M. and Ogawa, M.,** Studies on the artificial reproduction of *Tricholoma mastsutake* (S. Ito et Imai) Sing. V. Effects of some chemicals on the growth of mycorrhizal and non-mycorrhizal soil fungi, *Trans. Mycol. Soc. Jpn.,* 18, 391, 1977.
12. **Sobotka, A.,** Vliv aplikace nekterych biocidu v lesnich skolkach nakratke bocai koreny jednoletych semenacku borovice lesni, *Vysk. Ustav Lesn. Hospod. Mysliv.,* 36, 63, 1968.
13. **Tiefenbrunner, F.,** Mycelgewichtszunahme an Mycorrhizapilzen unter Einwirkung von Fungiziden in vitro, *Z. Pilzkd.,* 38, 105, 1972.
14. **Laiho, O. and Mikola, P.,** Studies on the effect of some eradicants on mycorrhizal development in forest nurseries, *Acta For. Fenn.,* 77, 1, 1964.
15. **Sobotka, A.,** Die Testung des Einflußes von Pestiziden auf die Mykorrhiza-Pilze in Waldboden, *Zentralbl. Bakteriol. Parasitenkd. Infektionskr. Hyg. Abt. 2,* 152, 723, 1970.
16. **Cudlin, P., Mejstrik, V., and Sasek, V.,** The effect of the fungicide dithane M-45 and the herbicide gramoxone on the growth of mycorrhizal fungi *in vitro, Ceska Mykol.,* 34, 191, 1980.
17. **Edgington, L. V., Khew, K. L., and Barron, G. L.,** Fungitoxic spectrum of benzimidazole compounds, *Phytopathology,* 61, 42, 1971.
18. **Kelley, W. D. and Snow, G. A.,** Non-target effect of benodanil on mycorrhizal fungi of loblolly pine, *Int. Congr. Plant Pathol.,* (Abstr.), Munich, 1978, 176.

19. **Peterson, G. W. and Smith, R. S.,** Forest Nursery Diseases in the United States, Agric. Handb. No. 470, U.S. Department of Agriculture, Washington, D.C., 1975.

20. **Bakshi, B. K. and Dobriyal, N. D.,** Effect of fungicides to control damping off on development of mycorrhiza, *Indian For.,* 96, 701, 1970.

21. **Hong, L. T.,** Mycorrhizal shoot roots development on *Pinus caribea* seedlings after fungicidal treatment, *Malays. For.,* 39, 147, 1976.

22. **Theodorou, C. and Skinner, M. F.,** Effect of fungicide on seed inocula of basidiospores of mycorrhizal fungi, *Austr. For. Res.,* 7, 53, 1976.

23. **Pawuk, W. H., Ruehle, J. L., and Marx, D. H.,** Fungicide drenches affect ectomycorrhizal development on container grown *Pinus palustris* seedlings, *Can. J. For. Res.,* 10, 61, 1980.

24. **Nesheim, O. N. and Linn, M. B.,** Deleterious effect of certain fungitoxicants on the formation of mycorrhiza on corn by *Endogone fasciculata* and on corn root development, *Phytopathology,* 59, 297, 1969.

25. **Bollen, G. J. and Fuchs, A.,** On the specificity of the in vitro and in vivo antifungal activity of benomyl, *Neth. J. Plant Pathol.,* 76, 299, 1970.

26. **Agnihotri, V. P.,** Persistence of captan and its effect on microflora respiration and nitrification of a forest nursery soil, *Can. J. Microbiol.,* 17, 377, 1971.

27. **de Bertoldi, M., Giovannetti, M., Griselli, M., and Rambelli, A.,** Effects of soil application of benomyl and captan on the growth of onions and the occurrence of endophytotic mycorrhizas and rhizosphere microbes, *Ann. Appl. Biol.,* 86, 111, 1977.

28. **Spokes, J. R., MacDonald, R. M., and Hauman, I. S.,** Effects of plant protection chemicals on vesicular-arbuscular mycorrhizae, *Pestic. Sci.,* 12, 346, 1981.

29. **Ivory, M. H.,** Mycorrhizal studies on exotic conifers in West Malaysia, *Malays. For.,* 38, 149, 1975.

30. **Kelley, W. D.,** Evaluation of systemic fungicides for the control of *Cronartium querocuum* f. sp. *fusiforme* on loblolly pine seedlings, *Plant Dis.,* 64, 773, 1980.

31. **Kelley, W. D.,** Control of *Cronartium fusiforme* on loblolly pine seedlings with the experimental systemic fungicide, benodanil, *Plant Dis. Rep.,* 62, 595, 1978.

32. **Moore-Parkhurst, S. M. and Englander, L.,** Influence of nursery practices soil amendments on Rhododendron mycorrhizae, 5th North American Conf. Mycorrhizae, Quebec, 1981, 43.

33. **Cline, M. L., France, R. C., Berry, S. G., and King, I. L.,** Growth performance of mycorrhizal loblolly pine seedlings treated with the fungicide beyleton, 5th North American Conf. Mycorrhizae, Quebec, 1981, 61.

34. **Mukerji, K. G., Sabharwal, A., Kochar, B., and Ardey, J.,** Vesicular-arbuscular mycorrhiza: concepts and advances, in *Progress in Microbial Ecology,* Mukerji, K. G., Agnihotri, V. P., and Singh, R. P., Eds., Print House, Lucknow, India, 1984, 489.

35. **Jalali, B. L. and Domsch, K. H.,** Effect of systemic fungitoxicants on the development of endotrophic mycorrhiza, in *Endomycorrhizae,* Sanders, F. E., Mosse, B., and Tinker, P. S., Eds., Academic Press, New York, 1975, 619.

36. **Anderson, J. R.,** Pesticides effects on nontarget soil microorganisms, in *Pesticides Microbiology,* Hill, I. R. and Wright, S. J. L., Eds., Academic Press, New York, 1978, 313.

37. **Burpee, L. L. and Cole, H.,** The influence of alachlor, trifluralin and diazinon on the development of endogenous mycorrhizal in soybeans, *Bull. Environ. Contam. Toxicol.,* 19, 191, 1978.

38. **Smith, T. F.,** Some effects of crop protection chemicals on the distribution and abundance of vesicular-arbuscular endomycorrhizae, *J. Austr. Inst. Agric. Sci.,* 44, 82, 1978.

39. **Domsch, K. H.,** Soil fungicides, *Annu. Rev. Phytopathol.,* 2, 293, 1964.

40. **Backman, P. A. and Clark, E. N.,** Peanut leaf spot research in Alabama, 1970—1976, *Auburn Univ. Agric. Exp. Stn. Bull.,* p. 1, 1977.

41. **El-Giahmi, A. A., Nicolson, T. H., and Daft, M. J.,** Effects of fungal toxicants on mycorrhizal maize, *Trans. Br. Mycol. Soc.,* 67, 172, 1976.

42. **Gray, L. E. and Gerdmann, J. W.,** Uptake of phosphorus-32 by vesicular-abuscular mycorrhizae, *Plant Soil,* 30, 415, 1969.

43. **Menge, J. A., Johnson, E. L. V., and Minassian, V.,** Effect of heat treatment and three pesticides upon the growth and reproduction of the mycorrhizal fungus *Glomus fasciculatus, New Phytol.,* 82, 473, 1979.

44. **Sutton, J. C. and Sheppard, B. R.,** Aggregation of sand-dune soil by endomycorrhizal fungi, *Can. J. Bot.,* 54, 326, 1976.

45. **Nemec, S.,** Effects of 11 fungicides on endo-mycorrhizal development in sour orange, *Can. J. Bot.,* 58, 522, 1980.

46. **Nemec, S. and Bannon, J. H. S.,** Response of *Citrus aurantium* to *Glomus etunicatus* and *G. mosseae* after soil treatment with selected fumigants, *Plant Soil,* 53(3), 351, 1979.

47. **Stewart, E. L. and Pfleger, F. L.,** Influence of fungicides on extinct *Glomus* species in association with field grown peas, *Trans. Br. Mycol. Soc.,* 69, 318, 1977.

48. **Harvey, I. C., Fletcher, L. R., and Emme, L. M.,** Effects of several fungicides on the Lolium endophytes in ryegrass plants, seeds and in culture, *N.Z. J. Agric. Res.,* 25, 601, 1982.

49. **Jalali, B. L. and Domsch, K. H.**, Effects of some fungitoxicants on the amino acid spectrum of wheat root exudates, *Phytopathol. Z.*, 90, 22, 1977.
50. **Ratanayake, M., Leonard, R. T., and Menge, J. A.**, Root exudation in relation to supply of phosphorus and its possible relevance to mycorrhizal formation, *New Phytol.*, 81(3), 543, 1978.
51. **Atilano, R. A. and van Gundy, S. D.**, Effects of some systemic, nonfumigant and fumigant nematicides on grape mycorrhizal fungi and citrus nematode, *Plant Dis. Rep.*, 63, 729, 1979.
52. **Boatman, N., Panget, D., Hayman, D. S., and Mosse, B.**, Effects of systemic fungicides on vesicular-arbuscular mycorrhizal infection and plant phosphate uptake, *Trans. Br. Mycol. Soc.*, 70(3), 443, 1978.
53. **Vyas, S. C.**, *Systemic Fungicides*, Tata McGraw-Hill, New Delhi, India, 1984.
54. **Rodriguez-Kabana, R. and Curl, E. A.**, Nontarget effects of pesticides on soilborne pathogens and disease, *Annu. Rev. Phytopathol.*, 18, 311, 1980.
55. **Trappe, J. M., Molina, R., and Castellano, M.**, Reactions of mycorrhizal fungi and mycorrhiza formation to pesticides, *Annu. Rev. Phytopathol.*, 22, 331, 1984.
56. **Menge, J. A.**, Effect of soil fumigants and fungicides on vesicular-arbuscular fungi, *Phytopathology*, 72, 1125, 1982.
57. **Nemec, S.**, Effects of five fungicides on endomycorrhizal development in sour orange, *Phytopathology*, 69(1), A8, 1979.
58. **Rhodes, L. H. and Larsen, P. O.**, Effect of fungicides on mycorrhizal development of creeping bentagrass, *Plant Dis.*, 65, 145, 1981.
59. **Powell, W. M., Hendrix, F. F., and Marx, D. H.**, Chemical control of feeder root necrosis of pecans caused by Pythium species and nematodes, *Plant Dis. Rep.*, 52, 577, 1968.
60. **Pitcher, R. S., Way, D. W., and Savory, B. M.**, Specific replant diseases of apple and cherry and their control by soil fumigation, *J. Hortic. Sci.*, 41, 379, 1966.
61. **Filer, T. H. and Toole, E. R.**, Effect of methyl bromide on mycorrhizae and growth of sweet gum seedlings, *Plant Dis. Rep.*, 52, 483, 1968.
62. **Martin, J. P., Baines, R. C., and Page, A. L.**, Observations on the occasions of temporary growth inhibition of citrus seedlings following heat fumigation treatment of soil, *Soil Sci.*, 85, 175, 1963.
63. **Kleinschmidt, G. D. and Gerdemann, J. W.**, Stunting of citrus seedlings in fumigated nursery soils related to the absence of endomycorrhizae, *Phytopathology*, 62, 1447, 1972.
64. **Tommerup, I. C. and Briggs, G. G.**, Influence of agricultural chemicals on germination of vesicular-arbuscular endophyte spores, *Trans. Br. Mycol. Soc.*, 76, 326, 1981.
65. **Kelley, W. D. and Rodriguez-Kabana, R.**, Effects of sodium azide and methyl bromide on soil bacterial populations, enzymic activities and other biological variables, *Pestic. Sci.*, 10, 207, 1979.
66. **Marx, D. H., Bryan, W. C., and Cordell, C. E.**, Growth and ectomycorrhizal development of pine seedlings in nursery soils infested with the fungal symbiont *Pisolithus tinctorius*, *For. Sci.*, 22, 91, 1976.
67. **Bird, G. W., Rich, J. R., and Glover, S. U.**, Increased endomycorrhizae of cotton roots in soil treated with nematicides, *Phytopathology*, 64, 48, 1974.
68. **O'Bannon, J. H. and Nemec, S**, Influence of soil pesticides on vesicular-arbuscular mycorrhizae in a citrus soil, *Nematropica*, 8, 56, 1978.
69. **Menge, J. A., Johnson, E. L. V., and Minassian, V.**, Effect of heat treatment and three pesticides upon the growth and reproduction of the mycorrhizal fungus *Glomus fasciculatus*, *New Phytol.*, 82, 473, 1976.
70. **Bailey, J. E. and Safir, G. R.**, Effect of benomyl on soybean endomycorrhizae, *Phytopathology*, 68, 1810, 1978.
71. **Bollen, G. J.**, A comparison of the in vitro antifungal spectra of thiophanates and benomyl, *Neth. J. Plant Pathol.*, 78, 55, 1972.
72. **Fuchs, A., van Den Berg, G. A., and Davidse, L. C.**, A comparison of benomyl and thiophanates with respect to some chemical and systemic fungitoxic characteristics, *Pestic. Biochem. Physiol.*, 2, 191, 1972.
73. **Austin, D. J. and Briggs, G. G.**, A new extraction method for benomyl residues in soil and its application in movement and persistence studies, *Pestic. Sci.*, 7, 201, 1976.
74. **Baude, F. J., Pease, H. L., and Holt, R. F.**, Fate of benomyl on field soil and turf, *J. Agric. Food Chem.*, 22, 413, 1974.
75. **Pfleger, F. L. and Stewart, E. L.**, The influence of fungicides on endemic population of *Glomus* spp. in association with field grown peas, *Proc. Phytopathol. Soc.*, 3, 274, 1976.
76. **Knauss, J. F.**, Pyroxychlor, a new systemic fungicide for control of *Phytophthora palmicora*, *Plant Dis. Rep.*, 58, 1100, 1974.
77. **Hayman, D. S.**, Endogone spore numbers in soil and vesicular-arbuscular mycorrhizae in wheat as influenced by season and soil treatment, *Trans. Br. Mycol. Soc.*, 54, 53, 1970.
78. **Redhead, J. F.**, Mycorrhizae association in some Nigerian forest trees, *Trans. Br. Mycol. Soc.*, 51, 377, 1968.
79. **Groth, D. E. and Martinson, C. A.**, Vesicular-arbuscular mycorrhizal infection of maize as affected by rotation, fertilization and metalaxyl, *Phytopathology*, 71, 221, 1981.

80. **Bruck, R. I., Kenerley, C. M., and Grand, L. F.,** The effects of metalaxyl on growth and mycorrhizal incidence of Fraser Fir, *Phytopathology*, 72, 335, 1982.
81. **Menge, J. A., Munnecke, D. E., Johnson, E. L. V., and Carnes, D. W.,** Dosage response of the vesicular-arbuscular mycorrhizal fungi *Glomus fasciculatus* and *G. constrictus* to methyl bromide, *Phytopathology*, 68, 1368, 1978.
82. **Hirrel, M. C. and Gerdmann, J. W.,** Enhanced carbon transfer between onion infected with vesicular-arbuscular mycorrhizal fungus, *New Phytol.*, 83, 731, 1979.
83. **Jalali, B. L.,** Effect of soil fungitoxicant on the development of VA-mycorrhiza and phosphate uptake in wheat, in *Soil-Borne Plant Pathogens*, Schippers, B. and Gams, W., Eds., Academic Press, New York, 1979, 525.
84. **Schonbeck, F.,** Endomycorrhiza in relation to plant disease, in *Soil-Borne Plant Pathogens*, Schippers, B. and Gams, W., Eds., Academic Press, New York, 1979, 271.

Chapter 13

ALGAE

I. INTRODUCTION

The role of microorganisms in the various component cycles contributing to the fertility of soil has been studied for more than a century and some aspects have been very well researched and documented. Algae are microscopic photosynthetic organisms; they range from unicellular to colonial and filamentous types and are part of the aquatic phytoplankton. Algae typically inhabit aquatic environments, soil surfaces, and other exposed locations. They also live in association with fungi (lichens) and plants (ferns and cycads); most algae are ecologically beneficial but some are pathogenic to plants. Blue-green algae (BGA) or cyanobacteria are important in soil and associated environments such as irrigation and drainage systems. They can colonize exposed or barren land and maintain and rejuvenate soil fertility through carbon fixation, synthesis of substances for plant growth, and binding of soil particles. Their most important activity is the fixation of atmospheric nitrogen. Inoculation of BGA in paddy fields is known to increase yield. These organisms are major contributors of photosynthesized organic materials to the carbon cycle. They aid in the binding of soil particles, thus reducing or preventing erosion, and also increase the moisture-holding capacity of soil. Some algae (principally BGA) are also able to fix atmospheric nitrogen to a sufficient extent to markedly increase the nitrogen content of soil. BGA comprise the groups of microorganisms having an oxygen-evolving photosynthetic system. Many genera are aerobic nitrogen fixers and some others are known to grow on molecular nitrogen in the anaerobic and symbiotic state. The combination of oxygen-evolving photosynthesis with the oxygen-sensitive nitrogen-fixing system is rare among microbes. In addition, BGA have the ability to invade otherwise uninhabited sections of the environment. The combined global nitrogen of BGA, both free-living and symbiotic forms (Azolla, Cycas), is important when future demands on protein in developing countries are considered. BGA, with both oxygenic photosynthesis and the oxygen-sensitive nitrogen-fixing system, have a division of labor: the vegetative cells carry out the photosynthetic process and the heterocysts are sites of nitrogen fixation in aerobic conditions. Heterocysts have a high reducing state and isotopic studies confirmed that they are the sites of nitrogen fixation of BGA.[1] The number of active heterocysts in BGA filaments can be taken as an index of the nitrogen fixation rate in media in which the available combined nitrogen is below the threshold level that allows the activity and differentiation of heterocysts.[2]

The extensive use of fungicides has greatly improved many facets of agriculture, but at the same time has created many unintended problems. The effect of fungicides on nontarget organisms, either direct or indirect, is a significant problem. A very important group of nontarget organisms is algae. Interference with the normal activities of algae in soil may be expected to have potentially serious consequences on the overall productivity of soil. Since algae inhabit the environment which fungicides enter directly or indirectly, the potential nontarget effects on the biological activity of algae must be considered. Several classical reviews have appeared on the interactions of fungicides and algae.[3-5]

II. NONSYSTEMIC FUNGICIDES

A. Copper

Copper oxychloride (Fytolon) was found to be nontoxic to *Aulosira* at 100 ppm, whereas the same concentration caused a 58.3 and 39.7% reduction in the growth of *Westiellopsis* and *Calothrix,* respectively. The growth of *Nostoc* and *Tolypothrix* was reduced by 18.3 and 6.25%, respectively, by 500 ppm of the fungicide.[6]

B. Quinones

Quinones are generally found to be more toxic to algae at lower concentrations than other fungicides in vitro as well as in vivo. Dichlone was a particularly potent inhibitor and appeared to be selectively toxic to bloom-forming species, all of which were killed by 5 ppm of the fungicide.[7] However, BGA, green algae, and diatoms were comparatively resistant to 15 ppm of this fungicide. In further studies, it was observed that dichlone was toxic to *Microcystis aeruginosa* and spray applications of a pond with dichlone, resulting in a water concentration of from 30 to 55 ppm, clumped and effectively killed BGA without any concomitant observable harmful effects on green algae.[8] Several quinone fungicides (menadione, benzoquinone, naphthoquinone, chloronil, and dichlone) were toxic to *Chlorella pyrenoidosa,* causing a reduction in cell number, chlorophyll concentration, and oxygen evolution.[9] However, menadione and benzoquinone had little effect on short-term oxygen evolution. In long-term experiments chloronil depressed oxygen production; dichlone, naphthoquinone, and chloronil also affected chlorophyll destruction in ways thought to be unrelated to the decreases in oxygen production. Cell numbers were found to be severely affected by dichlone and chloronil and only moderately so by other compounds.[10] Chloronil and dichlone checked the labeled carbon dioxide fixation of *C. pyrenoidosa* and caused an increase in the proportion of labeled carbon in sugar and glycine, accompanied by a reduction in labeled carbon in lipids and glutamic acid.[11] Sikka et al.[12] observed similar results. A reduction in acetate photometabolism accompanied by a decrease in labeled lipids with a corresponding increase in the proportion of labeled carbon in glutamic and citric acid was observed. Dichlone and chloronil application caused an increase in the efflux of labeled carbon in intracellular material from *C. pyrenoidosa* cells and the rate of loss of labeled carbon increased with the increase in fungicide concentration.[13] Rapid and irreversible changes in cell permeability were observed; the treated cells when exposed to light or under aerobic conditions released more labeled material than controls. Ultrastructural changes have also been observed in *Chlorella.* The cell membrane shrank from the cell wall and exhibited much invagination and the outer wall of the chloroplast appeared wavy. The cytoplasm in the treated cell was less structured and more homogeneous than the control cells.[14]

A range of studies were conducted on the effects of quinone fungicides on several algae.[15,16] Dichlone was toxic at 2 ppm to the growth of *Cylindrospermum licheniforme, M. aeruginosa, Gomphonema parvulum,* and *Nitzschia palea* but innocuous to *Chlorella pyrenoidosa.*[15,16] Later workers reported that dichlone was more toxic to BGA than green algae. Dichlone inhibited the growth of *Anabaena* sp. at 10 ppm.[16] Dichlone (0.1 to 0.5 ppm) had little or no effect on the filamentous algae and *Chara* sp.[17] The minimum toxic concentration of dichlone for *Anacystis nidulans* was 150 ppm.[18] Dichlone at 10 ppb inhibited the growth of *Anabaena* sp.[19] It was concluded that quinone fungicides caused a severe reduction in cell numbers; dichlone was most toxic and chloronil was moderately so. These fungicides caused irreversible damage to algal cells and resulted in an efflux of labeled carbon from the intercellular material of the treated cell. It was therefore concluded that quinones altered cell membrane permeability. The primary site of this was the cell membrane which might change the configuration of sulfhydryl groups.[9,13]

C. Dithiocarbamates

This group is represented by several important fungicides which are used for seed treatment (thiram), foliar application (zineb, ziram, and ferbam), soil treatment (vapam and nabam), and fruit application (ferbam). Zineb and ziram at 0.1 to 15 ppm affected the growth of *Scenedesmus quandricauda* and *Kirchveriella lumaris;* ziram was more toxic to these algae than zineb.[20,21] At 1.25 ppm zineb had little effect on the survival of *Euglena gracilis*.[21] A mixture of copper, zineb, and maneb was nontoxic to *Navicula osterearia* and *Phaeodactylum tricornutum*.[22,23] In the experiment, 30 different species of algae were treated with varying doses of ziram and it was found that the diatoms were most sensitive to ziram at all the concentrations used. Ferbam at 0.5 ppm eradicated 10% of the branched filamentous algae in two ponds; however, only 60 to 80% of *Cladospora* in the two ponds were killed by the fungicide at concentrations of 1.5 to 3 ppm.[17] Zinc dimethyldithiocarbamate (0.1 to 15 ppm) almost completely inhibited the growth of several algae at higher concentrations.[21] Similarly, a copper and ziram mixture inhibited the growth of two BGA.[24] Ziram was generally toxic to five unicellular algae, but not to *Scenedesmus* at concentrations of 0.25 to 2 ppm.[16] Among 30 species of algae the diatoms were affected at all the concentrations of ziram and BGA were more sensitive than green algae.[25] Ziram had a stimulatory effect on the growth of both *Cylindrospermum* and *Nostoc* at 0.01 ppm in carbon and nitrogen media; the sublethal range for *Cylindrospermum* was found to be 0.05 to 0.13 ppm, whereas for *Nostoc* it was 0.1 to 0.4 ppm in a carbon-nitrogen medium. The complete lethal dose in a carbon-nitrogen medium was 0.17 ppm for *Cylindrospermum* and 0.5 ppm for *Nostoc*.

Chlorella sorokiniana was sensitive to most of the fungicides tested on soil thin-layer plates;[27] 27 strains of BFA were treated with mancozeb and the maximum concentration allowing growth ranged from 5 to 50 ppm.[28,29] Zineb was lethal to some strains of *Anabaena* and *Nostoc* spp., even at the lowest concentrations used, but some strains grew well at 50 ppm, while *Tolypothrix tenuis* and *Aulosira fertilissima*, frequently introduced in paddy fields, tolerated high concentrations of zineb.[27] However, zineb did not inhibit the growth of 8 strains of *Anabaena* and 11 strains of *Nostoc* and *Aulosira* at 20 kg ha^{-1} and of *T. tenuis* and *Anacystis nidulans* at 200 kg ha^{-1}. Thus, the latter two genera are more resistant than the former three genera.[28] A reduction in the number of algal propagules was reported after soil application of mancozeb.[30] Zineb at 1 ppm inhibited the development of *Nostoc muscorum*[31] but not the growth. Zineb was found to have a sublethal range of 1 to 15 ppm with *Cylindrospermum* sp. and 5 to 25 ppm with *Nostoc muscorum* in a carbon-nitrogen medium. The complete toxic levels in this medium were 25 and 30 ppm for *Cylindrospermum* sp. and *Nostoc* sp., respectively. On an agar plate *Cylindrospermum* tolerated 5 ppm and *Nostoc* 15 ppm.[26]

Maneb was toxic to *E. gracilis* cells at 125 ppm and toxicity increased with the increase in the exposure time.[32] Similar toxic effects of maneb were also observed by Parachiv et al.[24] Maneb was more inhibitory than zineb to the growth of *E. gracilis*[32] but less so to fungi.[33]

Nabam and vapam were comparatively more toxic to fungi than other members of this group.[33] Palmer and Maloney[16] reported that nabam was nontoxic to algae and inhibited only flagellar movement. The fungicide affected the rate of photoassimilation of acetate of *N. muscorum* and completely checked the growth of a dozen strains of algae.[31,34] Later, Moore[32] exposed a culture of *E. gracilis* to 0.01 to 10 ppm and found that it was toxic at all concentrations tested; however, it was more toxic in light than in darkness. It was also observed that nabam gradually dissipated from the culture. The toxicity of nabam increased with the increase in the concentration and incubation time. Nabam at the concentration of 1 ppm adversely affected the growth of *Chlorella* and *Anabaena* spp.[35] Lazaroff[36] reported that nabam proved to be inhibitory to the growth of *Dunaliella*, *Phaeodactylum*, and *Monchrysis* and to *Chlorella* and *Proloccus* at 10 ppm in salt water. Resistant forms of freshwater algae were also reported.[36] Vapam, which may be used as a fungicide, nematicide, or herbicide, was found to have no influence on the growth of *N. muscorum*.[31]

Thiram, a broad-spectrum seed treatment fungicide, is used extensively for the control of externally seed-borne diseases and also to some extent for soil-borne diseases. Lindahl[37] conducted a series of experiments on the effects of thiram on the growth and photosynthesis of *Enteromorpha linza* and observed that the fungicide is toxic; however, the toxicity can be reversed by removing the fungicide.[37] Addition of metal ions or amino acids also caused a reduction in the amount of inhibition.[38] The inhibition was rapid when the fungicide was reapplied. *Enteromorpha* sp. degraded thiram and the degradation was greater in light than dark, indicating participation of the photosynthetic apparatus.[38] Lindahl[39] found that thiram was toxic to several marine and freshwater algae. Thiram had little or no effect on *S. quadricaude* or *C. pyrenoidosa*[20] up to 14 ppm. Moore[32] also found little effect of thiram on the survival of *Euglena gracilis* cells following a short-term exposure. Growth of two unicellular coccoid *(Aphanotheca* spp.) and two filamentous *(Anabaena variabilis* and *Nostoc* sp.) BGA was inhibited by thiram at concentrations above 50 ppm.[40]

D. Captan

Lazaroff[36] and Moore[31] reported that captan at 1 ppm inhibited the development but not the growth of freshwater algae and BGA; however, low concentrations of captan inhibited the growth of an axenic culture of green algae and BGA.[34] Captan strongly inhibited the photoautotrophic growth of several strains of *Chlorella* sp. at 10 to 100 ppm.[41] Only three strains showed resistance and two strains of *Scenedesmus* tolerated captan up to 50 ppm without being inhibited; however, resistance in *Chlorella* was not observed to be species-specific.[41] Captan at 0.01 to 8 ppm caused a reduction in growth of up to 40% in freshwater algae.[42]

E. Mercury Fungicides

Ethyl mercury phosphate was lethal to all marine phytoplankton species, even at 60 ppb in culture.[43] Mercury fungicides reduced growth and photosynthetic activity in the marine diatom *Nitzchia delicatissima* and also in the natural population of fresh water at 1 ppb.[44] Venkataraman and Rajyalakshmi[28,29] investigated the effects of fungicides on nitrogen-fixing BGA and found that 28 strains of *Anabaena* were inhibited at 0.01 to 100 ppm Ceresan; *T. tenuis* and *Aulosira fertilissima* tolerated high concentrations of Ceresan but one strain was inhibited by levels higher than 1 ppm. *C. sorokiniana* was sensitive to most of the fungicides tested on soil thin-layer chromatography plates.[27]

F. Miscellaneous

Pentachloronitrobenzene (PCNB) is used extensively for the control of *Sclerotium* and *Rhizoctonia* spp. responsible for many stem and root rot diseases in several economically important crops. *Westiellopsis* sp., *Aulosira* sp., *Nostoc* sp., *Tolypothrix* sp., and *Calothrix* sp. tolerated the highest concentration (1000 ppm) of PCNB; at 500 ppm the reduction in the growth of *Aulosira* and *Calothrix* was 17.1 and 10.2%, respectively, over controls; in *Westiellopsis* and *Nostoc* PCNB at 1000 ppm caused a 48.3 and 16.9% reduction in growth, respectively, over the control. *Tolypothrix* was not affected by 1000 ppm of PCNB.[6]

Captafol at 100, 500, and 1000 ppm caused a decrease in the growth of *Westiellopsis, Tolypothrix,* and *Nostoc* by 65, 35.9, and 1.3%, respectively.[45] Rovaral, a broad-spectrum fungicide, caused a reduction in the growth of *Westiellopsis, Aulosira,* and *Cylindrospermum* by 65, 4.2, and 2.5%, respectively, while at 1000 ppm it reduced the growth of *Nostoc* and *Tolypothrix* by 12.6 and 9.3%, respectively.[46]

When butachlor and HOE 2997 were incorporated into media and *Oscillatoria* and *Cylindrospermum* were cultured, reduction in the radial growth was observed. Butachlor was toxic to BGA, HOE 2997 completely inhibited the growth of *Oscillatoria* at 8 ppm, and 72% growth inhibition was noticed at 4 ppm whereas inhibition of *Cylindrospermum* was 100% at 2 ppm and 48% at 1 ppm.[47] Difolatan at a 100-ppm level caused a decrease in the growth of *Westiellopsis* by 65%; at 500 ppm it caused a decrease in the growth of *Tolypothrix* by 35.9% and at 1000 ppm it caused a decrease in the growth of *Nostoc* by 1.3%.[45]

III. SYSTEMIC FUNGICIDES

Benomyl, carbendazim, and thiophanates, members of the benzimidazole group, are broad-spectrum fungicides that are used extensively for the control of fungal diseases. Benomyl, carbendazim, and thiophanate-methyl were toxic to *Chlorella pyrenoidosa* at 0.34, 1.4, and 8.5 ppm, respectively.[48] *C. pyrenoidosa* tolerated concentrations of 100 to 1000 ppm of the fungicide ethirimol.[48] The algae degraded ethirimol in culture. Carboxin and its sulfone analog oxycarboxin were used for studying the effect on *C. pyrenoidosa* and it was observed that metabolism of ^{14}C-acetate was inhibited by 10^{-4} M carboxin, F 831, or oxycarboxin. However, photosynthesis was inhibited by 52% with 10^{-4} M carboxin but not with 10^{-4} M of F 831 or oxycarboxin; lower concentrations of carboxin were not inhibitory to photosynthesis.[49] The reason for the low sensitivity of most organisms to oxathiins is not known but may be related to the uptake of the compounds into cells and/or mitochondria.

Carbendazim at 500 ppm caused a decrease in the growth of *Nostoc* and *Calothrix* by 7.9 and 1.0%, respectively, while a 1000-ppm level caused a decrease in the growth of *Aulosira* and *Tolypothrix* by 5.7 and 3.2%, respectively.[45] Thiophanate-methyl at 50 ppm reduced the growth of *Calothrix* by 11.5% and 1000 ppm of this fungicide reduced by 31.4% the growth of *Westiellopsis*. At 300 ppm it reduced the growth of *Aulosira* and *Tolypothrix* by 12.8 and 14%, respectively; at 1000 ppm a 22.5% reduction in the growth of *Nostoc* was noted. Carbendazim was toxic to *Calothrix* sp. at 1 ppm, but at 50 ppm it caused a 5.8% increase in the growth of *Tolypothrix* and at 100 ppm a 5.6% increase in the growth of *Nostoc* over controls.[6] The growth of *Aulosira* was decreased by 21.4% at 100 ppm carbendazim. At 300 ppm carbendazim reduced the growth of *Westiellopsis*.[6]

IV. BIOLOGICAL NITROGEN FIXATION

Biological nitrogen fixation by prokaryotic organisms is an important step in the nitrogen cycle. The role of BGA, both free-living and symbiotic forms, should also be emphasized. According to one study, BGA fix 25 kg nitrogen per year per hectare, whereas other nonsymbiotic forms *(Azotobacter* and *Clostridium)* fix an insignificant amount of nitrogen — 0.026 to 0.22 kg nitrogen per year per hectare. The symbiont *Rhizobium* has a very high rate: 150 to 220 kg nitrogen per year per hectare.[50] Panigrahy[26] studied the nitrogen content in the medium where *Cylindrospermum* was incubated for 25 days in the presence of various dithiocarbamate fungicides. The medium had 1.0 mg N_2 at 0.1 ppm ziram, 0.6 mg N_2 at 15 ppm zineb, and 0.35 mg N_2 at 5 ppm mancozeb. In cultures of *Nostoc,* the N_2 content was 0.17 mg at 0.3 ppm ziram, 0.34 mg N_2 at 10 ppm zineb, and 1.4 mg N_2 at 10 ppm mancozeb.[51] The frequency of heterocysts of *Cylindrospermum* was not affected by any of the three fungicides (ziram, zineb, and mancozeb), but heterocyst activity was adversely affected.[26] The presence of the above-cited carbamates individually in a medium without a combined nitrogen source caused a decrease of the percentage of heterocysts in the filaments of *Nostoc*. However the number of heterocysts decreased along with the total cell number of both algae in the presence of fungicides.

V. FUTURE OUTLOOK

Generally, algae are not the target organisms of fungicide application, but they comprise an important segment of the ecosystem and thus may be affected. The lack of information on the possible synergistic effects of added fungicide surfactants is a matter of concern. The effects of fungicides on algae in soil may be considerably different from effects on the same algae grown in axenic cultures. Even if the fungicides are only 1% as effective in soil as in pure culture, they could still have rather dramatic effects on soil nutrition. In some instances, very low concentrations of fungicides would inhibit these beneficial organisms. Benomyl at low concentrations is known to significantly improve soil nutrition by stimulating BGA and at the same time reducing fungal contamination.

It has been established that algae show varied sensitivity towards systemic and nonsystemic fungicides. In order to fully understand the effects of various fungicides on the growth and reproduction of algae, detailed work and elaborate studies must be conducted in several areas. The effects of the combined application of various pesticides on artificially generated or natural communities of algae and other aquatic microorganisms require further study. Current research is limited to phytoplankton; future work should include the study of fungicidal effects on microscopic marine algae and the monitoring of the activity of algae, as well as on the population and its components. That fungicides/pesticides have an inhibitory effect on biological nitrogen fixation has been established through work with BGA. Fungicides, therefore, threaten the health of the soil. Soil organic matter would be greatly reduced in agricultural fields if the growth of cyanobacteria were limited as a result of agricultural fungicides. In addition, the possible effects on the stages in the reproductive cycle should be investigated, since gametes may be more susceptible than the purely metabolic, vegetative plant body. From available literature, it is apparent that the growth and activities of algae are adversely affected by commonly used agricultural fungicides.

REFERENCES

1. **Fogg, G. E., Stewart, W. D. P., Fry, P., and Walsby, A. E.,** The Blue Green Algae, *Academic Press,* New York, 1973.
2. **Tyagi, V. V. S.,** The heterocyst of blue green algae (Myxophyceae), *Biol. Rev.,* 50, 247, 1975.
3. **Ware, G. W. and Roan, G. G.,** Interaction of pesticides with aquatic microorganisms and plankton, *Residue Rev.,* 33, 15, 1970.
4. **Butler, G. L.,** Algae and pesticides, *Residue Rev.,* 66, 19, 1976.
5. **McCann, A. E. and Gullimore, D. R.,** Influence of pesticides on soil algal flora, *Residue Rev.,* 72, 11, 1979.
6. **Gangawane, L. V.,** Tolerance of nitrogen fixing blue green algae to Brassicol, Bavistin and Fytolon, *J. Indian Bot. Soc.,* 59, 157, 1980.
7. **Fitzgerald, G. P., Gorloff, G. C., and Skoog, F.,** Studies on chemicals with selective toxicity to blue-green algae, *Sewage Ind. Wastes,* 24, 888, 1952.
8. **Fitzgerald, G. P. and Skoog, F.,** Control of blue-green algae blooms with 2,3-dichloronaphthoquinone, *Sewage Ind. Wastes,* 26, 1136, 1954.
9. **Zweig, G., Hitta, J. E., and McMahon, R.,** Effect of certain quinones on $^{14}CO_2$ fixation by *Chlorella pyrenoidosa* Chick (Emerson strain), *Weed Sci.,* 16, 69, 1968.
10. **Frear, D. E. H.,** *Pesticides Index,* 4th ed., Pennsylvania State College, University Park, 1969.
11. **Zweig, G., Carroll, J., Tamas, I., and Sikka, H. C.,** Studies on effects of certain quinones. II. Photosynthetic incorporation of $^{14}CO_2$ by *Chlorella, Plant Physiol.,* 49, 385, 1972.
12. **Sikka, H. C., Carroll, J., and Zweig, G.,** Effect of certain quinone pesticides on acetate photometabolism and dark CO_2 fixation in *Chlorella, Pestic. Biochem. Physiol.,* 1, 381, 1972.
13. **Sikka, H. C., Saxena, J., and Zweig, G.,** Alteration in cell permeability as a mechanism of action of certain quinone pesticides, *Plant Physiol.,* 51, 363, 1973.
14. **Schwelitz, F. D., Sikka, H. C., Saxena, J., and Zweig, G.,** Ultrastructural changes in isolated spinach chloroplast and in *Chlorella pyrenoidosa* Chick (Emerson strain) treated with dichlone, *Pestic. Biochem. Physiol.,* 4, 379, 1974.
15. **Palmer, C. M.,** Evaluation of new algicides for water supply purpose, *J. Am. Water Works Assoc.,* 48, 1133, 1956.
16. **Palmer, C. M. and Maloney, T. E.,** Preliminary screening for potential algicides, *Ohio J. Sci.,* 55, 1, 1955.
17. **Eipper, A. W.,** Effect of five herbicides on farm pond plants and fish, *Game J.,* 6, 46, 1959.
18. **Whitton, B. A. and MacArthur, K.,** The action of two toxic quinones on *Anacystis nidulans, Arch. Mikrobiol.,* 57, 147, 1967.

19. **Bisiach, M.,** Algal infestation in Italian soils, *Riso,* 19, 129, 1970.
20. **Schulter, M.,** Untersuchungen über Algizide Eigenschaften einiger Fungizide und Herbizide, *Dtsch. Fisch. Ztg.,* 11, 247, 1964.
21. **Schulter, M.,** Untersuchungen über Algizide Eigenschaften von Fungiziden und Herbiziden, *Int. Rev. Gesamten Hydrobiol.,* 51, 521, 1966.
22. **Daste, P. and Neuville, D.,** Toxicité des pesticides agricoles on milieu marin, *Peche Marit.,* 55, 693, 1974.
23. **Neuville, D., Daste, P., and Longchamp, R.,** Toxicité comparé de divers pesticides le gard de deux especes de diatomes utiles a l'ostreiculture, *C. R. Acad. Sci. Ser. D,* 279, 675, 1974.
24. **Parachiv, M. E., Serbane, S., Popovici, G. H., and Djender, C.,** Algicide action of chemical substance on blue green algae, *Rev. Roum. Biol. Ser. Bot.,* 17, 195, 1972.
25. **Maloney, T. E. and Palmer, C. M.,** Toxicity of six chemical compounds to 30 cultures of algae, *Water Sewage Works,* 103, 509, 1965.
26. **Panigrahy, K. C.,** Physiological and Genetic Effect of Pesticides on Blue Green Algae: Effect of Carbamate Pesticides, Ph.D. dissertation, Berhampur University, Berhampur, India, 1984.
27. **Helling, C. S., Dennison, D. G., and Kaufmann, D. D.,** Fungicide movement in soil, *Phytopathology,* 64, 1091, 1974.
28. **Venkataraman, G. S. and Rajyalakshmi, B.,** Tolerance of blue-green algae to pesticides, *Curr. Sci.,* 40, 143, 1971.
29. **Venkaturaman, G. S. and Rajyalakshmi, B.,** Relative tolerance of nitrogen fixing blue-green algae to pesticides, *Indian J. Agric. Sci.,* 42, 119, 1972.
30. **Doneche, B.,** Effects du Mancozeb sur la microflora des sols du vignoble bordelais premiers resultats, *C. R. Acad. Sci. Ser. D,* 278, 3011, 1974.
31. **Moore, R. B.,** Algae as biological indicators of pesticides, *J. Phycol.,* 3, 4, 1967.
32. **Moore, R. B.,** Effects of pesticides on growth and survival of *Euglena gracilis, Bull. Environ. Contam. Toxicol.,* 5, 226, 1970.
33. **Nene, Y. L. and Thapliyal, P. N.,** Fungicides in Plant Disease Control, 2nd ed., Oxford and IBH Publishing, New Delhi, 1979.
34. **Moore, R. B. and Dorward, D. A.,** Accumulation and metabolisms of pesticides by algae, *J. Phycol.,* 4, 7, 1968.
35. **Audus, L. J.,** The action of herbicides on the microflora of the soil, Proc. 10th Br. Weed Conf., 1970, 1036.
36. **Lazaroff, N.,** Algal response to pesticide pollution, *Bacteriol. Proc.,* 149, 48, 1967.
37. **Lindahl, P. E. B.,** Inhibition of growth and photosynthesis in submerged plants with tetramethylthiuramdisulphide and sodium dimethyldithiocarbamate and reversal of this inhibition of photosynthesis, *Nature (London),* 191, 51, 1961.
38. **Lindahl, P. E. B.,** The inhibition of growth of *Enteromorpha linza* by sodium dimethyldithiocarbamate and tetramethylthiuramdisulphide, *Plant Physiol.,* 15, 607, 1982.
39. **Lindahl, P. E. B.,** On the reversal of the inhibition of photosynthesis induced by sodium dimethyldithiocarbamate and tetramethylthiuramdisulphide, *Plant Physiol.,* 19, 87, 1966.
40. **Gangawane, L. V.,** Tolerance of certain fungicides by nitrogen fixing blue green algae and their side effects on rice cultivars, *Pesticides,* 13, 37, 1979.
41. **Soeder, C. J., Liersch, R., and Trultzsch, U.,** Differential action of captan on the growth of some strains of *Chlorella* and *Scenedesmus, Arch. Mikrobiol.,* 67, 166, 1969.
42. **Butler, G. L., Deason, T. R., and O'Kelley, R. E.,** The effect of endrine, heptachlor, aldrin, dieldrine, captan, toxaphene and malathion on the growth of planktonic algae, *Bull. Environ. Contam. Toxicol.,* 13, 149, 1975.
43. **Ukeles, R.,** Growth of pure culture of marine phytoplankton in the presence of toxicants, *Appl. Microbiol.,* 10, 532, 1962.
44. **Harris, R. C., White, D. B., and MacFarlane, R. B.,** Mercury compounds reduce photosynthesis by plankton, *Science,* 170, 736, 1970.
45. **Gangawane, L. V. and Saler, R. S.,** Tolerance of certain fungicides by nitrogen fixing blue green algae, *Curr. Sci.,* 48, 306, 1979.
46. **Gangawane, L. V. and Kulkarni, L.,** Effect of pesticides on growth and heterocyst formation in *Nostoc* sp., *Marathwada Univ. J. Sci. Sect. B Biol. Sci.,* 19, 3, 1979.
47. **Ferrante, G. M. and Battino-Viterbo, A.,** Agar plate technique for potential algicide screening, *Riso,* p. 23, 1974.
48. **Teal, G.,** Metabolism of the Systemic Fungicide Ethirimol by Plants, Ph.D. thesis, University of Reading, Reading, England, 1974.
49. **Mathre, D. E.,** Effect of oxathiin systemic fungicides on various biological systems, *Bull. Environ. Contam. Toxicol.,* 8, 311, 1972.

50. **Mishushi, E. N. and Shil'nikova, V. K.,** *Biological Fixation of Atmospheric Nitrogen,* MacMillan, New York, 1971.
51. **Padhy, R. N.,** Cyanobacterium and pesticides, *Residue Rev.,* 95, 1, 1985.
52. **Shields, L. M. and Durell, I. W.,** Algae in relationship to soil fertility, *Bot. Rev.,* 30, 92, 1964.

Chapter 14

RHIZOBIA AND ROOT NODULATION

I. INTRODUCTION

The seeds of many legumes are attacked by a variety of plant pathogens, especially fungi. The destruction may be so extensive that yields are greatly reduced. In countries in which farmers earn enough to pay for fungicides, the chemicals have been widely used, even after it was realized that they are harmful to *Rhizobium*. Seed-applied fungicides are not specific for plant pathogens or even fungi and many kill or inactivate rhizobia and render useless the added inoculant. Thus, the number of *R. phaseoli* added to bean seeds declines as a result of the application of thiram, captan, and pentachloronitrobenzene (PCNB) to seeds.[1] Further, some of the best seed-protecting chemicals also reduce nodulation by *R. phaseoli*,[1] *R. japonicum,* and *R. leguminosarum.*[2] This decrease in nodulation of peanuts leads to reduced nitrogen fixation and pod yield.[3]

Studies conducted over the years on legume seed indicated that seed treatment with fungicides and rhizobial inoculation greatly increased emergence and nodulation, yet this crop is not usually treated with fungicides. The reason for this infrequent use of seed treatment is that many seed dressing fungicides have adverse effects on *Rhizobium* as indicated by biosensitivity and greenhouse and field studies.

The effect of fungicides on soil microflora has been widely investigated, but few detailed studies have been made on their interactions with specific microorganisms. The importance of the soil bacteria and *Rhizobium* spp. to world agriculture cannot be overemphasized. Biological fixation accounts for more than 90% of the terrestrial nitrogen turnover[4] and a large part of this nitrogen is fixed by *Rhizobium* spp. in association with legumes. Fungicides applied to leguminous plants, either as seed dressings or soil drenches, reach the soil and may affect this symbiotic relationship. Further, a fungicide applied to another crop may be sufficiently persistent to affect nitrogen fixation by legumes later grown in the same soil in order to enhance the nitrogen level.

Fungicides appear to differ in their effect on the growth of rhizobia in vitro and in vivo and there are differences in their effects in practice, depending on the strain of *Rhizobium* spp.[5] It is nonetheless believed that nodule numbers and nitrogen fixation may be adversely affected by fungicides applied as seed treatments or foliar application during the growing season. It is questionable whether harmful soil residues will ever accumulate from normal usage of fungicides but exceeding the recommended rate of application might cause undesirable effects on clover. One of the important attributes of systemic fungicides is translocation: acropetal (upward), basipetal (downward), and ambimobile (both directions). The ready translocation into plant roots of these systemic fungicides used against pathogens makes the effects of residues of such compounds in soil particularly worth investigating. Many legumes are grown in rotation involving cereals and other crops; nitrogen fixation may then be inhibited by soil residues of fungicides applied as sprays in seed treatments against pathogens. In general, systemic fungicides are not readily and rapidly deactivated[6] and this is advantageous for long-term plant protection. Fisher et al.[7] demonstrated that fungicides reduce nitrogen fixation in clover when present in soil at concentrations somewhat greater than those likely to be encountered in practice. Research on the interaction between fungicides and rhizobia has been summarized and the nontarget effects of fungicides are given in Table 1.

Table 1
COMPATIBILITY OF FUNGICIDES WITH *RHIZOBIUM* SPECIES

Crop/*Rhizobium* sp.	Fungicides	Compatibility (+)/ noncompatibility (−)	Ref.
Soybean, *Rhizobium japonicum*	Thiram	+	8
	Carboxin	+	
	PCNB and captan	−	
	Thiram	−	9
	Captan and PCNB	−	8
			50
	PCNB, carbendazim, captan, ziride, and agallol	+	21
	Zineb	+	83
	Benomyl, carbendazim, and pyracarbolid	−	
Peas, *R. leguminosarum*	Carbendazim, carboxin, benomyl, metalaxyl, tridemorph, triadimenol	+	104
	PCNB, captan, and thiram	−	
Peas and lentil, *R. leguminosarum*	Thiram	−	87
	Captan	−	
Groundnut, *Rhizobium* sp.	Thiram	−	36
Vetch, *Rhizobium* sp.	Thiram, Ceresan	−	91
Pigeonpea, *R. leguminosarum*	Carbendazim	+	94
Bean, *R. phaseoli*	Captan	−	51
	Thiophanate-methyl	+	96
	Ceresan	−	92
Clover, *R. trifolii*	Thiram and oxycarboxin	−	
	Benomyl, captan, thiophanate-methyl, Dodine, dimethirimol, tridemorph, triforine, and ethirimol	+	4
Mungbean, *Rhizobium* sp.	Carbendazim	+	90
Gram, *Cicer rhizobium*	Carbendazim	+	13, 93
			105

Note: Fungicidal seed treatment was at the recommended rate.

II. RHIZOBIAL INOCULATION TECHNIQUES AND APPLICATION OF FUNGICIDE

Inoculation of legume seeds with rhizobial cultures is one of the most effective processes for successful nodulation. Any slight deviation in the technique may result in the production of fewer or no nodules. Seed treatment and rhizobial inoculation on seed were carried out in one of three ways: (1) seeds were chemically treated and inoculated in one operation,[8] (2) seeds were treated first with fungicides and then inoculated with rhizobia,[9,10] and (3) seeds were first inoculated with rhizobia and then treated with fungicide.[11,12] Curly and Burton[8] demonstrated that applying fungicides before, after, or simultaneously with *Rhizobium* made no difference on the nodulation of soybeans. Their methods indicated that the nodulation efficiency of a fungicide-*Rhizobium* mixture on soybeans could best be determined by counting the number of nodules on the tap root 2 weeks after planting. There is very little information on this aspect on treated seeds.[8] However, the results that are available vary due to the workers' methods with regards to the quality and quantity of nodulation. Thomas[13] compared the above methods and found that first nodulation with rhizobia and then seed treatment gave satisfactory nodulation in vivo.

According to Mundade et al.,[11] fungicides should be applied to groundnut seeds after nodulation with *Rhizobium* rather than before, because the former method yielded a higher percentage of seed germination than the latter as compared to the control. However, Mukewari and Bhide[14,15] and Ostwal and Gaur[16] have also reported that there is no adverse effect in treating groundnut and soybean seeds, respectively, with protectants prior to bacterial inoculation.

III. COMPATIBILITY

Various methods have been developed for studying the compatibility of rhizobia with seed protectant chemicals. Some of them are briefly discussed here.

One of the earlier methods of studying the compatibility of rhizobia with seed protectant chemicals was the agar block method.[17] The fungicides were dispensed in a molten agar medium in a test quantity so as to give a final concentration equal to the concentration recommended for application on seeds. An 8-mm disk was cut from this plate and transferred to another empty sterilized plate; 10 mℓ of the medium preseeded with *Rhizobium* sp. was then poured evenly over the disk so that the medium covered the entire disk. After incubation at 30°C for 7 days the diameter of the zone of inhibition was recorded.[12] With this method, it was observed that the lupine group bacterium was highly sensitive to mercuric compounds while the alfalfa group bacterium was highly sensitive to arsenic compounds.

Another method of studying the compatibility of rhizobia with seed protectant chemicals is measurement of the zone of inhibition around a disk coated with a chemical and placed on the surface of a preseeded agar medium with *Rhizobium* spp.[8,18] All species of rhizobia were not equally sensitive to seed protectants; alfalfa, cowpea, lespediza, and lupine bacteria were less affected by thiram. Further, it was observed that chloronil had little effect on five species of rhizobia but was toxic to clover and cowpea organisms; dichlone was less toxic to *R. phaseoli* but was not toxic to other species used in the experiments.[18]

Hofer[19] used a gradient plate method for studying the compatibility of fungicides with rhizobia. In this method, petri plates were poured with a yeast-manitol-agar (YMA) medium with a given concentration of fungicide and tilted to produce a wedge-shaped layer of agar. An equal quantity of agar without fungicide was added to the surface of the hardened wedge-shaped original layer, thus providing a gradient plate. A preseeded YMA medium with rhizobia was poured over the second layer. Thus the thick edge of the first agar layer contained the larger amount of the fungicide and as the wedge thinned the amount decreased to a negligible quantity. By varying the quantity of fungicide in the first wedge of agar, it was possible to prepare a series of gradient plates. With the alfalfa bacterium, chloronil permitted the growth of colonies over half the plate in a dilution of 1:1600. However, with dichlone only half the place was covered with colonies and the critical dilution was 1:15,000. With Ceresan, the critial dilution was much higher than that for either of the compounds.[19]

In another in vitro study,[4] desired concentrations of fungicides were made in an aqueous slurry, dispensed in a molten medium, and shaken thoroughly, so that they were mixed uniformly. After the setting of the medium, a point inoculation with *R. trifolii* was made and incubated at 25°C. After 1 month, the radial growth of the bacterial colonies was measured.

Faizah et al.[20] described a technique for studying the biosensitivity of rhizobia with fungicides in vitro. Petri plates containing 20 mℓ YMA medium and 0.5% bromothymol blue were inoculated with 1 mℓ of a two-day-old culture of the appropriate *Rhizobium* in yeast extract broth. Four wells of 3 mm diameter were made in the medium and an aqueous solution of fungicide was poured into them. The plates were incubated at 30°C for 5 days, after which the diameters of the zone of inhibition of rhizobial growth were measured; captan, Ceresan, maneb, and karathane inhibited the growth of many rhizobial isolates.

Bandopadhyaya and co-workers[21] studied the in vitro sensitivity of *Rhizobium* spp. with fungicides using the methods of Runholff and Burton[18] with a slight modification. Instead of coating the paper disks with fungicide, they were dipped in fungicidal solutions and placed on the surface of the agar medium preseeded with rhizobia.

The effects of fungicides on rhizobia in soil or in sand culture are frequently measured in terms of the degree of nodulation, nodule weight, leghemoglobin content, or nitrogen-fixing efficiency;[22] the report of an international study group[23] recommended that any method of assessment should recognize the unique symbiotic relationship between the host and *Rhizobium*. It was proposed that plant growth and yield should be first assessed together, with an estimate of nodulation made later. Only where there is a critical effect would this be followed by tests to determine the fungicidal effects on the bacterium and nitrogen fixation (i.e., acetylene reduction). Results obtained from these criteria often do not wholly agree with the indications obtained from the growth in vitro of rhizobia in the presence of fungicides.

IV. RHIZOBIAL BIOSENSITIVITY AGAINST FUNGICIDES IN VITRO

In recent years, a large number of organic and systemic fungitoxicants have been used in plant disease control; their increasing use and accumulation in soil, water, and air has become a matter of concern to environmental biologists. Some of these organic fungicides are toxic to *Rhizobium* bacterium and therefore their accumulation in soils may be deleterious to agriculturally useful bacteria.

The effect of fungicides on *Rhizobium* may be divided into two types of action, which may be independent of each other: (1) action on the bacterium itself and on its growth and (2) action on the host plant, its infestation, root nodulation, and nitrogen fixation. The inhibitory effect of several fungicides on the growth of *Rhizobium* spp. in vitro has been reported; the bactericidal activity of dodine, a cationic surface-active agent, against Gram-negative organisms is well established.[24] The effect of Ethylan CP is less easy to explain. Nonionic surfactants are not necessarily bactericidal and factors other than surface activity may be responsible for growth inhibition. The antibacterial activity of captan has previously been reported.[25] The inhibition of growth of *Rhizobium* by oxycarboxin has been confirmed.[4,26] Mathre[27] reported that the rapid oxidation of carboxin to its corresponding sulfoxide may account for the lack of activity of this compound.

Many fungicides show toxicity towards rhizobia and/or nodulation. Toxicity is often associated more with some strains of rhizobia than others. Thus, strains associated with *Trifolium, Phaseolus,* and *Pisum* spp. may be highly susceptible, whereas those associated with *Melilotus* spp. may be either susceptible or resistant. Quinonoxime benzylhydrazine has been found to be highly toxic to strains from *Trifolium* and *Phaseolus,* but can also stimulate growth in some strains from *Melilotus*. The effects of fungicides on various *Rhizobium* spp. have been studied by measuring growth in vitro either on agar or in liquid cultures. Various reports on the biosensitivity of fungicides against *Rhizobium* spp. are discussed here.

Afifi et al.[28] found that at a concentration of 300 ppm, captan, thiram, Ceresan, dichlone, and methyl arsenic sulfide were toxic to some but not all rhizobial strains, whereas in the range of 200 to 30,000 ppm most strains were inhibited. The two mercurial fungicides methylmercurydicyandiamide (MMDD) and Ceresan were strongly toxic and lethal to seven strains of *R. leguminosarum,* but thiram, captan, dichlone, and chloronil were less inhibitory.[29] According to Daitloff,[30,31] the in vitro toxicity of fungicides to rhizobia was in the following order: Ceresan, thiram, captan, chloronil.

Captafol (10 to 1000 ppm) proved to be toxic to *R. trifolii* at 75 ppm and adversely affected growth at lower concentrations.[32] Captafol-tolerant mutants were isolated and all had lost the same plasmid and the ability to nodulate. Nodulation plasmids transferred from resistant donors to the tolerant mutants conferred the ability to nodulate and the rhizobia remained tolerant of captafol, indicating that genetic alteration leading to captafol tolerance was not necessarily detrimental to the ability of bacteria to form N-fixing nodules. The result of the investigation indicates that captafol may act as a plasmid curing agent in *R. trifolii*.[32]

In vitro experiments with 33 fungicides and 21 strains of *R. leguminosarum*[33] showed that captan, thiram, dichlone, and chloronil were moderately inhibitory whereas Ceresan and MMDD were strongly inhibitory. Makawi and Abdul-Ghaffar[34] found that captan, thiram, and benomyl were bactericidal towards a number of strains of rhizobia. Investigations with thiram and its degradation product, sodium dimethyl dithiocarbamate, showed that some rhizobia were sensitive but the degree of sensitivity depended upon the strain of *Rhizobium* spp. and the pH.[35]

Staphorst and Strijdom[36] investigated the in vitro effects of several fungicides on two rhizobial strains capable of nodulating in *Vigna anguiculata*. The fungicides were divided into three broad groups on the basis of their toxicity to the strains. The least toxic includes quintozene, benomyl, fenaminosulf, and 2-hydroxypropyl methane thiosulfonate (HPMTS); mixtures of dicloran and captafol and captan with phenylmercury acetate were intermediate in their effect, while carboxin, thiram, mancozeb, maneb, and 2-(thiocyanomethyl-thio)benzothiazole (TCMTB) were the most toxic.

Benomyl was bactericidal towards a number of strains of rhizobia, in contrast to other fungicides.[34] Benomyl, triforine, and tridemorph at concentrations of up to 200 ppm in an agar medium of pH 5.5 to 7.5 had little effect on strains of *R. leguminosarum* and *R. meliloti*. Using the purest grade unformulated fungicides, Fisher and Clifton[37] and Fisher[4] showed that the growth of *R. trifolii* in agar with 10, 50, 100, and 200 ppm of oxycarboxin markedly decreased, but benomyl, carboxin, dimethirimol, ethirimol, tridemorph, and thiophanate-methyl had little effect even at the highest concentrations. Triforine at 100 or 200 ppm was slightly inhibitory, but not at 10 or 50 ppm. In liquid culture, Fisher[4] confirmed the inhibitory effects of oxycarboxin at 10 to 1000 ppm. Fisher[4] and Fisher et al.[38] have also measured the effect of various protective, systemic fungicides required to reduce radial growth on agar by 50%: ED_{50} values were as follows: benodanil, benomyl, carbendazim, tridemorph, and ethirimol (5000); carboxin (2000); triforine (1000); oxycarboxin (162); fenarimol (26);[39] prochloraz and imazalil (10); propiconazole (24); triadimefon (110), and diclobutrazole (108 ppm).[39,40] Growth of *R. trifolii* on agar was strongly inhibited by fenarimol. Fungicides such as prochloraz, imazalil, diclobutrazole, triadimefon, and propiconazole inhibit sterol biosynthesis and are also comparatively more toxic than other fungicides;[41] however, sterols are absent from bacteria, so another mechanism must apply here. Sijpesteijn[42] has reported that the closely related fungicide triarimol also inhibits bacterial growth. The observed sensitivity of *Rhizobium* on agar to oxycarboxin confirms an earlier report of the inhibition of bacterial growth by oxathiins.[26] Although the ED_{50} value provides some comparison of the fungicides in vitro on pure cultures, it should not be considered the sole indicator of fungicidal side effects.[23] Fisher and Hayes[39] also demonstrated this and have shown that results of in vitro tests are not necessarily reflected in the effects on symbiosis.

A recent study indicated (Table 2) that all *Rhizobium* spp. were insensitive to carbendazim. Only *R. japonicum* was sensitive to triforine and metalaxyl. However, triadimefon and captafol had an inhibitory effect on *R. leguminosarum, Rhizobium* sp. (groundnut), and *R. phaseoli. R. japonicum* was found to be sensitive to captafol also. Captan was inhibitory to all the rhizobial species except *Rhizobium* sp. (groundnut). However, thiram was inhibitory to all the rhizobial species except *Cicer rhizobium. Rhizobium* sp. (lentil) was sensitive to captafol, captan, and thiram.[13] *R. meliloti*, a strain of rhizobium responsible for nodulation in chickpea *(C. arietinum* L.), was 90% inhibited in the presence of oxycarboxin and carboxin in the liquid yeast-manitol medium as determined by spectrophotometry.[43]

Legume nodulation and nitrogen fixation may be affected directly by seed dressing, by fungicides applied during the growing season, or by residual fungicides from previous

Table 2

BIOSENSITIVITY OF VARIOUS *RHIZOBIUM* SPECIES WITH DIFFERENT FUNGICIDES[13]

Rhizobium species	Carbendazim			Triforine			Metalaxyl			Triadimefon			Captafol			Captan			Thiram	
	1	2	3[a]	1	2	3	1	2	3	1	2	3	1	2	3	1	2	3	1	2
							Diameter of Inhibition Zone (mm)[a]													
Rhizobium japonicum	0	0	0	0	0	15	0	0	14	0	0	0	0	0	21	0	0	20	0	0
Cicer rhizobium	0	0	0	0	0	0	0	0	0	0	0	0	0	0	0	0	0	15	0	0
R. leguminosarum	0	0	0	0	0	0	0	0	0	0	0	14	0	14	15	0	0	14	0	14
Rhizobium sp. (groundnut)	0	0	0	0	0	0	0	0	0	0	0	15	0	0	16	0	0	0	0	18
R. phaseoli	0	0	0	0	0	0	0	0	0	0	0	14	0	0	14	0	0	16	0	0

[a]   Diameter measured by the mean of three replications; 1 = 500, 2 = 1000, and 3 = 2000 ppm concentrations of the fungicide.

application. Daitloff[30,31] has suggested that the toxic effects of fungicides may be prevented by coating legume seeds with a layer of polyvinylacetate after fungicide treatment but before inoculation with rhizobia.

V. VIABILITY OF RHIZOBIUM ON FUNGICIDE-TREATED SEED

Very few reports are available on rhizobial viability on seeds treated with fungicides. *Rhizobium* counts on the fungicide-treated and inoculated seeds were estimated by the serial dilution and planting technique on a congo red yeast extract manitol agar medium. Curly and Burton[8] found that pentachloronitrobenzene (PCNB) and captan (0.89 g kg^{-1} seed) significantly reduced *R. japonicum* survival on soybean seeds while carboxin-treated seeds maintained viability up to 8 hr after treatment. They further observed that thiram (0.69 g kg^{-1} seed) had no adverse effect on rhizobia even when seeds were treated 24 hr before planting. Captan (2.0 g kg^{-1}) and thiram (0.93 g kg^{-1}) reduced nodule mass on soybean,[44-47] clover,[48] alfalfa,[49] and lentil.[50] Graham et al.[51] studied the survival of *R. phaseoli* on fungicide-treated seeds and found that captan rapidly affected the viability of rhizobia while PCNB-treated seeds maintained a count of 1000 rhizobia per gram of seed until almost 48 hr after treatment. The results suggest that fungicides and bacteria interact in a positive or negative manner, which requires further study.

VI. MODE OF ACTION

Fisher,[4] investigating the mode of action of fungicides on *R. trifolii,* found that captan, ethirimol, and dodine all lowered respiratory activity to varying degrees. The severe inhibition caused by dodine and captan reflects the drastic decrease in lateral growth on agar; the action is probably bactericidal rather than bacteriostatic. Ethirimol reduced oxygen uptake but had no effect on the growth on agar, which suggests that its action must have been on the general metabolic activity of the bacterium rather than on the reproductive process; in contrast, oxycarboxin caused a considerable reduction in the growth of the bacterium but had no effect on oxygen uptake: the action is bacteriostatic. Low levels of tridemorph and triarimol stimulated oxygen uptake.[4] The oxygen uptake of plant and fungal protoplasts was also stimulated by tridemorph.[52] In eukaryotic cells, such effects have been attributed to changes in the permeability of intracellular membranes confining endogenous reserves. Such an explanation is less likely with the prokaryotic *Rhizobium* cell, although electron microscopic studies of soybean nodule bacterioids reveal an inclusion probably containing lipid material.[4] Among fungicides, sterol inhibitors are more toxic to *R. trifolii.*[39,40] Sterols are absent from bacteria; therefore another mechanism must apply here. Other compounds which inhibit sterol biosynthesis in fungi also inhibit bacterial growth.[39,42,53] As mentioned above, the ED$_{50}$ values for fungicides benodanil, benomyl, carbendazim, pyracarbolid, and ethirimol were 5000 ppm; for carboxin 2000 ppm and oxycarboxin 162 ppm. These fungicides are mostly inhibitors of proteins, nucleic acid, or the respiratory process. However, fungicides inhibiting sterol biosynthesis have very low ED$_{50}$ values, such as prochloraz and imazilil (ED$_{50}$ 10 ppm), diclobutrazole (ED$_{50}$ 108 ppm), triadimefon (ED$_{50}$ 110 ppm), propiconazole (ED$_{50}$ 24 ppm), fenarimol (ED$_{50}$ 26 ppm), and triforine (ED$_{50}$ 1000 ppm).[40]

Acetylene-reducing activity (ARA) is indicative of nitrogen-fixing activity and as such is useful for ranking qualitatively the effects of fungicides on nitrogenase. Nitrogen fixation can be conveniently measured by the reduction of acetylene to ethylene and the superiority of this technique over measurement of nodule leghemoglobin content or nodule fresh weight as a criterion of nitrogen-fixing capacity of legumes has been demonstrated. Nitrogen fixation by excised clover roots was affected only by thiram, Ethylan CP, and oxycarboxin. Thiram had little effect on the growth of *Rhizobium* on agar plates and Richardson[54] has reported that thiram treatment of soil resulted in an increase in the leghemoglobin content of root nodules. Therefore, it probably caused a direct inhibition of nitrogenase enzymes. Nitrogenase, as measured by nitrogen fixation per gram fresh weight or per gram dry weight, was significantly inhibited by the highest concentrations of carboxin, oxycarboxin, and tridemorph, although only the latter compound reduced nodulation in white clover where a *R. trifolii* is a symbiotic nitrogen fixer.[40] Earlier, Fisher and Hayes[39] demonstrated that benodanil, pyracarbolid, and tridemorph were toxic and phytotoxic at the highest concentrations used. This reduction in plant size is reflected in the significant decrease in nitrogen fixation by pyracarbolid. With benodanil, growth inhibition was evidently insufficient to affect significantly the site of fixation.[39] Nitrogen fixation was also decreased by diclobutrazol but, despite the reduction in plant size, the effect of the lowest concentration in soil was not significant.[40] However, the fungicide markedly reduced the rate of fixation whether expressed on the basis of the whole plant or the root weight. It was observed that diclobutrazol (1.0 and 5.0 mg kg^{-1} of soil) reduced nodulation of clover roots and the effect on nitrogen fixation expressed on the basis of dry root weight suggests that nitrogenase may be directly affected. Rennie and Dubetz[55] reported that thiram and captan applied at 1.0 g kg^{-1} seed along with rhizobium inoculation reduced shoot nitrogen yield at anthesis in soybean, but a lower rate of thiram (0.54 g kg^{-1}) increased nitrogen yield; soybean plants assimilated more soil and/or fertilizer nitrogen to compensate for lower nitrogen fixation rates.[56] This would explain why the effects of fungicides on ARA and presumably on nitrogenase activity were not reflected in shoot nitrogen or seed yield.

Captan and/or thiram have been shown to interfere with nodulation and nitrogen fixation in field beans,[1] lentil,[50] alfalfa,[49] and clover.[48,51] Captan and thiram are compatible with *R. leguminosarum* inoculation for peas[57,58] and fababean[59] and general inoculation technique guidelines,[60] but others[61] suggest that some seed-applied rhizobia can be harmed by these fungicides. Captan (2.0 g kg^{-1} seed) reduced nodule mass and ARA in pot experiments when either seed-applied or granular *R. japonicum* inoculant was applied.[44] Thiram (0.93 g kg^{-1}) inhibited ARA over the entire 7 weeks of a pot experiment on soybean.[45-47] Rennie et al.[62] made a detailed study on the effect of seed-applied fungicides on growth and nitrogen fixation in peas, lentil, and fababean: captan reduced nitrogen fixation and ARA in all three legumes; thiram and metalaxyl stimulated ARA in all three legumes. However, thiram reduced nodulation in peas by 46% whereas metalaxyl reduced nodulation in fababean by 35%.[62] It was also observed that captan-containing fungicides showed varied effects: B-3 and Evershield containing 33.5 and 29.5% captan also reduced nodulation or nitrogenase activity in some of the legumes, but DL-PLUS containing 15% captan was not toxic. This suggests that compounds containing over 29.5% captan may not be compatible with seed-applied *R. leguminosarum;* therefore, caution must be exercised in recommending such compounds.

Fisher and Hayes[39] measured the residual toxic effects of fungicides applied to cereal on the rhizobia and the nitrogen-fixing capacity of clover. They observed that concentrations of benodanil in the top 5 cm of soil following a single application are unlikely to exceed 3 μg g^{-1} dry soil; corresponding figures for fenarimol, pyracarbolid, and tridemorph are 0.15, 0.1, and 1.5 μg g^{-1}, respectively. Benodanil at 5 μg g^{-1} had no significant effect on plant weight or nitrogenase activity; 15 μg g^{-1} significantly reduced plant weight but had no effect on nitrogenase. Fenarimol had no effect at concentrations of up to 5 μg g^{-1}. Pyracarbolid at 1 μg g^{-1} had no effect, but 5 μg g^{-1} reduced plant weight, although enzyme activity was unaffected. Tridemorph at 2 μg g^{-1} had no effect; 5 μg g^{-1} reduced plant weight whereas 15 μg g^{-1} almost completely inhibited nitrogenase.[39] Therefore, it is unlikely that any of these fungicides would seriously inhibit symbiotic nitrogen fixation by legumes in the field.

VII. RESISTANCE OF RHIZOBIAL BACTERIUM TOWARDS FUNGICIDES

Effects of fungicides on rhizobia are not easy to quantify because there are variations in susceptibility among different strains. Reports are available on the development of resistance to fungicides in rhizobia. Some wild mutants resistant to fungicides are encountered. *R. meliloti* strains generally tolerated higher rates of captan and a mixture of benzimidazole in vitro than did strains of *R. trifolii* and *R. leguminosarum*. All strains were inhibited by 50 ppm of captan, but spontaneous mutants resistant to this concentration were isolated.[63]

Staphorst and Strijdom[36] unsuccessfully attempted to isolate a thiram-resistant strain, but isolates resistant to chloronil were not observed. Certain isolates of *R. phaseoli* were resistant to spergon and some strains of *R. meliloti* resistant to thiram were isolated from the natural rhizobial population.[64] Strains of *R. meliloti* resistant to thiram following exposure to high concentration have also been developed in the laboratory.[65] The resting cells of the bacteria formed large quantities of dimethyldithiocarbamate (DMDTC), dimethylamine, and CS_2 from thiram. The resistant population of this bacterium to thiram and spergon showed cross resistance to each other,[65] indicating a common site for the mode of action of these compounds in bacterium. However, the resistant strains possessed the same nitrogen-fixing capacity as wild rhizobia. The plants derived from alfalfa seeds treated with thiram and then inoculated with the parent rhizobia were stunted, chlorotic, and weak N_2-fixers.[64] In contrast, alfalfa plants derived from seeds fortified with thiram and inoculated with the resistant strains of rhizobia grew well and fixed nitrogen as well as the legume derived from fungicide-free seeds. When resistant strains were inoculated into cowpea and alfalfa, growth and N-fixation were improved compared to plants in which inoculation of nonresistant strains had been used. These strains would tolerate seed treatment with the appropriate fungicide.

Lennox and Alexander[66] reported that the number and weight of pods and weight and nitrogen content of tops of beans harvested from the seeds inoculated with thiram-resistant strains of *R. phaseoli* increased, if seeds were treated with thiram before planting in soil. The resistant strains produced a greater percentage of nodules and were more effective with fungicide than without it.[66] Thiram treatment protects the rhizobia from the attack of pathogenic microorganisms and the enhanced growth accounts for the better nodulation. If thiram-resistant strains were present in the soil, then thiram seed treatment would also be beneficial. Gupta and Shirkot[67] reported that three rhizobia isolated from soybean, mungbean, and cowpea were adapted to thiram after a series of transfers (5-19) on a medium containing a higher amount of thiram. The developed resistance was stable as was evident from reversion experiments.

Colonization of the rhizosphere by *Rhizobium* spp. is probably important in determining the extent of nodulation of legumes. The proliferation of rhizobia as a necessary step for successful nodulation has been stressed[66] and a correlation between the failure to nodulate and low rhizobial numbers in the rhizosphere has been found. Microbial antagonism may also contribute to the extent of the colonization of the rhizosphere and the nodulation of the host. Application of fungicides also enhanced nodulation in several crops by suppressing microbial activity. Lennox and Alexander[66] reported that the application of thiram to bean seeds suppresses protozoa and allows for more extensive development of thiram-tolerant *R. phaseoli*; treatment of seeds with mancozeb allows for the establishment of a mancozeb-resistant bacterium on corn roots.[68] In order to increase the colonization of the rhizosphere of soybean by *R. japonicum*, a strain of this bacterium resistant to benomyl was used.[69] The number of rhizobia rose quickly in the first 2 days after soybean seeds were planted in soil and then rapidly fell. The decline was slower if the seeds were coated with benomyl. This fungicide reduced the number of bacteria and protozoa in the rhizosphere, but the effect lessened or disappeared as the plants grew. Nodulation and plant yield were increased by the addition of benomyl to soybean seeds sown in sterile soil inoculated with *R. japonicum* and a mixture of microorganisms.[69] Inoculation of thiram-treated seed with a thiram-resistant strain of *R. phaseoli* resulted in higher numbers of *R. phaseoli* in the spermosphere and rhizosphere of kidney beans as compared with seedlings that arose from untreated seeds.[69,70] The favorable effect of benomyl on colonization by *R. japonicum* probably results from its effects on microorganisms antagonistic to the rhizobium.

Rhizobium developed resistance to thiram under natural conditions and also in the laboratory.[66-71] It was observed that developed resistance was stable, but with the development of resistance, dehydrogenase activity and glycolipids decreased, total lipids and phospholipids increased, and neutral lipids showed fluctuations. The presence of compounds which stimulated lipid production by rhizobia increased with their resistance to thiram;[67] moreover, the nodulation capacity of thiram-resistant strains was similar to that of parent wild strains. The mutants were able to infect host plants, suggesting that the use of rhizobia mutants may be a means of ensuring that the fungicide will not adversely affect these bacteria in practice.

Despite the close chemical relationship between the fungicides captafol and captan, the former was found to be much less toxic to *R. trifolii*.[72] The viability of the rhizobium treated with captafol (175 μg mℓ^{-1}) decreased approximately tenfold every 24 hr of incubation, whereas the same loss of viability of the bacterium occurred in about 25 min with captan (110 μg mℓ^{-1}); however, captafol seems to be more stable and its activity continues longer.[72]

As captafol is a fungicide of wide agricultural use, studies were conducted to develop resistant strains of *R. trifolii*.[72] Over 100 captafol-resistant strains have been developed, but none was able to form nitrogen-fixing nodules on *Trifolium alexandrum*. Thus, it appeared that resistance to captafol was directly related to the loss of the ability to nodulate. This was tested by Ruiz-Sainz and co-workers.[72] Plasmids were first examined to determine if a particular plasmid which might carry host range genes was lost; in a second experiment, host range plasmids from other strains of *Rhizobium* were introduced to determine if the captafol-resistant strains could then form nodules and if they were still captafol-resistant. Agar gel electrophoresis showed that the wild-type strain had three plasmids and that in the four captafol-resistant derivatives assayed the smallest plasmid was missing.[72] This suggests that this plasmid may carry symbiotically important genes. It has a molecular weight of approximately 180 $\times$ 10^6 because it has the same mobility in agar gel as the host range plasmid of the *R. trifolii* strain.[73] The other two resistant plasmids are probably larger than 30 $\times$ 10^6. Captafol is a mutagenic agent[74] and thus there is still the possibility that Nod$^-$ phenotypes are a consequence of captafol-induced mutations. It was reported that when an *R. trifolii* strain is treated with acridine orange it becomes noninfective and loses its smallest plasmid. Although acridine orange is also known to be a mutagenic agent, it seems clear that the Nod$^-$ phenotype in the strain is associated with the curing of the same plasmid.

The observation that the captafol-resistant strains carrying symbiotic plasmids were able to form nitrogen-fixing nodules on peas and clovers, respectively, and that isolations from these nodules were still captafol-resistant shows that captafol resistance was unlikely to be related directly to the ability of the bacteria to form normal nodules.[72] It seems that captafol is a plasmid-curing agent as all the strains selected on this basis were captafol-resistant and had, coincidentally, lost a plasmid. It has been suggested that further work is needed to determine if captafol is a general plasmid-curing agent for *Rhizobium* and other bacterial species. It is likely that many of the problems caused by fungicides and pesticides on *Rhizobium* inoculants can be overcome by genetic manipulations. These fungicide-resistant mutants were able to destroy the fungicides, were fully effective on their host plants, and were better inoculants than their parent strains when applied to legume seeds that had been treated with these fungicides.

Studies on the natural resistance or sensitivity of rhizobia to pesticides have been made.[28,35,75] In agriculture, it is important that efficient N$_2$-fixing strains of rhizobia be resistant to adverse environmental conditions such as pesticide applications. It is also desirable that adapted strains should not lose their efficiency of N_2-fixation. The adapted rhizobial strains were stable and developed a normal symbiotic relationship with soybean, cowpea, and beans;[67] however, Schwinghamer[76] showed that following the development of resistance in *R. leguminosarum*, *R. trifolii*, and *R. meliloti* to viomycin and neomycin, the normal symbiotic relationship with legumes is lost. No loss of effectiveness is observed in streptomycin-resistant strains.

VIII. FUNGICIDAL EFFECTS

A. Seed Treatment

Chemicals used for protective purposes, those which contain mercury, copper, or zinc, are extremely toxic to rhizobia; on the other hand, thiram, dichlone, and chloronil, which do not contain heavy metals and are insoluble in water, are less toxic to rhizobia.[77] The results of some workers[78-80] showed that nodulation was not affected or increased as a result of seed treatment with thiram, chloronil, or dichlone. Hofer[19] found that mercurial fungicides were extremely toxic to the root nodule bacteria of *Pisum*, *Trifolium*, and *Vigna*. Seed treatment of lupines and fenugreek with Alberton 52T (ethyl mercury chloride, Hg equivalent 2.0%) and Rhizoctol combi (methyl arsenic sulfide 10%) markedly reduced the number of nodules, total nitrogen, and the dry weight, which were reduced to a lesser extent by Ceresan, while captan and thiram were less toxic. Chloronil had no effect on nodulation.[12]

In field studies, thiram reduced nodulation of soybean when grown in noninoculated soil, but nodulation was not affected by *R. japonicum*.[9] Seed treatment with PCNB and captan reduced tap root nodulation while treatment with carboxin and thiram had little effect on tap root nodulation even when seeds were sown 24 hr after treatment.[8] Nodulation, dry weight, and nitrogen content of plants were increased by treating the inoculated seed of soybean with benomyl, but nodulation was decreased by treatment with oxycarboxin and carboxin.[81] Seed dressing of soybean seed with carboxin at 200 to 300 g kg^{-1} gave high field germination and led to the formation of a large number of nodules on roots, ensuring high yield.[82] Zineb stimulated nodulation whereas benomyl, carbendazim, and pyracarbolid reduced the soybean plant dry weight, nodulation, and total nitrogen content.[83] Plants raised from groundnut seeds treated and inoculated with Ceresan and *Rhizobium* sp. had the highest dry matter and nitrogen content as compared to plants raised from untreated seeds.[84] Captan was inhibitory to the nodulation of inoculated groundnut grown in the greenhouse, whereas thiram was not.[36] Nodulation, dry weight, and nitrogen content of plants were increased by treating the inoculated seed of groundnut with benomyl.[81] Captan, Ceresan, and PCNB do not adversely affect nodulation in groundnut plants inoculated with *Rhizobium*.[85] Seed treatment of inoculated pea seed with benomyl, oxycarboxin, and carboxin increased the nodulation, dry weight, and nitrogen content, but treatment with carbendazim decreased nodulation.[81] Seed treatment of pea seed with carbendazim, PCNB, captan, Ceresan dry, mancozeb, thiram, and BASF 380 F in combination with *Rhizobium* gave good germination and nodulation. However, carbendazim, thiram, and Ceresan gave the highest germination whereas carbendazim gave the highest nodulation.[86] Captan inhibited the nodulation of peas where the soil contained a very low number of *R. leguminosarum*, but thiram did not.[87] Field studies by Graham et al.[51] indicated that a delicate balance exists between inoculants and seed protectants in beans. Nodule number per plant was not affected when thiram- or PCNB-treated bean seeds were inoculated and promptly planted in a soil which contained an appropriate number of *R. leguminosarum*.[87] Captan seed treatment did not affect seed yield, seed protein content, or nodulation of lentil where the soil was populated by the appropriate strain of *R. leguminosarum*. Captan was inhibitory to *Rhizobium* when inoculated and treated seeds were planted in virgin soil. However, this inhibitory effect may not occur if a higher rate of inoculum is used on the seeds and if the appropriate *Rhizobium* strain is present in the soil.[50] Manikasundaram and Rambadam[88] have established that seed treatment with benomyl (0.15%), Ceresan M (0.1%), and zineb (0.3%), along with rhizobial inoculation, did not cause any deleterious effects on nodulation.

Spergon, phygon, and Ceresan M are used extensively for seed treatment and they have varied toxic effects on different species of *Rhizobium.* A study conducted by Hofer[19] indicated that the bacterium for alfalfa was least injured while that for clover was highly sensitive. The mercurial fungicide Ceresan M was more toxic than spergon and phygon; a dilution of 0.5 ppm was highly critical for growth of the clover bacterium. These results suggest the advisability of in vitro testing of fungicides proposed for use on leguminous seeds to determine the effects of their interaction with inoculants.[19]

Bandopadhyaya and co-workers[89] reported that PCNB, captafol, zineb, and agallol, although not carbendazim, were inhibitory to one or other species of *Rhizobium* in terms of numbers of nodule per plant. Carbendazim showed a slight inhibition to peanut but increased nodulation in soybean and berseem. The fresh weight of nodules was significantly low in all the fungicides used. Similarly, Chandra et al.[90] also reported that carbendazim (0.05, 0.1, and 0.2%) seed treatment had no adverse effects on the nodulation of *V. mungo.* Four seed dressing fungicides (thiram, captan, PCNB, and Dithane M-45) were tested for their relative effectiveness against charcoal rot *(Macrophomina phaseolina)* on soybean variety Clark-63, with special reference to seed treated with rhizobial inoculants. Results showed that Dithane M-45 alone or in combination with thiram gave the best yield and seed emergence in *M. phaseolina*-infested soil when the seeds were also inoculated with *R. leguminosarum.* No adverse effects of fungicides on nodulation were observed (Tables 4 and 5).

Thiram and Ceresan can inhibit rhizobia in culture, but in pot experiments Elek and Kecskes[91] found that both compounds improve yield and nodule formation in vetch *(Vicia sativa)* plants grown from seeds inoculated with rhizobia in an arid forest soil and were without effect in a calcareous soil. However, Taha et al.[92] observed that Ceresan can inhibit broad bean and lentil nodulation and growth.

With the introduction of several new systemic compounds which translocate upward as well as downward in treated plants, studies were undertaken to evaluate the effect of these fungicides on nodulation.[13] The data in Table 3 indicate that rhizobial inoculation and carbendazim (0.15%) seed treatment gave good nodulation in soybean *(Glycine max),* gram *(Cicer arietinum),* pigeonpea *(Cajanus cajan),* black gram *(Vigna mungo),* green gram *(V. radiata),* and groundnut *(Arachis hypogaea);* in soybean, gram, and pigeonpea, triforine, metalaxyl, triadimefon, captafol, captan, and thiram gave significant nodulation as compared to controls. However, in green gram and black gram more significant nodulation was observed in thiram-treated seeds than with other treatments. These results indicate that the effects of rhizobial inoculation and subsequent seed treatment with fungicides vary; hence their compatibility should be evaluated prior to recommendation for seed treatment in inoculated legume seeds.[13,93,94]

Chemicals such as PCNB[95] and captan and Ceresan[89] seem to stimulate legume growth. Thiram and carboxin had little or no effect on the rhizobium-legume symbiosis of pigeonpea, soybean, peanut, bengal gram, and green gram when added directly to seeds but improved germination, nodulation, and yield.[89] The number and weight of nodules per plant increased when the seeds were treated with thiophanate-methyl at 0.175% active ingredient (a.i.), whereas higher doses resulted in decreases in both number and weight of nodules.[96]

Narayan and co-workers[83] reported that carbendazim, benomyl, and pyracarbolid adversely affected the nature of *Rhizobium,* whereas zineb showed a stimulatory effect. A similar effect was also recorded in inoculated *Rhizobium.*[97] Thiram inhibited the growth of rhizobia and delayed plant growth, but no drastic effects were observed after 8 weeks. Manikasundaram and Rambadam[88] found no adverse effect of zineb on groundnut seeds but observed stimulatory effects that enhanced the survival of rhizobia. The noninjurious nature of fun-

Table 3

EFFECT OF RHIZOBIAL INOCULATION AND FUNGICIDAL SEED TREATMENT ON NODULE NUMBER AND DRY WEIGHT PER PLANT OF VARIOUS LEGUMES[13]

Fungicides	Green gram		Black gram		Groundnut		Soybean		Gram		Pigeonpea	
	No. of nodules per plant	Nodule dry wt per plant (mg)	No. of nodules per plant	Nodule dry wt per plant (mg)	No. of nodules per plant	Nodule dry wt per plant (mg)	No. of nodules per plant	Nodule dry wt per plant (mg)	No. of nodules per plant	Nodule dry wt per plant (mg)	No. of nodules per plant	Nodule dry wt per plant (mg)
Carbendazim	20.1	48.5	64.0	50.7	46.2	107.5	7.37	196.00	13.7	51.6	4.05	35.8
Triforine	15.2	37.6	48.5	55.2	35.4	95.2	4.37	90.8	11.2	49.9	3.47	30.4
Metalaxyl	16.3	43.5	56.7	50.2	34.3	116.2	4.50	81.5	11.5	42.0	4.10	29.7
Triadimefon	13.8	32.2	20.3	30.5	12.9	54.6	1.56	54.7	10.7	47.3	2.25	16.8
Captafol	18.9	45.0	51.6	40.5	32.5	102.9	4.00	71.2	10.0	40.4	2.80	29.8
Captan	17.2	34.3	54.2	42.5	28.2	95.4	5.10	73.0	5.5	25.9	3.25	25.1
Thiram	13.3	51.5	57.3	60.0	33.5	126.1	3.62	79.4	11.5	47.2	0.97	7.7
Control	13.8	29.5	36.0	31.0	25.5	68.6	1.43	48.6	5.37	35.9	2.32	22.9
Critical difference at 5%	3.95	0.66	16.72	13.63	8.08	28.15	2.42	57.59	2.99	5.76	0.92	3.76

gicides on nodulation was also reported by Fisher.[4] Wrobel[98] reported that organic mercury and thiram fungicides delay nodule formation and plant growth in beans and peas.

Captan is the most common seed-applied fungicide recommended for soybeans. Curly and Burton[8] found that captan at 0.8 g kg^{-1} seed significantly reduced the number of rhizobia after a 24-hr incubation. In a pot experiment, Chambers and Mantas[44] found that captan at 2.0 g kg^{-1} seed reduced nodule mass and ARA when either seed-applied or granular inoculants were applied. Although captan did not affect seed yield, it lowered protein yields, particularly with the seed-applied inoculant. *R. japonicum* strain USDA 110 was less sensitive to captan than CB 1809. Curly and Burton[8] found that thiram at 0.6 g kg^{-1} seed had no adverse effect on the rhizobial number of seed, but at 0.93 g kg^{-1} seed it inhibited growth of *R. japonicum* in pure culture and reduced nodule mass and ARA over the entire 7 weeks of the pot experiment.[45,47] Curly and Burton[8] with soybean and Graham et al.[1] with field beans noted that immediate inoculation and planting of fungicide-treated seed decreased the inhibitory effects. However, this is not common practice in a normal farming operation. Rennie and Dubetz[55] reported that the seed-applied fungicides thiram and captan at 0.54 g kg^{-1} had no adverse effects on nodulation and nitrogen fixation under experimental conditions when a granular inoculant is applied. The inhibitory effects of captan and thiram noted on rhizobia in soybean and bean may be attributed to the closer contact of fungicide and inoculant and the higher dosage used by the authors. Thus, the use of granular inoculants may alleviate the problem of seed-applied fungicides.[1,44]

Seed treatment of chickpea with carboxin and oxycarboxin (0.05%) resulted in increased nodulation over controls, which had an influence on the growth of the plant.[43] The fungicides increased the growth of soybeans and beans raised after seed treatment.[99] The process of nodulation occurs only after seeds have rooted and become established. The seed coats carrying the residual fungitoxicants remain at the surface of the soil. Carboxin and oxycarboxin move into the treated seeds within 24 hr of germination and subsequently these chemicals become distributed to the roots, stem, and leaves. The maximum accumulation, however, occurs in the roots as seen in soybean and mungbean.[100,101] The concentrations of fungicides that accumulate in chickpea may not be toxic,[43] but studies should be conducted on the fungitoxicity of other compounds used for seed treatment.

Generally, nodulation failure in legumes is attributed to the inhibition of rhizobia caused by the toxicity of fungicides; however, Karanth and Vasantharajan[102] observed that the phytotoxicity of fenaminosulf was responsible for the failure to nodulate. Fenaminosulf is a soil fungicide used for the control of damping off caused by *Pythium* sp. and is applied either as a seed or soil dressing. In pot culture studies fenaminosulf was found to be toxic to sunhemp *(Crotalaria juncea),* a green manure plant, and also to affect nodulation.[102] It was also observed that the growth of *Rhizobium* in culture was not affected by 200 μg mℓ$^{-1}$ and *Rhizobium* spp. capable of nodulating in sunhemp, soybean, chickpea, and sesbania were also resistant to this fungicide. However, the fungicide is toxic to sunhemp even at 100 μg g^{-1} soil; tap root elongation was particularly affected.[102] The inhibition was irreversible by external addition of either 3-indole acetic acid (IAA) or gibberellic acid at the concentrations used (0.1 to 1.0 μg). One of the essential steps in successful nodulation is the curling of root hair, which is caused by IAA produced from tryptophan excreted by the plant.[103] Fenaminosulf inhibits the conversion of tryptophan to IAA in soil; thus, it is concluded that the observed phytotoxicity may well be due to its effect on phytohormone metabolism.

Peas and pea seedlings are exposed to several seed-borne pathogens when they are planted; hence, the seeds are treated with several fungicides.[104] In order to establish nodulation they are also inoculated with the rhizobial bacterium. Studies conducted[104] showed that a mixture of thiram and PCNB at 0.25% (in a ratio of 1:1) yielded a higher germination percentage and nodule number (Table 4). Bhattacharya and Sengupta[105] also reported similar results for

Table 4
**EFFECT OF SYSTEMIC AND NONSYSTEMIC FUNGICIDE SEED TREATMENT
ON EMERGENCE AND NODULATION OF PEA SEEDS INOCULATED WITH
RHIZOBIUM LEGUMINOSARUM AFTER 1, 4, AND 24 hr TREATMENT (VAR. T-
163)[104]**

| | Time after treatment | | | | | |
| | 1 hr | | 4 hr | | 24 hr | |
Fungicide	Emergence	Nodulation	Emergence	Nodulation	Emergence	Nodulation
Systemic						
Carboxin	80	16	80	15	85	10
Carbendazim	75	35	85	45	89	40
Benomyl	60	12	70	20	76	10
Metalaxyl	75	45	79	50	84	30
Tridemorph	60	32	56	37	56	12
Triadimenol	80	40	78	45	67	15
Nonsystemic						
PCNB	90	45	80	40	80	5
Thiram	85	40	85	38	85	10
Captan	65	15	65	40	65	9
None						
Control 1[a]	50	30	50	30	50	30
Control 2[b]	80	10	80	10	80	10
Control 3[c]	50	10	50	10	50	10

Note: The data are the mean of four replications; emergence counts were made after 7 days of germination; nodulation counts were made after 21 days of germination.

[a] Seeds only inoculated.
[b] Seeds only treated.
[c] Seeds neither treated nor inoculated.

chickpeas. When planning seed treatment for the control of seed-borne disease, all possible interactions should be taken into account, as it was observed that pea cultivars respond differently to seed-treatment fungicides (Table 5).[104]

B. Soil Application

Systemic and nonsystemic fungicides are applied to soil for the control of phytopathogenic fungi. Some are very specific whereas others are broad spectrum. Fungicides present in soil may originate from soil treatments, seed dressing, or runoff from foliar sprays. The effects of fungicides on rhizobia in soil or in sand culture are frequently measured by the degree of nodulation, nodule weight, and nitrogen-fixing capacity.

Early work on the effects on soil microflora has been reviewed by Kreutze.[106] Few investigators have studied the influence on specific microorganisms. Comparatively little information is available on the persistence of many of the recently developed fungicides, particularly systemic ones, in soil; among the protective fungicides, only captan[107] and thiram[54] are slowly degraded. The systemic fungicides benomyl[108] and thiophanate-methyl[109] are converted into carbendazim, the long persistence of which has been demonstrated by Dimond.[110] Ethirimol and dimethirimol are extensively adsorbed by soil[111] and former compounds have been seen to persist under these conditions.[112] The adsorption of soil particles to the surface of *Rhizobium* cells has been demonstrated. Graham-Bryce[6] considered that systemics are not usually rapidly degraded in soil. The presence of fungicides in the soil affects rhizobial symbiosis with legumes. Thiram was found to inhibit *R. leguminosarum;*[113]

Table 5

EFFECT OF SYSTEMIC AND NONSYSTEMIC FUNGICIDAL SEED TREATMENT ON EMERGENCE AND NODULATION OF PEA CULTIVARS INOCULATED WITH *RHIZOBIUM LEGUMINOSARUM*[104]

Emergence (%) and nodule counts after seed treatment[a] with

Fungicide	Carbendazim E	N	Metalaxyl E	N	Triadimenol E	N	Thiram E	N	PCNB E	N
T-163	92	+ + +	80	+ + +	80	+ + +	90	+ + +	75	+ + +
Pant Pea	100	+ + +	80	+ +	90	+ +	80	+ + +	80	+ + +
388-1	75	+	65	+	70	+	60	+	65	+ +
L-116	80	+	70	+	60	+	60	+	75	+
Rachna	100	+	100	+	90	+	65	+	80	+
HFP-5	100	+	80	+	80	+	80	+	70	+
Pant Pea-4	100	+ + +	100	+ + +	90	+ + +	70	+ +	60	+ +
RAV-21	80	+ +	75	+	80	+ + +	80	+ +	75	+ +
RP-21	70	+ +	75	+ +	80	+ +	70	+ +	75	+ +
RP-1	100	+ +	70	+ +	80	+ +	80	+ +	75	+ +
RAV-18	100	+ + +	80	+ + +	80	+ + +	70	+ +	65	+ +
PG-3	100	+ +	100	+ + +	100	+	80	+ + +	65	+ + +

Note: Data are the mean of four replications. E = emergence: counted after 7 days of germination. N = nodulation: counted after 21 days of germination.

[a] Seed treatment with systemic fungicides at 0.1%; nonsystemic fungicides at 0.25%.

it had no effect on the nodulation of vetch in sand culture or in various soils under greenhouse and light-chambered conditions.[29] It was further observed that Ceresan strongly inhibited nodulation; MMDD was found to be unstable and harmful to vetch seeds and chloronil, dichlone, captan, and copper oxychloride had an intermediate effect on nodulation.[29] Kratka and Ujevic[114] found that at rates equivalent to 0.3 kg^{-1} captan had no adverse effect on the nodulation of vetch. The effects of fungicides on the nodulation of legumes vary according to the strain of rhizobium.[36] It was shown that quintozene was less toxic in vitro than in vivo; thiram was not toxic in vitro, and had relatively no effect on nodulation or dry mass of plants. Furthermore, while captan had an intermediate effect in vitro, it inhibited nodulation only for the first 25 days after treated weeds were planted but then showed no further harmful effects. Benomyl allowed nodulation of plants by one strain of *Rhizobium* but prevented nodulation by another strain, whereas in vitro tests had shown it to be among the least toxic. Fenarimol, inhibitory to *R. trifolii* on agar, did not reduce nodulation or nitrogen fixation; the amount of fungicide reaching the bacteria within the nodule was probably too small to have a detectable effect.[36] A characteristic lack of tap root nodulation in all plants from seeds treated with fungicide at the rate of 2 g kg^{-1} has been observed. This may be due to the leaching of fungicides in the lower layer of the soil. Fisher[4] observed that with root drenches of carbendazim and ethirimol at 25 and 50 kg ha^{-1}, in sterile and unsterile soil, the dry weight and nitrogen content of *R. trifolii*-inoculated white clover plants were not affected. Oxycarboxin (2.5, 5.0, and 15 kg ha^{-1}) significantly decreased nitrogen fixation, whereas ethirimol did not. Wainwright and Pugh,[115] in a survey of the effects of various fungicides, concluded that soil bacteria are tolerant to triarimol in vivo.

C. Foliar Application

Fungicidal sprays used for controlling crop diseases have far-reaching effects that influence more than just the target microorganisms. Prasad and Ramani[116] reported that zineb (0.3%)

and copper oxychloride (0.3%) sprays did not affect the dry weight, total nitrogen, nodulation, or leghemoglobin content. Foliar application of captan and zineb (0.3%) on chickpea adversely affected the weight, number, and leghemoglobin content of nodules.[117]

IX. EFFECT OF SURFACTANT

Surfactants are used as a dormant spray for eradicating overwintering fungal pathogens. Sprays can be cationic, anionic, or nonionic compounds. Sprays containing as much as 4.5% a.i. are necessary and a considerable buildup of surfactants in soil might be expected. Some surfactant toxicity to soil microorganisms in vitro has been observed.[118] Fisher and co-workers[38] have studied the effects of surfactants on nitrogen-fixating bacteria *Rhizobium* spp. They reported that the growth of *R. trifolii* was most inhibited by the cationic spray Hyamine 3500. The antibacterial activity of cationic surfactants against Gram-negative organisms is well established.[119] Fisher[4] reported that the cationic protective fungicide dodine is also inhibitory to bacteria. Fisher et al.[38] observed that anionic Manoxol OT had the highest degree of toxicity. A similar inhibition was caused by the two Aromox compounds; the charge characteristics of these lie between those of cationic and neutral surfactants. Nonionic compounds are less growth inhibiting.[38] These results agree with those of Swisher,[120] who ranked surfactant toxicity as cationic > anionic > nonionic. Above certain concentrations, the nonionics ALK 3, PP 222, and Triton-X45 caused little further growth inhibition; micelle formation is a possible explanation.[38] It was further observed that the surfactant had less effect on the respiration of *R. trifolii;* Hyamine 3500 was again less inhibitory, Manoxol OT and the Aromox compound even less so, whereas ALK 3, PP 222, and Triton X45 caused moderate inhibition at concentrations as high as 2000 $\mu g\ m\ell^{-1}$. Low concentrations of ALK 3 caused a slight stimulation at pH 7.5; this could be due to changes in membrane permeability and consequently increased substrate availability. Although the primary effects of surfactants almost certainly involve changes to the cell membrane, inhibition of respiration is probably a secondary mechanism of toxicity. Surfactants have similar effects on fungal spores.[121]

Fisher and co-workers[38] reported that young clover plants were sensitive to surfactants at very high concentrations (5000 $\mu g\ g^{-1}$) in soil and soil pH was not a significant factor in any of the interactions. Similarly, no significant alteration was observed in *R. trifolii* nodulation and nitrogen fixation by red clover due to surfactants. Fisher et al.[38] reported that the correlation between tests in vitro and in vivo was poor. Hyamine 3500, the most inhibitory compound in vitro, had no effect on fixation in vivo; the cationic surfactant was probably entirely absorbed into negatively charged sites in the soil. The anionic surfactant Manoxol OT, also particularly inhibitory to growth and respiration in vitro, showed a corresponding effect on fixation in vivo. The other compounds affecting fixation, Triton X45 and PP 222, were among those with the least effect in vitro; both are essentially alkyl phenol ethoxylates that are difficult to degrade biologically.[120] Factors such as ease of penetration and partition characteristics, which favor their efficiency as mildew eradicants, are probably also responsible for their inhibitory action within the clover root nodule. Ethylan CP, similar in composition, had no deleterious effect.[4]

Micrographs of nodules from clover plants grown in soil containing Triton X45 and PP 222 showed that the bacterial cell wall contracted so that it no longer filled the space enclosed by the membrane envelope of the host.[38] Degenerated nodules have a similar appearance.[122] This effect has also been associated with the loss of symbiotic effectiveness of nodules caused by autolysis.[123] Membrane damage by surfactants is probably selective; thus, membrane surfaces which are of host origin remain intact after the bacteria within them are destroyed. Although Manoxol OT did not cause cell shrinkage, membrane damage was apparent; changes to nuclear material are consistent with the inhibition of cell division in vivo.[38] Nitrogenase is probably localized in the bacterioid cytoplasm,[124] but interference with membrane function probably results in the inhibition of many intracellular enzymes and the eventual death of the cell. It was further observed that none of the compounds used had any effect on adhesion of the bacterial cells to root hairs. It is evident that the effect of surfactants depends on alterations to the bacterioid membrane rather than the failure of the initial infection process.[38]

However, because of the large number of variables involved, it is difficult to make accurate measurements of effects under field conditions. Until this can be accomplished, the use of eradicant surfactants should proceed with caution.

X. FUTURE OUTLOOK

To meet the imminent crisis in the world food supply, it is important that the resources of the earth be mobilized as rapidly and effectively as possible. Basic to such mobilization is a knowledge of the magnitude of the dynamic processes in the biosphere which affect the availability of N_2, the one element most often lacking in byproducts of foodstuffs. Of paramount importance is the process of biological nitrogen fixation: photosynthesis utilizes freely available CO_2 and draws on the unlimited supply of atmospheric N_2. Its potential role in increasing N_2 availability has long been recognized.

Several methods have been proposed or are used to minimize the nontarget effect of fungicides on rhizobia, because chemicals vary greatly in their effect on rhizobia.[8,30,31] A simple approach is to use those seed treatment protectants that are least deleterious; unfortunately, these are not always the most desirable for control of pathogens. A second approach involves covering the fungicide-treated seeds with a coating such as polyvinyl acetate.[31] A third approach is based on obtaining fungicide-resistant mutants derived from effective rhizobia and applying these to the seeds together with the chemicals.[65] While it may be possible to circumvent the toxic effects of fungicides on seed-applied rhizobia by inoculating with a granular inoculant, as has been done in field bean[1] and soybean,[55] this is not at present an economically viable proposition. Granular inoculation of legumes planted in narrow interrow spacings requires over 28 kg of rhizobial inoculant per hectare and this is too expensive. Accordingly, seed-applied rhizobial inoculants will remain the choice for seed. Thus, it will be necessary to select superior N_2-fixing strains of *Rhizobium* that are tolerant of, if not resistant to, fungicides, as has been reported by Odeyemi and Alexander.[65]

Each procedure is useful in permitting nodulation or increasing yield as compared to farmers' conventional practices. Nevertheless, because of dissimilar farming practices in many regions of the world and the continued introduction of new fungicides and different rhizobia, fungicide toxicity remains an issue.

REFERENCES

1. **Graham, P. H., Ocampo, G., Ruiz, L. D., and Duque, A.,** Survival of *Rhizobium phaseoli* in contact with chemical seed protectants, *Agron. J.,* 72, 625, 1980.
2. **Stovald, G. E. and Evans, J.,** Fungicide seed dressings: their effects on emergence of soybean and nodulation of pea and soybean, *Aust. J. Exp. Agric. Anim. Husb.,* 20, 497, 1980.
3. **Chendrayan, K. and Prasad, N. N.,** Effect of seed treatment with Tillex and Ceresan on Rhizobium groundnut symbiosis, *Madras Agric. J.,* 63, 520, 1976.
4. **Fisher, D. D.,** Effects of some fungicides on *Rhizobium trifolii* and its symbiotic relationship with white clover, *Pestic. Sci.,* 7, 10, 1976.
5. **Anderson, J. R.,** Pesticides effect on nontarget soil microorganisms, in *Pesticides Microbiology*, Hill, I. R. and Wright, J. S. L., Eds., Academic Press, New York, 1978, 333.

6. **Graham-Bryce, I. J.,** Cereal seed treatments: problems and progress, *Proc. 7th Br. Insectic. Fungic. Conf.,* 3, 921, 1973.

7. **Fisher, D. J., Pickard, J. A., and McKenzie, C. M.,** Uptake of the systemic fungicide triadimefon by clover and its effects on symbiotic nitrogen fixation, *Pestic. Sci.,* 10, 75, 1979.

8. **Curly, R. L. and Burton, J. C.,** Compatability of *Rhizobium japonicum* with chemical seed protectants, *Agron. J.,* 67, 807, 1975.

9. **Brinkerhoff, L. A., Fine, G., Karsen, L., and Sinoff, R. A.,** Further studies on the effect of chemical seed treatment on nodulation of legumes, *Plant Dis. Rep.,* 39, 393, 1955.

10. **Peterson, G. W. and Bucholtzmets, W. F.,** Relative toxicity of legume seed protectants to two species of *Rhizobium, Iowa State Coll. J. Sci.,* 29, 95, 1975.

11. **Mundade, D. L., Kande, B. K., and More, S. B.,** Effect of different pesticides on growth of *Rhizobium* sp. and on germination of groundnut seed, *Pesticides,* 14(9), 24, 1980.

12. **Mickail, K. V., Moniband, M., and Abd-El Malik, Y.,** Effects of seed dressing fungicides on Rhizobium and symbiotic nitrogen fixation, *Agric. Res. Rev.,* 57(12), 115, 1980.

13. **Thomas, M.,** Studies on the Effect of Rhizobial Inoculation and Seed Treatment with Fungicides on Nodulation of Legumes, M.Sc. (Ag) thesis, J. N. Agricultural University, Jabalpur, India, 1983.

14. **Mukewari, P. M. and Bhide, B. P.,** Effect of seed treatment with fungicide and antibiotic aureofungin on the efficacy of nodulation by rhizobium strain of groundnut, *Hind. Antibiot. Bull.,* 11, 172, 1969.

15. **Mukewari, P. M. and Bhide, B. P.,** Effect of seed treatment with fungicide and antibiotic aureofungin on the efficacy of nodulation by rhizobium strain of groundnut, *Hind. Antibiot. Bull.,* 12, 75, 1970.

16. **Ostwal, K. P. and Gaur, A. C.,** Effect of seed dressing fungicide and aureofungin on soybean inoculated with *Rhizobium japonicum, Hind. Antibiot. Bull.,* 13, 74, 1971.

17. **Allyn, W. P. and Baldwin, A.,** The effect of oxidation reduction character of the medium on the growth of an aerobic form of bacterium, *J. Bacteriol.,* 21, 417, 1930.

18. **Runholff, M. and Burton, J. C.,** Compatability of rhizobia with seed protectants, *Soil Sci.,* 72, 283, 1951.

19. **Hofer, A. W.,** Selective action of fungicides on *Rhizobium, Soil Sci.,* 86, 282, 1957.

20. **Faizah, A. W., Broughton, W. J., and John, C. K.,** Rhizobia in tropical legumes, survival in the seed environment, *Soil Biol. Biochem.,* 12, 219, 1980.

21. **Bandopadhyaya, S., Bhattacharya, P., and Mukerjee, N.,** In vitro sensitivity of *Rhizobium* spp. of some fungicides and insecticides, *Pesticides,* 13(1), 22, 1979.

22. U.S. Environmental Protection Agency, Registration of pesticides in the United States, proposed guidlines, *Fed. Regist.,* 43(132), Part II, 29696-29741, 1978.

23. **Greaves, M. P., Poole, N. J., Domsch, K. H., Jagnow, G., and Verteaete, W.,** Recommended tests for assessing the side effects of pesticides on the soil microflora, *Tech. Rep. Agric. Res. Counc. Weed Res. Organ.,* 59, 15, 1980.

24. **Baker, Z., Harrison, R. W., and Miller, B. F.,** Action of synthetic detergents on the metabolism of bacteria, *Gen. Exp. Med.,* 73, 249, 1941.

25. **Kittleson, A. R.,** A new class of organic fungicides, *Science,* 115, 84, 1952.

26. **Mathre, D. E.,** Effects of oxathiin systemic fungicides on various biological systemic, *Bull. Environ. Contam. Toxicol.,* 8, 311, 1972.

27. **Mathre, D. E.,** Mode of action of oxathiin systemic fungicides, *J. Agric. Food Chem.,* 19, 872, 1971.

28. **Afifi, N. M., Moharram, A. A., and Hamid, Y. A.,** Sensitivity of *Rhizobium* spp. to certain fungicides, *Arch. Mikrobiol.,* 66, 121, 1969.

29. **Kecskes, M. and Vincent, M. J.,** The effect of some fungicides on *Rhizobium leguminosarum.* II. Investigations in the ligjht chamber and glasshouse, *Agrokem Talajtan,* 18, 461, 1969.

30. **Daitloff, A.,** The effects of some pesticides on root nodule bacteria and subsequent nodulation, *Aust. J. Exp. Agric. Anim. Husb.,* 10, 56, 1978a.

31. **Daitloff, A.,** Overcoming fungicide toxicity to rhizobia by insulating with a polyvinyl acetate layer, *J. Aust. Inst. Agric. Sci.,* 36, 293, 1970.

32. **Ruiz-Saniz, J. E., Beringer, J. E., and Gutierrez-Navarro, A. M.,** Effect of the fungicide captafol on the survival and symbiotic properties of *Rhizobium trifolii, J. Appl. Bacteriol.,* 57, 361, 1984.

33. **Kecskes, M.,** Comparative investigation of the action of fungicides on *Rhizobium leguminosarum* Frank and its symbiosis with *Vicia sativa* L., *Meded. Fac. Landbouwwet. Rijksuniv. Gent,* 35, 505, 1970.

34. **Makawi, A. A. M. and Abdul-Ghaffar, A. S.,** The effect of some pesticides on the growth of root-nodule bacteria, *U.A.R. J. Microbiol.,* 5, 109, 1970.

35. **Sud, R. K. and Gupta, K. G.,** On the sensitivity of isolates of *Rhizobium* and *Azobacter chroococcum* to TMTD and its degradation product NADDC, *Arch. Mikrobiol.,* 85, 19, 1973.

36. **Staphorst, J. L. and Strijdom, B. W.,** Effects of rhizobia of fungicides applied to legume seed, *Phytophylactica,* 8, 47, 1976.

37. **Fisher, D. J. and Clifton, G.,** Effect of fungicides on Rhizobium, *Long Ashton Res. Stn. Univ. Bristol Rep. for 1975,* p. 126, 1976.

38. **Fisher, D. J., Hayes, L. N., and Jones, C. A.,** Effects of some surfactant fungicides on *Rhizobium trifolii* and its symbiotic relationship with white clover, *Ann. Appl. Biol.,* 90, 73, 1978.

39. **Fisher, D. J. and Hayes, L. N.,** Effects of some fungicides used against cereal pathogens on the growth of *Rhizobium trifolii* and its capacity to fix nitrogen in white clover, *Ann. Appl. Biol.,* 28, 101, 1981.

40. **Fisher, D. J. and Hayes, L. N.,** Effects of some systemic imidazole and triazole fungicides on white clover and symbiotic nitrogen fixation by *Rhizobium trifolii, Ann. Appl. Biol.,* 101, 19, 1982.

41. **Buchenauer, H.,** Mode of action of triadimefon in *Ustilago avenae, Pestic. Biochem. Physiol.,* 7, 309, 1977.

42. **Sijpesteijn, A. K.,** Effect on fungal pathogens, in *Systemic Fungicides,* Marsh, R. W., Ed., Longman, London, 1977, 131.

43. **Garg, F. C., Tauro, P., and Grover, R. K.,** Effect of fungitoxicants on rhizobia, *Indian J. Microbiol.,* 13, 179, 1977.

44. **Chamber, M. A. and Mantas, F. J.,** Effects of some seed disinfectants and methods of rhizobial inoculation of soybeans *(Glycine max* L. Merrill), *Plant Soil,* 66, 353, 1982.

45. **Tu, C. M.,** Effect of fungicides on growth of *Rhizobium japonicum* in vitro, *Bull. Environ. Contam. Toxicol.,* 25, 364, 1980.

46. **Tu, C. M.,** Influence of pesticide seed treatments on *Rhizobium japonicum* and symbiotically grown soybean in soil under laboratory conditions, *Prot. Ecol.,* 3, 41, 1981.

47. **Tu, C. M.,** Effects of some pesticides on *Rhizobium japonicum* and on the seed germination and pathogens of soybean, *Chemosphere,* 11, 1027, 1982.

48. **Heinonen-Tanski, H., Oros, G., and Kecskes, M.,** The effect of soil pesticides on the growth of red clover rhizobia, *Acta Agric. Scand.,* 32, 283, 1982.

49. **Tu, C. M.,** Effect of fungicidal seed treatments on alfalfa growth and nodulation by *Rhizobium melitoti, Chemosphere,* 10, 127, 1981.

50. **Duczek, L. J. and Buchan, J. A.,** The effect of captan seed treatment on emergence, nodulation, seed yield and seed protein content of lentils, *Can. J. Plant Sci.,* 61, 727, 1981.

51. **Hill, I. R. and Wright, S. J. L.,** *Pesticides Microbiology,* Academic Press, 1978, 844.

52. **Fisher, D. J.,** A note on the mode of action of the systemic fungicide tridemorph, *Pestic. Sci.,* 5, 219, 1974.

53. **Borgers, M.,** Mechanism of action of antifungal drugs, with special reference to the imidazole derivatives, *Infect. Dis. Rev.,* 5, 520, 1980.

54. **Richardson, L. T.,** The persistence of thiram in soil and its relationship to the microbiological balance and damping off control, *Can. J. Bot.,* 32, 335, 1954.

55. **Rennie, R. J. and Dubetz, S.,** Effect of fungicides and herbicides on nodulation and nitrogen fixation in soybean lacking indigenous *Rhizobium japonicum, Agron. J.,* 76, 451, 1984.

56. **Rennie, R. J., Dubetz, S., Bole, J. B., and Muendal, H. F.,** Dinitrogen fixation measured by ^{15}N isotope dilution in two Canadian soybean cultivars, *Agron. J.,* 74, 725, 1982.

57. **Alikhan, S. T. and Zimmer, R. C.,** Production of Field Peas in Canada, Publ. 1710, Agriculture Canada, Ottawa, Ontario, 1980.

58. **Slinkard, A. E., Drew, B. N., and Kirkland, K. J.,** Dry Pea Production in Saskatchewan, Publ. 225, University of Saskatchewan, Saskatoon, 1977.

59. **Rowland, G. G. and Drew, B. N.,** Fababean Production in Saskatoon, Publ. 416, University of Saskatchewan, Saskatoon, 1982.

60. **Alikhan, S. T.,** Seed Inoculation in Pulse Crops, Canadex 255.24, Agriculture Canada, Ottawa, Ontario, 1980.

61. **Slinkard, A. E. and Buchan, J.,** Inoculation of Pulse Crops, Agedex 149.23, Publ. 381, University of Saskatchewan, Saskatoon, 1982.

62. **Rennie, R. J., Howard, R. J., Swanson, T. A., and Flores, G. H. A.,** The effect of seed-applied pesticides on growth and N_2-fixation in pea, lentil and fababean, *Can. J. Plant Sci.,* 65, 23, 1985.

63. **Gillbarg, B. O.,** On the effects of some pesticides on *Rhizobium* and isolation of pesticides resistance mutants, *Arch. Mikrobiol.,* 75, 203, 1971.

64. **Odeyemi, O. and Alexander, M.,** Resistance of *Rhizobium* strains to Phygon, Spergon and thiram, *Appl. Environ. Microbiol.,* 33, 784, 1977.

65. **Odeyemi, O. and Alexander, M.,** Use of fungicide resistant rhizobia for legume inoculation, *Soil Biol. Biochem.,* 9, 247, 1977.

66. **Lennox, L. B. and Alexander, M.,** Fungicide enhancement of nitrogen fixation and colonization of *Phaseolus vulgaris* by *Rhizobium* phaseoli, *Appl. Environ. Microbiol.,* 41, 404, 1981.

67. **Gupta, K. G. and Shirkot, C. K.,** Tetramethylthiuramdisulphide (TMTD) resistance in slow growing rhizobia, *Soil Biol. Biochem.,* 60, 399, 1981.

68. **Mendez-Castro, F. A. and Alexander, M.,** Method for establishing a bacterial inoculum on corn roots, *Appl. Environ. Microbiol.,* 45, 248, 1983.

69. **Hossain, A. K. M. and Alexander, M.,** Enhancing soybean rhizosphere colonization by *Rhizobium japonicum, Appl. Environ. Microbiol.,* 48, 468, 1984.

70. **Ramirez, C. and Alexander, M.,** Evidence suggesting protozoan predation on rhizobium associated with germinating seeds and in the rhizosphere of beans *(Phaseolus vulgaris* L.), *Appl. Environ. Microbiol.,* 40, 492, 1980.

71. **Ingham, E. R.,** Review of the effects of 12 selected biocides on target and nontarget soild microorganisms, *Crop. Prot.,* 4(1), 3, 1985.

72. **Grag, F. C., Tauro, P., and Grover, R. K.,** Effect of fungitoxicants on rhizobia, *Indian J. Microbiol.,* 13, 179, 1973.

73. **Hooykaas, P. J. J., van Brussel, A. A. N., Den DulkRas, H., van Slogteren, G. M. S., and Schilperoort, R. A.,** Sym plasmid of *R. trifolii* expressed in different rhizobial species and *A. tumefaciens, Nature (London),* 291, 351, 1981.

74. **Ruiz-Vazquez, R., Pueyo, C., and Cerda-Olmedo, E.,** A mutagen assay detecting forward mutations in an arabinose-sensitive strain of *S. typhimurium, Mutat. Res.,* 54, 121, 1978.

75. **MacKenzie, K. A. and MacRae, I. C.,** Tolerance of nitrogen fixing system of *Azotobacter vinelandii* for four commonly used pesticides, *Antonie van Leeuwenhoek J. Microbiol. Serol.,* 38, 529, 1972.

76. **Schwinghamer, E. A.,** Association between antibiotic resistance and ineffectiveness in mutant strains of *Rhizobium* spp., *Can. J. Microbiol.,* 10, 221, 1964.

77. **Allington, W. B.,** Results of the uniform soybean seed treatment test in 1944, *Plant Dis. Rep.,* 29, 219, 1945.

78. **Allison, J. L. and Torrie, J. H.,** Effect of several seed protectants on germination and stands on various forage legumes, *Phytopathology,* 34, 799, 1944.

79. **Miller, L. I.,** Root nodulation of Holland Jumbo strain of peanuts grown from seed treated with a fungicide, *Phytopathology,* 38, 48, 1948.

80. **Vlitos, A. J. and Preson, P. A.,** Seed treatment for field legumes, *Okla. Agric. Exp. Stn. Bull.,* 332, 3, 1949.

81. **Nedelchev, N.,** Effect of the date of soybean seed dressing with fungicides on number of nodules and plant productivity, *Rastenievud. Nauki.,* 18(3), 110, 1981.

82. **Furtode, A.,** Effects of systemic fungicide on soil microbial population and nodulation of *Rhizobium* spp., *Thesis Abstr.,* 3(4), 286, 1977.

83. **Narayan, Y. D., Radhakrishna, D., and Rai, P. V.,** Effect of certain fungicides on nodulation of cowpea *(Vigna unguiculata* Walp.), *Curr. Res.,* 10(3), 23, 1981.

84. **Balaraman, K. and Prasad, N. N.,** Influence of wet Ceresan seed treatment on nitrogen fixation by *Rhizobium* in groundnut, *Pesticides,* 7(1), 11, 1973.

85. **Sardeshpande, J. S., Kulkarni, J. H., and Bagyaraj, P. J.,** Influence of fungicidal seed treatment on the nodulation of groundnut inoculated with *Rhizobium, Pesticides,* 7(4), 11, 1973.

86. **Singh, D. V. and Singh, R. R.,** Search of safer fungicides for treating the pea seeds in combination with Rhizobium, *Pesticides,* 14(1), 34, 1980.

87. **Duczek, L. J.,** The effect of captan, thiram and MBC seed treatment on emergence, nodulation and yield of lentil and peas, Pest. Res. ECPUAA Con., Ottawa, Ontario, 1979, 404.

88. **Manikasundaram, G. and Rambadam, R.,** Effect of different fungicidal application on nodulations of groundnut varieties, *Madras Agric. J.,* 63, 644, 1976.

89. **Bandopadhyaya, S., Bhattacharya, P., and Mukerjee, N.,** Effect of some seed treating fungicides on nodulation of five legumes, *Pesticides,* 17(3), 27, 1983.

90. **Chandra, G., Shrivastava, R. C., and Mathur, S. N.,** Responses of *Vigna munga* (L.) Hepper to seed treatment of malathion and carbendazim and effect on growth, nodulation and yield, *Pestology,* 7, 22, 1983.

91. **Elek, E. and Kecskes, M.,** The effect of some seed treatments using micro elements and fungicides on vetch plants grown from seeds inoculated with *Rhizobia,* in Symp. Biologica Hungarica, Proc. Symp. Soil Microbiology, Vol. 2, Szegi, J., Ed., Budapest, Hungary, 1972, 431.

92. **Taha, S. M., Mahmood, S. Z. Z., and Saleem, S. H.,** Effect of pesticides on *Rhizobium* inoculation and symbiotic nitrogen fixation of some leguminous plants, in Symp. Biologica Hungarica, Proc. Symp. Soil Microbiology, Szegi, J., Ed., Budapest, Hungary, 1972, 423.

93. **Thomas, M. and Vyas, S. C.,** Nodulation and yield of chickpea treated with fungicides at sowing, *Int. Chickpea Newsl.,* 11, 37, 1984.

94. **Thomas, M. and Vyas, S. C.,** Effect of fungicides seed treatment on pigeonpea nodulation, *Int. Pigeonpea Newsl.,* p. 4, 1985.

95. **Pande, J. S. S., Kulkarni, J. H., and Bagyaraj, P. J.,** Influence of fungicidal seed treatment on the nodulation of groundnut inoculated with *Rhizobium, Pesticides,* 7, 11, 1973.

96. **Virk, K. S.,** Effect of thiophanate-methyl on leaf pigments and nodulation of few crop plants, *Haryana Agric. Univ. J. Res.,* 11(3), 408, 1981.

97. **Tu, C. M.,** Effects of pesticides seed treatments on *Rhizobium japonicum* and its symbiotic relationship with soybean, *Bull. Environ. Contam. Toxicol.,* 18, 190, 1977.

98. **Wrobel, T.,** Influence of fungicides on symbiosis of *Rhizobium* with pea and lupines, *Acta Microbiol. Pol.,* 12, 203, 1963.

99. **von Schemeling, B. and Clark, M.,** Oxathiin induced plant growth stimulation, 7th Int. Congr. Plant Protection, Paris, Abstr., 1970.

100. **Thapliyal, P. N. and Sinclair, J. B.,** Translocation of benomyl, carboxin and chloroneb in soybean seedlings, *Phytopathology,* 61, 130, 1971.

101. **Vyas, S. C.,** *Systemic Fungicides,* Tata McGraw-Hill, New Delhi, India, 1984.

102. **Karanth, N. G. K. and Vasantharajan, V. N.,** Phytotoxicity of Dexon (p-dimethylaminobenzenediazosodium sulfonate) towards the legume *(Crotalaria juncea), Proc. Indian Natl. Acad. Sci. Part B,* 40, 576, 1975.

103. **Dixon, R. O.,** Rhizobia, *Annu. Rev. Microbiol.,* 23, 137, 1969.

104. **Vyas, S. C., Thomas, M., and Gajendragadkar, G. R.,** Effect of fungicide seed treatment on nodulation of peas and pea cultivars, *Pesticides,* 19(8), 53, 1985.

105. **Bhattacharya, P. and Sengupta, K.,** Effect of seed dressing fungicides on nodulation and grain yield of chickpea, *Int. Chickpea Newsl.,* 11, 41, 1984.

106. **Kreutzer, W. A.,** Selective toxicity of chemicals to soil microorganisms, *Annu. Rev. Phytopathol.,* 1, 101, 1963.

107. **Munnecke, D. E.,** A biological assay of nonvolatile diffusible fungicides in soil, *Phytopathology,* 48, 499, 1958.

108. **Clemons, G. P. and Sisler, H. D.,** Formation of fungitoxic benomyl derivatives, *Phytopathology,* 59, 705, 1969.

109. **Vonk, J. W. and Sijpesteijn, A. K.,** Methyl benzimidazole-2yl carbamate, the fungitoxic principal of thiophanate-methyl, *Pestic. Sci.,* 2, 160, 1971.

110. **Dimond, A. E.,** Effects of physiology of the host and on host/pathogen interactions, in *Systemic Fungicides,* Marsh, R. W., Ed., Longman, London, 1972, 116.

111. **Graham-Bryce, I. J. and Coutts, J.,** Interactions of pyrimidine fungicides with soil and their influence on uptake by plants, *Proc. 6th Br. Insectic. Fungic. Conf.,* 2, 419, 1971.

112. **Hill, I. R. and Arnold, D. J.,** Degradation of ethirimol in soil, *Proc. 7th Br. Insectic. Fungic. Conf.,* 1, 47, 1973.

113. **Kecskes, M. and Vincent, M. J.,** The effect of some fungicides on *Rhizobium leguminosarum.* I. Laboratory investigations, *Agrokem. Talajtan,* 18, 57, 1969.

114. **Kratka, J. and Ujevic, M.,** Seed treatment of winter vetch with fungicides and their effect on formation of root nodules, *Pol'nohospodarstvo,* 16, 203, 1970.

115. **Wainwright, M. and Pugh, G. J. F.,** Effect of fungicides on the numbers of microorganisms and frequency of cellulytic fungi in soil, *Plant Soil,* 43, 561, 1975.

116. **Prasad, N. N. and Ramani, C. E.,** Effect of certain pesticides on nodulation and nitrogen fixation by *Rhizobium* sp., *Madras Agric. J.,* 63, 646, 1976.

117. **Annapurna, T. and Rao, P. R.,** Effect of foliar spray of captan and Dithane Z-78 on the nodulation of *Cicer arietinum, Madras Agric. J.,* 67, 831, 1980.

118. **Hislop, E. C., Barnaby, V. M., and Burchill, R. T.,** Aspect of the biological activity of surfactants that are potential eradicants of powdery mildew, *Ann. Appl. Biol.,* 87, 29, 1977.

119. **Baker, Z., Harrison, R. W., and Miller, B. F.,** Action of synthetic detergents on the metabolism of bacteria, *J. Exp. Medicine,* 73, 249, 1941.

120. **Swisher, R. D.,** *Surfactant Biodegradation,* Marcel Dekker, New York, 1970.

121. **Forsyth, F. R.,** Surfactants as fungicides, *Can. J. Bot.,* 42, 1335, 1964.

122. **Mosse, B.,** Electron-microscope studies of nodule development in some clover species, *J. Gen. Microbiol.,* 36, 49, 1964.

123. **Bassett, B., Goodman, R. N., and Novackya, A.,** Ultrastructure of soybean nodules. II. Deterioration of symbiosis in effective nodules, *Can. J. Microbiol.,* 23, 873, 1977.

124. **Planque, K., Kennedy, I. R., Vries, G. E., Quispel, A., and Van Brussel, A. A. N.,** Location of nitrogenase and ammonia-assimilatory enzymes in bacteriods of *Rhizobium leguminosarum* and *Rhizobium lupini, J. Gen Microbiol.,* 102, 95, 1977.

Chapter 15

BIOCONTROL AGENTS

I. INTRODUCTION

Extensive research has been conducted over the last several decades to evolve methods of biological control on a range of plant pathogens, both foliar and soil-borne, and, most recently, seed-borne. Despite this, practical methods of biocontrol have not yet materialized. This may be due to several handicaps: not enough effort to study biocontrol per se, build theories, or reveal the principle of production systems; inattention to the importance of the ecological complexity continuum principle, which can lead scientists to begin their studies with the more complex ecosystems rather than with the simpler ones; economic feasibility; and lastly, ignorance of modern tools for biotechnology and bioengineering. Perhaps success in reducing losses due to the most difficult to control soil-borne pathogens most likely depends on a clever blending of biocontrol technology[1] with other methods of control.

An integration of several control technologies may be used to reduce pathogenic activities to a tolerable or permissible threshold, with chemicals applied only when absolutely necessary. If blending involves pesticides, such an approach can be successful only if the biocontrol agents are compatible with such pesticides and also possess tolerance/resistance.

Research conducted in the last decades on the resistance developed by plant pathogenic fungi as a result of exposure to modern fungicides has been reviewed.[2-4] Delp,[5] who used the words "resistance" and "tolerance" interchangeably, suggested several mechanisms of resistance to fungitoxicants. Possible roles of genetically active fungicides in causing the formation of new fungal biotypes have been discussed.[6] Despite the abundance of research work conducted on the tolerance of plant pathogens to fungicides, virtually no research has been done on the tolerance of biocontrol fungi to fungicides or on the purposeful induction (nontarget effect) of tolerance in biocontrol agents for use in the integrated control of plant diseases. Papavizas et al.[7] were the first to report tolerance induction in *Trichoderma harzianum* to benomyl by UV light irradiation and selection. They observed that tolerance was a stable characteristic of the new biotypes. Biological control agents must be effective in suppressing a particular disease and must be compatible with fungicides used to control other diseases. Results of research on the biological control of plant pathogens using antagonists and fungicides have been presented in Table 1. Tolerance to carbendazim fungicides in biotypes of *Trichoderma* or *Talaromyces* developed by UV radiation or chemical mutagens would be an important characteristic and a first step in making them good candidates for blending technologies. Formulations of such biotypes could be used in tank mixtures of fungicides or fertilizers in spray programs or in soil together with fungicides (e.g., use of the benomyl-resistant biotype T-1-g[9] of *Trichoderma viride* together with benomyl to control soil-borne Fusarium wilt of chrysanthemum).[1] The use of some tolerant strains of fungi together with selective pesticides may offer a distinct possibility of control of soil-borne diseases where no single components are currently effective.[1]

Table 1

BIOLOGICAL CONTROL OF PLANT PATHOGENS AFTER DEVELOPMENT OF FUNGICIDAL RESISTANCE IN ANTAGONISTS AND BLENDING WITH FUNGICIDES

Host/pathogen systems	Antagonist(s)	Fungicide(s)	Ref.
Mustard/*Alternaria brassicicola*	*Trichoderma harzianum, Pencillium earylophilium, P. Oxalicum, Gliocladium virens*	Benomyl, delan, iprodione, metiram	51
Iris/*Rhizoctonia solani, Sclerotium rolfsii*	*T. harzianum*	Carbendazim	39
Beans/ *S. rolfsii*	*T. viride*	Carbendazim	31
Peas/*Pythium ultimum*	*T. viride*	Carbendazim	31
Melon/*Fusarium oxysporum* f. sp. *melonis*	*T. viride*	Carbendazim	31
Onions/*S. cepivorum*	*T. harzianum*	Chlorothalonil, procymidone, iprodione, vinclozolin	30
Potato/*Verticillium dahliae*	*Talaromyces flavus*	Benomyl, captan, mancozeb, maneb, metiram	32
Peas/*P. ultimum, F. solani* f. sp. *pisi*	*Trichoderma* spp.	Benomyl, metalaxyl	12, 13
Pineapple and avocado/*Phytophthora cinnamomi*	Soil biota	Metalaxyl	13
Apricot/*Eutypa armeniacae*	*F. lateritium*	Benomyl	47, 48
Sugar beet/*Pythium aphanidermatum*	*T. harzianum*	Metalaxyl	28
Soybean/*R. bataticola*	*T. harzianum*	Carbendazim	29
Cyclamen/*Botrytis cinerea*	*P. brevicompactum*	Benomyl	44
Peas/*F. oxysporum, Aphanomyces euteiches, Pythium* spp., *R. bataticola*	*P. oxalicum*	Captan	38

Blending control technology needs to be stressed as a part of integrated control in that the approach may be directed against more than one pathogen. Such an approach would seek to develop resistance in the antagonist to the fungicide to be used and to use a selective fungicide not toxic to the antagonist, but at the same time effective against the pathogen(s). According to Todd,[8] a multipurpose treatment might simply be defined as one that provides control for two or more diseases. It is possible that plant pathologists and soil microbiologists, involved in the intriguing prospects of biological and cultural control of soil-borne plant pathogens, have neglected the possibility of integrated control of plant disease. Integrated control is defined by Irving[9] as a compatible system of control in which various methods are used in the proper sequence and timed so as to cause the least hazard to man and the environment and to permit the maximum assistance to natural control. The term is a common phrase for a practical approach conceived by modern environmentalists who designed an approach in which all methods of biological control may be brought into operation to reduce pathogenic activities to a tolerable level. It is obvious that integrated control is no more than blending of chemical and biological control (including disease resistance), with the intent of shifting emphasis from chemical to biological control. Papavizas[1,10] has comprehensively reviewed the status of biological control of soil-borne pathogens and strongly recommended that the blending technique should be adopted to fight soil-borne diseases by integrated control. Several proposals have been tested. The one with the most potential is the combination of chemical seed treatments and *Trichoderma,* which increases the level of disease control more than resistance germ plasm.[7,11-14] Biological control of plant diseases does not involve the use of fungicides, but an integrated pest management system advocates that no components of plant disease control be excluded. Hence, an attempt has been made to review the nontarget effects of agricultural fungicides in the ecology of fungi related to the biological control of plant pathogens.

II. SOIL-BORNE DISEASES

The nontarget effects of fungicides in the biological control of plant diseases perhaps offers greater potential of manipulating the microbial balance than other methods. A major objection to many chemical controls currently in use is their indiscriminate effect on organisms other than the pathogen; the intent of fungicides in the biological control of plant pathogen is not to kill the pathogen directly but to weaken it and make it more vulnerable to the antagonism of the associated microflora. Alternately, fungicides and fungistats directed at some purely saprophytic elements of the flora to shift the microbial balance may indirectly control the nontarget microorganisms.

A few reports are available on fungicides which controlled a disease even though the materials or rates used were not considered directly fungitoxic. One of the earlier examples of simple integrated control of a soil-borne pathogen is the suppression of *Armillaria mellea* mediated indirectly by fumigating the soil with nontoxic concentrations of carbon disulfide.[15] The fumigant stimulated the multiplication in soil of the antagonistic fungus *Trichoderma*. *A. mellea* was destroyed in soil by the antibiotic action of *Trichoderma* by the end of the 7-day fumigation period, but was not complete until after 24 days. Control of *Aphanomyces cochloides* and *Pythium aphanidermatum* of sugar beets with thiram was partially the result of direct fungicidal action on the pathogen, but due more to nontarget effects caused by a shift in the microbial balance.[16] The control of damping-off with captan at rates not considered directly fungicidal to the pathogen suggested that the biological control initiated by the chemical was involved.[17] Although *Helminthosporium sativum* conidia were not inhibited by 5 ppm of methylmercurydicyandiamide (MMDD), this rate controlled root rot of wheat caused by the fungus.[18] Certain microorganisms, especially *Penicillium* spp., were observed to have increased profusely in the treated soil. *Pythium ultimum* was difficult to establish in soil treated with thiram, even after the fungicide was largely dissipated 2 months after treatment. Even though residual amounts of the compound were low and probably sublethal, the antagonistic mycoflora prevented the natural buildup of the pathogen in soils recropped to peas.[19]

Pincard[20] found that field application of monosodium salt of hexachlorophene at 280 g ha^{-1} to cotton plants resulted in a significant decrease in disease control. Altering the microbial flora of the soil with very small dosages of certain selective chemicals may help to stimulate another type of microbial flora of substantial antagonism to troublesome root pathogens. The evidence suggested that the observed field control of several cotton diseases with amounts of a chemical nontoxic to the pathogens *(Rhizoctonia, Pythium, Phytophthora, and Fusarium)* is indirect and is related to the ability of the hexachlorophene to alter the ecological balance of the soil.

A parallel phenomenon was reported by Stankova-Opocenska and Dekker.[21] *Pythium debaryanum,* which causes damping-off in many plants, is insensitive in vitro to the systemic fungicide 6-azauracil, even at 1000 ppm. They obtained protection of cucumber seedlings against this pathogen by soaking the seed in a solution of 1 ppm 6-azauracil, but not when the seed was soaked in an aqueous solution of 10 or 100 ppm. The microflora in the rhizosphere of developing seedlings was changed by 6-azauracil: bacteria increased and fungi decreased. The authors suggested that the control was obtained via the increased microflora at 1 ppm but not at 10 or 100 ppm.[21]

Fumigation of the soil with chloropicrin controlled pineapple root rot caused by *Phytophthora cinnamomi* in Hawaii for a period of 3 years, although the pathogen could be isolated from soil as early as 2 weeks after treatment.[22] Fumigation resulted in an increase in the population of *Trichoderma viride.* On the other hand, pentachloronitrobenzene (PCNB) increased disease severity over that in nontreated plots and reduced counts of *Trichoderma, Pencillium,* and actinomycetes. Application of ethyl mercury phosphate at 56 kg ha^{-1} applied 10 cm deep also increased the disease, but 168 kg ha^{-1} controlled it. As the *Trichoderma* population decreased in these trials, the disease increased, and vice versa. The fungus was an active antagonist in moist soil, but was suppressed under wet conditions; it was also inhibited by a soil pH above 5.4. Application of coral sand to temporary sugarcane fields greatly increased soil pH and heart rot caused by *Phytophthora parasitica* in subsequent plantings of pineapple, due to the suppression of *T. viride.*[11] Disease control was thus thought to be indirect, through the effect of treatments on organisms that ordinarily inhibit *Trichoderma* or other antagonists of *Phytophthora.*

All fungicides do not bring about disease control indirectly. Where the fungicide selectively inhibits antagonists more than the pathogen, or where competing pathogens are affected differently by the fungicide, the pathogen affected least by the fungicide may cause considerably more damage than without the fungicide. Fenaminosulf is a fungicide specific for the control of *Pythium* spp.; application of this fungicide increased the severity of dry rot due to *Rhizoctonia* spp. on several crops.[23] In still other instances, control of both *R. solani* and *Pythium* increased seedlings damage caused by *Fusarium.* Fenaminosulf and PCNB destroyed the antagonists of *Rhizoctonia* and *Pythium,*[24] respectively. Increased damping-off of conifer seedlings caused by *Pythium* sp. has been observed. The pathogen made better growth through soil incubated in the presence of PCNB vapors than through untreated soil. *Penicillium paxieli* was common in untreated soil, but not in PCNB-amended soil. Moreover, the antagonistic fungus inhibited the growth of *Pythium* in vitro in the absence of PCNB. If *P. paxieli* is in fact the cause of the partial suppression of *Pythium* in untreated soil, then fungicides would be selected on the basis of their preferential effect on the pathogen over the antagonist.[24] PCNB is a broad-spectrum fungicide but has a narrow spectrum of activity against bacteria. The suppression of PCNB-sensitive fungi reduces the competition for nutrients of more tolerant pathogens like species of *Fusarium* and *Pythium,* which then became more active under such situations.[25] Glucose is utilized more slowly in soil with PCNB than without PCNB, but without glucose as an energy source for microbial growth, PCNB had no effect on soil respiration or microbial number; hence PCNB selectively affected growth response to nutrients.

Many soil fungicides alter antagonistic flora and thus control disease; without this effect the fungicide would only be partially effective.[26] Fenaminosulf alone gave significant control of damping-off of conifers, but the stand was poor when fenaminosulf was mixed with anilazine or PCNB. It was suggested that mixing these fungicides with fenaminosulf hampered biological control. This effect on the soil microflora can explain the great variation in disease control from year to year that workers obtain in fungicide trials.

Unexpected interactions have been observed among systemic fungicides and microflora in the control of pathogenic fungi. While controlling *Phytophthora cinnamomi* by metalaxyl soil applications, hyphal lysis was more stimulated in leachates from unsterile soil than in leachates from sterile soil with equivalent concentrations of fungicide.[13]

Damping-off and root rot caused by *Pythium ultimum* and *Fusarium solani* f. sp. *pisi* of peas are serious diseases that can be controlled by a combination of seed treatments with metalaxyl or benomyl (0.6 active ingredient (a.i.) kg^{-1}) and *Trichoderma viride* spores at a concentration of 3.8 $\times$ 10^{10} ppm/500 g seed plus 1 g methyl cellulose to act as sticker.[13] The fungicide was infused into the seed by acetone before applying *Trichoderma* spores.[12] An interesting observation was that populations of *Trichoderma* were not reduced as much when seeds were coated with metalaxyl and *Trichoderma* as with *Trichoderma* alone. Papavizas also reported similar results.[14] Earlier Papavizas and Papavizas et al.[7] developed

Table 2

**BIOLOGICAL CONTROL OF DRY ROOT ROT OF
SOYBEAN CAUSED BY *RHIZOCTONIA BATATICOLA* BY
SEED INOCULATION WITH TOLERANT (T) AND WILD
(W) STRAINS OF *TRICHODERMA HARZIANUM* (TH) TO
CARBENDAZIM AND SEED TREATMENT WITH
CARBENDAZIM ALONE AND IN COMBINATION WITH
THE ANTAGONIST**

Treatment	Concentration	Infected plants (%)
Seed alone (control)	—	60.00
Seed + TH W	5 g kg^{-1}	32.68
Seed + TH T	5 g kg^{-1}	31.00
Seed + carbendazim	1000 ppm	29.80
Seed + carbendazim + TH W	1000 ppm + 5 g kg^{-1}	25.40
Seed + carbendazim + TH T	1000 ppm + 5 g kg^{-1}	10.40
Critical difference at 5%		7.90

and used strains of *Trichoderma* resistant to metalaxyl and benomyl up to 100 μg mℓ$^{-1}$ and these strains were used in this study. A recent trend in research is to develop resistance in effective antagonists in order to make them compatible with fungicides in the integrated control of soil-borne diseases. Strains of *T. harzianum* were obtained that are resistant to soil fungicides such as quintozene, thiram, and captan.[7]

Damping-off of sugar beet and tobacco caused by *P. aphanidermatum* is most destructive in the seedling stage and its control using *T. harzianum* has been studied.[28] It was reported that the control of damping-off was augmented by blending the antagonist and metalaxyl seed treatment (0.1%); however, the antagonist or metalaxyl alone did not yield the same degree of control.

The unavailability of commercially viable, agronomically suitable, and disease-resistant cultivars and the expense of pesticides and cultural control measures prompted a search for alternative methods of plant disease control. Very recently, Papavizas[1] reported on the blending technique of plant disease control based on the utilization of the antagonist and the fungicide simultaneously (blending) for the control of soil and foliar plant diseases. In this way, low doses of fungicides can be used to save cost. Excessive application of pesticides which pollute the atmosphere may also be avoided. Vyas and Khare[29] showed that the antagonist *T. harzianum* showed tolerance towards increasing concentrations of carbendazim in successive transfers from lower to higher. In the first series, the antagonist was exposed initially to 100, 200, 500, and 1000 ppm of carbendazim on potato dextrose agar (PDA); growth was observed on 100 and 200 ppm while concentrations of 500 and 1000 ppm were highly toxic. For the second series, the inoculant disks were taken from the culture that initially grew in 200 ppm. Thus, during the fifth series, a strain was obtained by gradual transfers that could grow normally at carbendazim concentrations of 500 and 1000 ppm as if on plain PDA. This strain was maintained on 1000 ppm for several days to stabilize the tolerance. Similar results were also obtained by Abd-El Moity et al.[30] and Papavizas and Lewis.[31] This tolerant strain was used for seed inoculation.

For the control of soil-borne diseases for which other methods of plant disease control fail, biological control is the only viable proposition.[1] The data presented in Table 2 indicate that the pathogen *Rhizoctonia bataticola,* which causes dry root rot in soybean, results in a 60% reduction in the crop stand. This seems to accompany a significant loss in plant density, particularly in Kharif (July through October) and in light soils where the activity of *R.*

Table 3

**BIOLOGICAL CONTROL OF DRY ROOT ROT OF
SOYBEAN CAUSED BY *RHIZOCTONIA BATATICOLA* BY
AMENDING INFESTED SOIL WITH THIS FUNGUS AND
LATER BY ANTAGONIST *TRICHODERMA HARZIANUM*
(TH) WILD (W) AND TOLERANT (T) STRAINS AND
CARBENDAZIM ALONE AND IN COMBINATION**

Treatment	Concentration	Infected plants (%)
Seed (control)	—	86.00
TH W	5 g kg^{-1}	29.20
TH W + carbendazim	5 g kg^{-1} + 0.1%	15.20
TH T	5 g kg^{-1}	25.00
TH T + carbendazim	5 g kg^{-1} + 0.1%	3.00
Critical Difference at 5%		6.90

bataticola and *Sclerotium rolfsii* prevails. Seed inoculation with wild and tolerant strains of *T. harzianum* significantly reduced the incidence of infected plants. There was no significant difference between the strains with fungicide alone, but they were significantly superior to the control. However, seed inoculation and carbendazim treatment were significantly superior to other treatments (Table 2). Also, the infected plant stands raised from seeds treated with tolerant strains were significantly superior to those inoculated with wild strains. Carbendazim, a systemic fungicide, possesses high effectivity towards *S. rolfsii* and *Fusarium* spp., which are the other important pathogens of soybean. Similar results were also reported by Kraft and Papavizas[13] for peas and by Abd-El Moity et al.[30] for onions with different pathogen host systems.

Another experiment was conducted to test the suitability of this method by amending the soil with carbendazim and tolerant and wild strains of *T. harzianum,* which were earlier amended with *R. bataticola*. In this case, the treatment combination of the *T. harzianum*-tolerant strain and amended carbendazim caused a significant reduction in infected plants as compared to the control and other treatments (Table 3). Although significant reduction in mortality has been observed under both conditions, taking the cost and other factors into consideration, it is concluded that seed inoculation and fungicidal seed treatment have potential for soil-borne disease control. Seed-treatment fungicides such as carbendazim and soil treatment yielded identical results, but in this situation reinfestation often occurs.[29]

III. SEED-BORNE DISEASES

Treatment of seeds and seed pieces is a common method of applying fungicides to control plant pathogens and is an effective and efficient way to introduce biological control agents. Fungal preparations have been applied to the seeds of a variety of crops in an attempt to control a number of seed-borne and soil-borne plant pathogens.[32] Biological control agents must be effective and compatible with current production practices in order to become readily integrated into production systems. Biological control agents which are generally used to suppress a particular disease must be compatible with the fungicides which are used to control other diseases. Ferries and Mitchell[33] reported the effect of caladium bulbs treated with fungicides on the surrounding soil biota. *Pseudomonas fluorescens* and *P. putida*, which are antagonists to *Erwinia carotovora*, were compatible in vitro with fungicides applied to potato seed pieces.[34] Biotypes of *T. viride* and *T. harzianum* from an irradiated wild type were differentiated by spore germination in vitro at high concentrations of benomyl, thiabendazole, and thiophanate-methyl.[7,31]

In the greenhouse and field, *Talaromyces flavus* (TFl) is an effective biocontrol agent of Verticillium wilt of eggplant caused by *Verticillium dahliae*.[35] *T. flavus* becomes established in the rhizosphere and on potato seed pieces in the soil when applied as a drench.[36] In potato production, fungicides are routinely applied to control seed piece decay caused by *Fusarium* spp. and other fungi. The compatibility of conidia and ascospores of a wild and a benomyl-resistant (TFl-1) strain of TFl with five fungicides recommended for use on potato seed pieces was investigated in vivo and in vitro. The fungicides used were captan, mancozeb, maneb, metiram, and thiabendazole.[32] The wild isolate TFl, the ascospore inoculum form, and thiabendazole and metiram fungicides were the most significant factors allowing maximum survival of TFl applied to potato seed pieces in field experiments. Studies on comparative responses of ascospores and conidia to fungicides and their comparative survival when introduced to the soil indicated that the form of the antagonist inoculum may affect the potential of microbial agents to survive in natural soils in the presence of certain fungicides.[32] It was further demonstrated that the ascospore inoculum of both isolates survived better than the conidial inoculum in the presence of thiabendazole. Both benomyl and thiabendazole bind with fungal tubulin.[37] If cross-resistance usually exists, as stated by these workers, one would expect that the resistance of TFl-1 to benomyl would result in resistance to thiabendazole. Although TFl-1 was 86 times more tolerant than thiabendazole in vitro, in the field test both isolates were similar in their responses to the fungicide. Also, although in vitro both isolates were more tolerant to mancozeb and metiram than to thiabendazole, in the field this difference was not evident.[32] It was observed that the interactions between fungi and fungicides in vitro were not consistent indicators of their interactions in the field. Ascospores and conidia were applied with fungicides to potato seed pieces in the field and growing mycelium was placed on agar amended with fungicides in vitro. Thiabendazole was the fungicide most compatible with both spore types of both isolates in the field, whereas in the laboratory test, thiabendazole was the most toxic fungicide to the TFl and was second most toxic to TFl-1.[32] There are several obvious differences between the soil and petri dish system.[11] In the petri dish, the fungicide is uniformly distributed throughout the substrate which supports the fungus, whereas when the fungus and fungicide are both on the surface of the seed piece there is a gradient of fungicidal activity away from the seed piece. Secondary interactions also may be important; the fungicide may affect populations of nontarget soil microorganisms interacting with the antagonists or the seed piece may serve as a food source, affecting the antagonist and microorganisms in the vicinity of the seed piece.[32]

Both the isolates TFl and TFl-1 survived better in the field and were more tolerant of more fungicides when applied to the potato seed pieces as ascospores than as conidia. In the field, both fungi were most tolerant of thiabendazole and metiram and least tolerant of captan.[32]

Pea cultivars suffer heavily from the wilt fungi *F. oxysporum*, *Aphanomyces euteiches*, *Pythium* spp., and *R. solani*. Seeds treated with *Penicillium oxalicum* gave greater seedling stand than untreated seeds in four cultivars; however, seed inoculation with the antagonist fungus and captan seed treatment gave significantly better germination than inoculating seed alone with the antagonists.[38] Chet et al.[39] reported that coating iris bulbs with *Trichoderma harzianum* and quintozene gave significant protection from soil-borne and bulb-borne *R. solani* and *S. rolfsii*, which involve problems in production in iris. Also, as shown earlier,[40] treatments combining a biological method with chemicals were most successful.

IV. FOLIAR DISEASES

Many pathogens affecting aerial plant parts, especially perennial and annual plants, apparently are inhibited by competitors, antagonists, or hyperparasites.[11,41-43] Where such biological control is operative, as in soil, it is a reasonable expectation that fungicides will, at least at times, create problems rather than solve them.[34] Botrytis rot of cyclamen corms was initially controlled effectively by benomyl (500 ppm dip); however, continuous use resulted in greater disease severity on treated plants than on controls. The reason for this was the development of resistance to benomyl by a strain of *Botrytis cinerea*. Bollen[44] later observed that the disease on treated plants was no worse than on those untreated; on these, two strains of *Penicillium brevicompactum* were found which were not only resistant to benomyl but also antagonistic to the fungus. *P. brevicompactum* controlled the infection of corms by *B. cinerea*, although it did not completely suppress it. The use of benomyl effectively controlled both fungi; however, the emergence of benomyl-tolerant strains of *B. cinerea*, in the absence of the naturally occurring anatagonists, resulted in a more severe infection of *Botrytis*. Subsequent adaptation of *P. brevicompactum* to benomyl returned the situation effectively to its original state.[44]

Dieback or gummosis of apricot caused by *Eutypa armeniacae* is a major disease of mature apricot trees in south Australia. It was first recorded in 1932, but its emergence as an important disease followed the introduction of routine sprays of copper fungicides for control of shot hole disease *Clasterosporium carpophilum*. Copper spraying depleted tree-surface microflora containing species capable of rapidly colonizing pruning wounds.[45] These wounds are now known to be prone to infection by *E. armeniacae*.[46] Dieback, which earlier proved to be difficult to control, is now seen to have its origins in a disease-control practice for a less important disease. Carter[45] found that pruned shoots which are not infected with *E. ameniacae* frequently yielded the saprophytic fungus *F. lateritium*. It can be demonstrated that preinoculation of cut shoots with this species effectively blocks infection by *E. armeniacae*.

Carter and Price[47,48] reported an interesting example of integrated disease control involving benomyl and *F. lateritium*, now known to be an antagonist of *E. armeniacae* (a vascular pathogen of pruned apricot sapwood). The antagonist is ten times less sensitive to benomyl than the pathogen so that the action of benomyl and the antagonist on the pathogen was complementary. *F. lateritium* strains produce a nonvolatile, diffusible metabolite in vitro that is toxic to the ascospores and mycelium of *E. armeniacae*.[45] In vitro studies revealed tenfold difference between *E. armeniacae* and *F. lateritium* in sensitivity to benzimidazoles benomyl and thiabendazole. An integrated approach was suggested to control the pathogen, based upon initial transient chemical protection which would then be enhanced by increasing biological protection resulting from the establishment of an antagonistic competitor at the infection sites. The protection of pruned apricot sapwood by benomyl. *F. lateritium*, and a combination of benomyl and *F. lateritium* against invasion by *E. armeniacae* was evaluated in vitro and in the field.[48] The concentration of benomyl should not be less than 2.2 kg ha^{-1} which gives a concentration of 12 µg/cm^2; if the concentration of benomyl in the sapwood is lowered, the desired control is not achieved. The spore load of *F. lateritium* should be 4.4×10^3 to 4.4×10^5 mℓ$^{-1}$. All the treatments in the fields effected significant control of the pathogen. Price and Carter[49] established the pattern of persistence of this chemical in apricot sapwood. It is clear that treatments which deliver a minimum of 12 µg/cm^2 to a large proportion of the pruned sites effectively protect trees against invasion of *E. armeniacae*.

Although recognition of dieback as an iatrogenic disease came late, it stimulated a detailed appraisal of its etiology and this in turn has provided a basis for the potential integration of biological and chemical methods. The practical problem with both biological and chemical control is the large amount of material which would have to be used in conventional spray application. This can now be overcome by the use of an applicator coupled to the pruning shears.[50]

The pyrimidine fungicides ethirimol and dimethirimol, highly specific for the control of cereals and cucurbits powdery mildews, respectively, act without any effect on all other groups of saprophytes and parasites. Frequent application, however, induced resistance in the powdery mildew fungi.[2,43] This specificity has been effectively utilized by integrating fungicides and biological control agents as a means of reducing losses caused by pathogens of glasshouse crops. Until the introduction of dimethirimol, it was obligatory to use sulfur, quinomethionate, or dinocap for the control of cucurbits powdery mildew. These materials, though effective, were sometimes phytotoxic and acaricidal and were not truly compatible with biological control of red spider mites and white fly predators. When dimethirimol was substituted by benomyl (because of its resistance to the pyrimidine), biological control was less satisfactory because the benzimidazole reduced the fecundity and egg-hatch of the mite, thus altering the food supply of the predator.[19]

Biological control of foliar pathogens has not to date been considered a practical proposition. There is, however, some evidence that phylloplane populations can interefere with the development of some necrotrophic pathogens, especially toward the end of the growing season. Crop senescence seems most affected by a number of filamentous fungi which are ubiquitous on aerial plant tissues. However, their significance in hastening senescence appears to be markedly affected by environmental conditions and possibly the cultivar. Hence there are at least two possibly conflicting considerations concerning the use of fungicides and their possible effects on phylloplane microflora. Sprays applied late or of persistent compounds may decrease the natural biological control, but may also control those fungi which are otherwise responsible for senescence. In the future, decisions to spray crops with broad-spectrum fungicides may have to take both effects into account.

V. FUTURE OUTLOOK

Although specific antagonists may be involved in the indirect control or the accentuation of disease by fungicides, as suggested by some of the examples above, such an effect is largely nonspecific and may be related to the total microbiota.[11] Probably the greatest role of fungicides in biological control will involve their effect on the relative competititve advantage of pathogen vs. antagonist. When any two microorganisms that are in competition are exposed to a differentially harmful factor, other conditions being equally favorable, the microorganisms least affected will be most successful. Many instances of control with sublethal doses of a fungicide undoubtedly involve this mechanism. Sublethal is not synonymous with harmless; even mild dosages are restrictive in some way to the pathogen, even if the effect is not measurable by reduced conidial germination or colony growth rate. Where an increase in dosage obviously inhibits the pathogen, it may be assumed that the milder dosage is also harmful, but in a more subtle way. On the other hand, it is common for a toxic compound to be stimulatory at very low dosages, which could also influence the relative competitive advantage of a pathogen or its antagonist.

The potential for selective inhibition or stimulation of microorganisms or groups of microorganisms with chemicals clearly exists, but awaits man's ingenuity to ascertain which compounds or rates should be used. Indirect benefits of fungicides have been found more or less by accident in tests in which the direct killing of the pathogen by the chemical was the objective. Far more progress would be made if research were directed specifically at indirect control, enlisting the aid of antagonistic microorganisms.

REFERENCES

1. **Papavizas, G. C.**, Soil-borne plant pathogens: new opportunities for biological control, *Br. Crop Prot. Conf. Pests and Diseases*, 1, 371, 1984.
2. **Dekker, J.**, Acquired resistance to fungicides, *Annu. Rev. Phytopathol.*, 14, 405, 1976.
3. **Georgopoulos, S. G.**, Development of fungal resistance to fungicides, in *Antifungal Compounds*, Vol. 1, Siegel, M. R. and Sisler, H. D., Eds., Marcel Dekker, New York, 1977, 439.

4. **Vyas, S. C.,** *Systemic Fungicides,* Tata McGraw-Hill, New Delhi, India, 1984.

5. **Delp, G. J.,** Coping with resistance to plant disease control agents, *Plant Dis.,* 64, 652, 1980.

6. **Hastie, A. V.,** Genetic effects of fungicides, Proc. 9th Int. Congr. Plant Protection, (Abstr.) Washington, D.C., August 5 to 12, 1979, 351.

7. **Papavizas, G. C., Lewis, J. A., and Abd-El Moity, T. H.,** Evaluation of a new biotype of *Trichoderma harzianum* for tolerance to benomyl and enhanced biocontrol capabilities, *Phytopathology,* 72, 126, 1982.

8. **Todd, F. A.,** Telone C soil fumigant, an effective multipurpose disease control treatment for tobacco, *Down Earth,* 26, 5, 1971.

9. **Irving, G.,** Agricultural pest control and the environment, *Science,* 168, 1419, 1970.

10. **Papavizas, G. C.,** Status of applied biological control of soil-borne plant pathogens, *Soil Biol. Biochem.,* 5, 709, 1973.

11. **Baker, K, F., and Cook, R. J.,** Biological control of plant pathogens, Freeman and Company, San Francisco, 1974.

12. **Kraft, J. M.,** Field and greenhouse studies on pea seed treatment, *Plant Dis.,* 66, 798, 1982.

13. **Kraft, J. M. and Papavizas, G. C.,** Use of host resistance *Trichoderma* and fungicides to control soilborne diseases and increase seed yields of peas, *Plant Dis.,* 67, 1234, 1983.

14. **Papavizas, G. C.,** Survival of *Trichoderma harzianum* in soil and in pea and bean rhizosphere, *Phytopathology,* 72, 121, 1982.

15. **Bliss, D. E.,** The destruction of *Armillaris mellea* citrus soils, *Phytopathology,* 41, 665, 1951.

16. **McKeen, W. E.,** A study of sugar beet root rot in south Ontario, *Can. J. Res.,* 27, 284, 1949.

17. **Domsch, K.,** Untersuchunger zur Wirkung einiger Bodenfungizide, *Mitt. Biol. Bundesanst. Land Forstwirtsch. Berlin-Dahlem,* 107, 1, 1959.

18. **Chinn, S. H. F.,** Biological effect of Panogen-PX in soil on common root rot and growth response of wheat seedlings, *Phytopathology,* 61, 98, 1971.

19. **Richardson, L. T.,** The persistence of thiram and its relationship to the microbiological balance and damping off control, *Can. J. Bot.,* 32, 335, 1954.

20. **Pincard, J. A.,** Microbial antagonisms encouraged by the monosodium salt of hexachlorophene, *Phytopathology,* 60, 1308, 1970.

21. **Stankova-Opocenska, E. and Dekker, J.,** Indirect effect of 6-azauracil on *Pythium debaryanum, Neth. J. Plant Pathol.,* 76, 152, 1970.

22. **Anderson, E. J.,** Indirect effects of agricultural chemicals in soil: long term effects of soil fungicides, *Proc. Annu. Conf. Control Soil Fungi,* 9, 17, 1962.

23. **Garren, K. H.,** Evidence for two different pathogens of peanut pod rot, *Phytopathology,* 53, 746, 1963.

24. **Gibson, I. A. S., Ledger, M., and Boehm, E.,** An anomalous effect of pentachloronitrobenzene on the incidence of damping off caused by a *Pythium* sp., *Phytopathology,* 51, 531, 1961.

25. **Farley, J. D. and Lockwood, J. L.,** Reduced nutrient competition by soil microorganisms as a possible mechanism for pentachloronitrobenzene induced disease accentuation, *Phytopathology,* 59, 718, 1969.

26. **Vaartaja, O., Pitblado, R. E., Buzzall, R. I., and Crowford, L. G.,** Chemical and biological control of Phytophthora root rot and stalk rot of soybean, *Can. J. Plant Sci.,* 59, 307, 1979.

27. **Nasbaum, C. J. and Todd, F. A.,** The role of chemical soil treatment in the control of nematode-disease complexes of tobacco, *Phytopathology,* 50, 516, 1960.

28. **Mukhopadhyaya, A. N. and Chandra, I.,** Biocontrol of sugar beet and tobacco damping off by *Trichoderma harzianum,* (Abstr), Seminar on Management of Soil-borne Diseases of Crop Plants, Tamil Nadu Agricultural University, Coimbatore, India, January 8 to 10, 1986, 34.

29. **Vyas, S. C. and Kharem, M. N.,** Biological control of dry root rot of soybean caused by *Rhizoctonia bataticola* by carbendazim and antagonist, (Abstr.), Seminar on Management of Soil borne Diseases of Crop Plants, Tamil Nadu Agricultural University, Coimbatore, India, January 8 to 10, 1986, 29.

30. **Abd-El Moity, T. H., Papavizas, G. C., and Shatla, M. N.,** Induction of new isolates of *Trichoderma harzianum* tolerant to fungicides and their experimental use for the control of white rot of onion, *Phytopathology,* 72, 396, 1982.

31. **Papavizas, G. C. and Lewis, J. A.,** Physiological and biocontrol characteristics of stable mutants of *Trichoderma viride* resistant to MBC fungicides, *Phytopathology,* 73, 407, 1983.

32. **Fravel, D. R., Marois, J. J., Dunn, M. T., and Papavizas, G. C.,** Compatibility of *Talaromyces flavus* with potato seed pieces fungicides, *Soil Biol. Biochem.,* 17, 163, 1985.

33. **Ferries, R. S. and Mitchell, D. J.,** Population dynamics of soil microorganisms associated with fungicide-dusted caladium seed pieces, *Soil Biol. Biochem.,* 13, 57, 1981.

34. **Burr, T. J., Schroth, M. N., and Suslow, T.,** Increased potato yields by treatment of seed pieces with specific strains *Pseudomonas fluorescens* and *P. putida, Phytopathology,* 68, 1377, 1978.

35. **Marois, J. J., Johnstone, S. A., Dunn, M. T., and Papavizas, G. C.,** Biological control of Verticillium wilt of eggplant in the field, *Plant Dis. Rep.,* 66, 1166, 1982.

36. **Marois, J. J., Fravel, D. R., and Papavizas, G. C.,** Ability of *Talaromyces flavus* to occupy the rhizosphere and its interactions with *Verticillium dahliae, Soil Biol. Biochem.,* 16, 387, 1984.

37. **Davidse, L. C. and Flach, W.,** Interaction of thiabendazole with fungal tubulin, *Biochim. Biophys. Acta,* 543, 82, 1978.
38. **Windels, C. E. and Kommedahl, T.,** Pea cultivar effect on seed treatment with *Pencillium oxalicum* in the field, *Phytopathology,* 72, 541, 1982.
39. **Chet, I., Elad, Y., Kalfon, A., Hadar, Y., and Katan, J.,** Integrated control of soil-borne and bulb-borne pathogens in iris, *Phytoparasitica,* 10, 229, 1982.
40. **Elad, Y., Katan, J., and Chet, I.,** Physical, biological and chemical control integrated for soil-borne diseases in potatoes, *Phytopathology,* 70, 418, 1980.
41. **Dickinson, C. H.,** Effects of ethirimol and zineb on phyllosphere microflora of barley, *Trans. Br. Mycol. Soc.,* 60, 423, 1973.
42. **Strashnow, Y., Elad, Y., Sivan, A., and Chet, I.,** Integrated control of *Rhizoctonia solani* by methyl bromide and *Trichoderma harzianum, Plant Pathol.,* 34, 146, 1985.
43. **Griffiths, E.,** Lastrogenic plant diseases, *Annu. Rev. Phytopathol.,* 19, 69, 1981.
44. **Bollen, G. J.,** Resistance to benomyl and some chemically related compounds in strains of *Pencillium* species, *Neth. J. Plant Pathol.,* 77, 187, 1971.
45. **Carter, M. V.,** Biological control of *Eutypa armeniacae, Aust. J. Exp. Agric. Anim. Husb.,* 11, 687, 1971.
46. **Carter, M. V.,** Further studies on *Eutypa armeniacae, Aust. J. Agric. Res.,* 11, 498, 1960.
47. **Carter, M. V. and Price, T. V.,** Biological control of *Eutypa armeniacae.* II. Studies on the interaction between *E. armeniacae* and *Fusarium lateritium* and their relative sensitivities to benzimidazole chemicals, *Aust. J. Exp. Agric. Anim. Husb.,* 25, 105, 1974.
48. **Carter, M. V. and Price, T. V.,** Biological control of *Eutypa armeniacae.* III. A comparison of chemical, biological and integrated control, *Aust. J. Exp. Agric. Anim. Husb.,* 26, 537, 1975.
49. **Price, T. V. and Carter, M. V.,** Persistence of benomyl in pruned apricot sapwood, *Aust. J. Agric. Sci.,* 26, 529, 1975.
50. **Carter, M. V. and Mullet, L. F.,** Biological control of *Eutypa armeniacae.* IV. Design and performance of an applicator for metered delivery of protection aerosols during pruning, *Aust. J. Exp. Agric. Anim. Husb.,* 18, 287, 1978.
51. **Wen, W. S. and Lu, J. H.,** Seed treatment with antagonists and chemical to control *Alternaria brassicola, Seed Sci. Technol.,* 12, 851, 1984.

Chapter 16

FUNGICIDE-INDUCED HOST RESISTANCE

I. INTRODUCTION

With regard to fungicide-induced host resistance, fourth-generation compounds show very little or no fungitoxicity in vitro, but decrease infection or reduce disease intensity when applied to plants. This indicates that the fungicides are metabolized to fungitoxicants in the host plant or that they affect the pathogenicity of the fungus or the resistance of the host. A compound of this type may interfere with factors critical to pathogenesis, such as the production or activity of fungal toxins and enzymes or elicitors and repressors of host resistance. On the other hand, the primary action may be on the host rather than the pathogen. A fungicide may increase passive resistance of the host by affecting cell wall composition, or it might accentuate active resistance mechanisms that are triggered by an attack of the pathogen.

A wide range of chemicals can elicit responses that resemble defense reactions of plants and can also stimulate systemic resistance to fungi in plants.[1] In screening tests designed to identify antifungal chemicals, it was observed that a surprising proportion of the compounds that cause visible damage to test plants also induce some measure of systemic resistance. The results were obtained when 11-day-old rice plants growing in compost were treated by root drenches with suspensions (50 μg mℓ^{-1}) of 854 diverse compounds. About 1% of the compounds caused some kind of visible damage to the rice plants. They reduced by at least 95% the area of leaf that was covered by lesions following inoculation with *Pyricularia oryzae*. From these results it would be expected that between eight and nine compounds would both cause damage to the plants and reduce their susceptibility to *P. oryzae* if there were no connection between the two effects. However, the observed number of such compounds was 37, a highly significant deviation from the predicted figure. These observations demonstrate that a wide range of substances can cause disease resistance by altering the metabolism of the plants significantly enough to cause visible damage. Any substance that cause a change in the metabolism of the plant to which it is applied, whether or not there is visible damage, may interfere with the susceptibility of the plant to a pathogen.[1] There is a need for compounds that alter the plant in a suitable way such that, on attack by a pathogen, it would behave in the same way as a genetically resistant plant, so that host resistance mechanisms would be activated only at the precise times and places where needed. This enhanced host resistance is an interesting and promising approach to plant disease chemotherapy. The action of the most sophisticated compound of this type would be expected to cause toxicity to the pathogen only in its parasitic phase. The literature on this subject is reviewed in this chapter on the use of alternative chemical agents for controlling plant disease. The subject has been reviewed very recently by Sisler and Ragsdale,[2] Langcake,[3] Wade,[4] and Dekker.[5] Some of the recent researches on this role of fungitoxicants which increase host resistance in plants and important achievements made in this field are listed in Table 1.

II. TRIARIMOL

Historically, triarimol seems to be the first compound which showed indirect action against fungal pathogens. It has been shown to have no effect on spore respiration or germination of *Asperigillus niger* at 20 ppm or *Cladosporium cucumerinum*. Even when 50 ppm of triarimol was applied to leaves, normal germ tubes were produced by *Erysiphe polygoni*

Table 1
SOME IMPORTANT ACHIEVEMENTS IN FUNGICIDE-INDUCED HOST-MEDIATED DEFENSE REACTIONS

Year	Achievement	Ref.
1969	Triarimol was developed	6
1971	Indirect mode of action of triarimol identified	7
1975	Development of WL-28325 which stimulates disease resistance in rice	8
1976	Tricyclozole was developed with indirect mode of action	16
1977	Probenazole seen to control disease by host-mediated defense reaction	22
1979	Production of phytoalexin by fosetyl-Al in plant	37
1980	Acylalalines (metalaxyl) control plant disease by producing phytoalexin	55
1981	Many reviews appear on host defense accentuation by fungicides	3, 4
1984	Practical utility gaining foothold	5

STRUCTURE 1

spores,[6] although 5 to 10 ppm was sufficient to prevent macroscopic signs of disease. Development of mildew beyond the formation of haustoria was prevented by 10-ppm sprays of triarimol for up to 4 days before or after inoculation, or by 0.5 g/500 g soil. The possibility, however, that this compound produced systemic disease control by altering host metabolism is shown both by the phytotoxic effects seen on barley seedlings, strawberry plants, and rooted apple shoots and tomato plants[7] and by the failure of tomato leaf sap to inhibit the growth of *C. fulvum* and *Botrytis cinerea* on agar after the application of sufficient triarimol. The fungicide was introduced into the soil to prevent the development of leaf mold and the spread of *B. cinerea* lesions in vivo.[7]

III. WL-28325

Langcake and Wickins,[8] Langcake et al.,[9] and Cartwright et al.[10] reported the development of a compound, 2,2-dichloro-3,3-dimethyl cyclopropane carboxylic acid (DDCC) (WL-28325, Structure 1) which is specific for the control of *P. oryzae* with low intrinsic fungitoxic activity, but is inactive against all other host-parasite combinations tested, including *P. sasakii* on rice and even *P. oryzae* on other hosts such as rye grass and barley.

WL-28325 is a systemic prophylactic compound highly specific for rice blast disease. In this respect it resembles tricyclozole. It was found that WL-28325 is not sufficiently fungitoxic to account for disease control and it does not appear to give rise to fungitoxic metabolites within the plant. Since WL-28325 is not converted in rice plants to a derivative of higher intrinsic fungitoxicity, it has been suggested that the compound may act by stimulating the disease resistance mechanism of rice.[8]

WL-28325 does not prevent epidermal penetration of rice plants by *P. oryzae*, but markedly accentuates host cellular responses once penetration has occurred.[10] Histological examination showed that in rice plants host cell necrosis preceded visible damage to the infection hyphae of *P. oryzae* inoculated into leaves treated with WL-28325. In untreated leaves, the growth of infection hyphae proceeded with little host response. Thus, treated plant cells respond in a rapid, hypersensitive fashion to penetration by *P. oryzae,* which is in contrast to a delayed, mild reaction by cells of the untreated plants, but hyphal development is halted soon after penetration in cells of treated plants. It was suggested that WL-28325-induced resistance to blast might involve a direct suppression of fungal growth through the production of inhibitory compounds by the host. This suppression of hyphal growth apparently results from the accumulation of phytoalexins — two known 9-B-pyrimidine diterpenes and momilactones A and B — in the tissue surrounding the infection centers and formation of brown pigments around the infection sites in WL-28325-treated plants. The accumulation of these fungitoxic substances is more rapid and far greater in magnitude in WL-28325-treated plants than in untreated plants.[10,11] Phytoalexins do not occur in uninfected leaves, but accumulate both in and around infection sites of treated leaves to concentrations which are well above minimal inhibitory values in vitro. While accumulation of momilactones is regarded as a major factor in the suppression of *P. oryzae* by WL-28325, the biochemical basis of marked enhancement of phytoalexins in rice and host defense mechanisms by which this is triggered remains unclear.[12] The mode of action of the compound is presumably based on interference with certain determinants of susceptibility and resistance in the rice-*P. oryzae* relationship.[12] Such determinants might involve the production of compounds by the fungus which either suppress hypersensitivity[13,14] or elicit resistance.[15]

IV. TRICYCLOZOLE (TCZ)

TCZ (Structure 2), which is used commercially for control of *P. oryzae,* was also thought to have this mode of action. It is 25 to 35 times more active against the development of the fungus in vivo than against mycelial growth, spore germination, or germ tube elongation in vitro.[16] It has been shown to be a melanin biosynthesis inhibitor[17] that blocks the conversion of 1,3,8-trihydroxynaphthalene to vermelone and the conversion of 1, 3, 6, 8-trihydroxy-naphthalene to scytalone in the polyketide pathway to melanin.[18] Other fungicides act in a similar or identical manner to TCZ.[16,19-21] The systemic activity of TCZ in rice plants has been shown indirectly by the control of foliar disease symptoms after seed, root, and soil application and directly by the presence of the compound in the leaves of plants grown from TCZ-treated seed. Full disease control is obtained at concentrations in the leaves that are 25 to 50 times lower than those required to inhibit fungal growth in vitro. Hence, it has been suggested that TCZ is altered in the plant or that it affects host resistance, pathogenicity of *P. oryzae,* or host-parasite interaction.[16,17]

V. PROBENAZOLE

Probenazole (Oryzemate) (Structure 3), a relative of saccharin, controls blast and bacterial leaf blight caused by *P. oryzae* and *Xanthomonas oryzae,* respectively, at relatively low doses as foliar spray and after root application.[22,23] Probenazole is converted to two active principles: *N*-D-glucopyranosyl-benzisothiazole-1,1-dioxide and benzisothiazole-1,1-dioxide.[24] Its in vitro toxicity is weak and not sufficient to explain its activity in vivo. Hyphal growth of *P. oryzae* in vitro is not inhibited at concentrations up to 800 μg mℓ$^{-1}$. Also, conidial germination was not checked on slides. However, conidial germination on rice leaf

$$\left[\begin{array}{c} H_5C_2O \diagdown \quad \diagup O \\ \quad P \\ H \diagup \quad \diagdown O \end{array} \right]_3 Al^{3+}$$

STRUCTURE 4

sheath was quite sensitive to the compound;[23] when conidia suspended in a 10-μg mℓ$^{-1}$ solution were injected into rice leaf sheaths, appressorium formation, penetration, and mycelial growth were inhibited. While this effect suggests that probenazole may act prior to penetration, other tests show that the compound suppresses lesion spread after the fungus has penetrated. Blast lesions on treated rice are similar to hypersensitive lesions of resistant cultivars. Analysis of the fungicide-treated plants inoculated with *P. oryzae* revealed an accumulation of 1-linoleic acid and three other fungitoxic materials with similar properties. These are believed to form a chemical barrier against the pathogen. Moreover, there is an increase in treated and inoculated leaves in the activity of peroxidase, phenylalanine ammonia lyase, and catechol-*O*-methyltransferase in inoculated plants treated with probenazole. In addition, there is a marked increase in the production of antimicrobial substances. These increases in the enzymatic activity and antimicrobial substances are believed to reflect an enhanced lignoid barrier formation around the invading pathogen.[25,26] This indicates that the disease-controlling effect of probenazole can be attributed to a host-mediated defense reaction[23,27] involving the formation of a physical and chemical barrier in and around the invaded cells.[26] A similar sequence of events can be initiated by the application of plant hormones such as auxins and ethylene, which also results in the suppression of rice blast. Abscissic acid counteracts the activity of the growth hormones as well as that of probenazole. The available evidence suggests that probenazole interferes with the initial processes of pathogenesis.[27]

Some of the most interesting and effective compounds used against pathogens in their parasitic phase have been developed for the control of rice blast disease caused by *P. oryzae*. The dichlorocyclopropane carboxylic acids are a class of compounds which enhance resistance of rice plants to blast disease.[10,11,28] Although the pathogen penetrates the cuticle and epidermal cell walls of plants treated with these chemicals, the subsequent spread within the tissue is halted. This is apparently due to accentuated production of momilactone phytoalexins.[10]

The mode of action of probenazole in controlling rice blast disease is less clear. Therapeutic effectiveness for control of the disease is not related to its toxicity to saprophytic growth of the pathogen on agar media. Spore germination on rice sheaths is relatively sensitive to probenazole. The compound accentuates several enzyme activities in inoculated rice plants and increases the level of fungitoxic components.[22,29,30] It prevents lesion spread, which indicates that it increases host resistance or suppresses pathogenicity of the fungus within the tissue.

VI. ALUMINUM TRIETHYL PHOSPHONATE (ALUMINUM-*O*-TRIS)

Aliette (aluminum triethyl phosphonate, Structure 4) is a novel xenobiotic and is not a fungicide (at least it is not toxic in vitro), but against the diseases caused by Oomycetes certain instances of excellent preventive and curative activity have been reported. Aliette, like many Oomycete fungicides, is very soluble in water.[31] Some control of *Phytophthora cinnamomi* on avocado roots has been achieved from a foliar spray. Even more noteworthy is a report by Laville[32] that foliar sprays at intervals of 20, 40, or even 60 days with 2000 μg mℓ$^{-1}$ will arrest cankers on large branches caused by *Phytophthora* on citrus. However, concentrations that control *Phytophthora* disease showed little toxicity to the mycelial growth of the pathogens in vitro.[33,34] The ED_{50} values for inhibition of radial mycelial growth of *P. parasitica* and *P. citrophthora* often range from 300 to 1000 μg mℓ^{-1}. However, sporangial formation on the host is strongly inhibited at 5 μg mℓ^{-1}, which is a relatively low concentration.[33-35] In view of the low fungitoxicity, it has been suggested that the indirect action of the fungicide via the plant be utilized.[36] These hypotheses are more or less substantiated by the finding that the chemical enhances the production of phenolic compounds and phytoalexins in tomato leaves after inoculation with *P. capsici;* however, treatment of uninoculated plants does not have such an effect.[37] The compound is degraded in buffers or plant tissue to phosphorus acid, a product which is involved in the productive activity of the plant.[38,39] According to the workers, the chemical triggers a necrotic blocking (defense) reaction, which increases phenol accumulation in plant tissues around infection sites: as a consequence, further advance of the pathogen into the tissue is halted.

When applied to cure tomato leaves infected by *P. capsici,* the enzyme phenylalanine ammonia lyase, which plays a key role in secondary metabolism, appears to be stimulated.[39] Similarly, tomato leaves produced more phenolic compounds following inoculation with the pathogen when the leaves were floated on a solution (100 μg mℓ^{-1}) of aluminium-tris than when floated on distilled water. The defense reaction operates in the tissue of pepper and French bean, but not in the tissue of potato. Induction of defense reactions is suppressed by phosphate ions.[40] Although substantial evidence indicates that in the presence of aluminium-tris the general defense mechanism is mediated through host systems, not normally evoked by pathogens, that are induced and become effective, the primary target that is responsible for induction or accentuation of these systems has not been identified.

VII. ACYLALANINES

Some fungicides that have potent antifungal effects in vitro also induce responses in the host plant identical to the natural resistance reponse. It is difficult to exclude the possibility that these responses contribute to the effectiveness of some fungicides. There are several example of this type of effect.

The antifungal activity of acylalanines was first recognized in 1972 when curative and systemic properties of various lead compounds, derived from anilide herbicides, were observed against *P. infestans.*[40] One of these compounds, CGA 29212, showed systemic, curative, and protective activity at low dosages against plant pathogens in the Peronosporales, although it was unfortunately still phytotoxic. This phytotoxicity was overcome by substituting for the chloroacetyl group other acyl compounds, metalaxyl, and furalaxyl.[41,42]

The fungitoxicity of the acylalanines and acylalanine fungicides is dependent on chirality and the activity of the R-enantiomers, whereas the S-enantiomers are almost devoid of any fungitoxicity.[40] This contrasts with the chirality dependency of the weed-killing activity of compounds such as S-enantiomers.[43] Thus, the mechanism of action of these fungicides and herbicides in plants may differ from its primary mode of action in fungi. With the development of metalaxyl and furalaxyl, several other related compounds have been commercially formulated. Benalaxyl[44] still carries the alanine methylester but in ofurace[45] and cyprofuram[46] this compound is in the cyclic butyrolactone form. In oxadixyl,[47] the latter form is substituted by an oxazolilinone compound. The last three compounds are not acylalanines: because of their structural and biological properties, they can best be described as acylalanine-type fungicides. The fungicides of this group are almost all selective and inhibit the Peronosporales; fungi of the order Saprolegnials are not sensitive[48] and, apart from Oomycetes, only members of the Hyphochytridiomycetes are inhibited.[49] The mode of action of these compounds in fungi has been studied in detail.[50,51] They inhibit the incorporation of uridine in RNA and also the activity of RNA polymerase.

Metalaxyl (Structure 5) and furalaxyl (Structure 6) have almost revolutionized downy

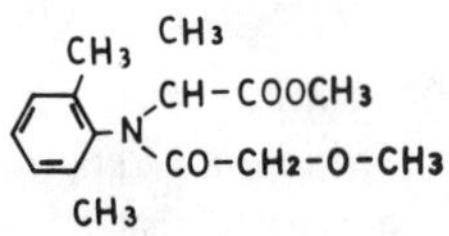

STRUCTURE 5 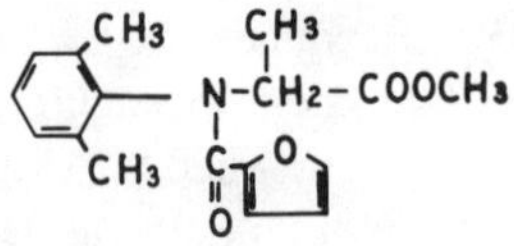   STRUCTURE 6 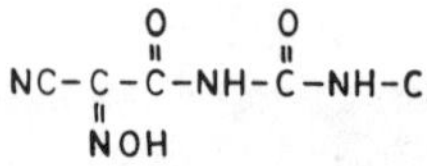   STRUCTURE 7

mildew control programs. They show potent activity in vitro; in plants, however, they seem to be even more active against the fungus than in vitro. However, no discernible effects are seen unless the fungus has penetrated the host and has started to form haustoria.[52-54] Ward et al.[55] studied phytoalexin production in soybean plants by metalaxyl in two situations; lesions were found on treated plants which are indistinguishable from lesions on plants inoculated with an avirulent (naturally resistant) race of the pathogen. The levels of phytoalexins that accumulated were very similar in the two cases. Metalaxyl per se did not promote the formation of phytoalexins.[55] Two possibilities were postulated: phytoalexins may accumulate because the fungicide kills the fungus, releasing fungal metabolites that nonspecifically stimulate production of defense compounds, or the fungicide may interfere with either the pathogenicity of the fungus or the resistance mechanisms of the host. It is believed that phytoalexins would accumulate in either case. Since the fungicides are active against the fungus directly it is likely that the effects on host resistance mechanisms are incidental or secondary. It is possible, however, that they make a signficiant contribution to the effectiveness of the fungicide.

Another oomycetes-effective fungicide, cymoxanil (Structure 7), following infection by downy mildew fungus in grape vines, produces hypersensitive necrosis with enhanced synthesis of phytoalexins.[4] Neither metalaxyl[55] nor cymoxanil, in the absence of infection, induced phytoalexin production in soybean or vines, respectively.

The phytoalexin produced by metalaxyl in soybean in the presence of *P. megasperma* has been identified as glyceollin.[55] It has been suggested that cell wall polymers, released by the fungus killed by metalaxyl, elicit phytoalexin synthesis. The ability to cause the release of such elicitors from fungal cell walls could even provide a basis for the development of successful systemic fungicides for plant disease control.[55] The production of glyceollin was confirmed by Borner et al.[56] It was proposed that, under conditions where metalaxyl concentrations were not fully inhibitory, glyceollin could play a critical role in preventing the spread of the pathogen. However, in a subsequent examination of this possibility it was found that even in partially inhibitory metalaxyl treatments the glyceollin concentrations in lesions still exceeded the in vitro ED_{50} but failed to prevent the spread of the pathogen.[57] Thus, it appeared that as long as the in vivo activity of glyceollin was comparable to that in vitro, glyceollin did not contribute to the control of *P. megasperma* f. sp. *glycinea* in metalaxyl-treated seedlings.

Keen et al.[58] reported that the herbicide glyphosate suppresses glyceollin production and induces susceptibility to avirulent races of *P. megasperma* f. sp. *glycinea* in treated soybean seedlings. Glyphosate has been reported to block the shikimate pathway by inhibiting 5-enolpyruvylshikimate-3-phosphate synthetase; hence, it could restrict glyceollin production by decreasing the supply of phenylalanine. Consequently, if glyceollin contributes to the control of *P. megasperma* f. sp. *glycinea* on metalaxyl-treated seedlings, it would be expected that inhibition of glyceollin production by glyphosate should reduce the effectiveness of metalaxyl treatments. Ward[59] reported that glyphosate appreciably reduced the ability of marginally inhibitory concentration of metalaxyl to restrict the spread of *P. megasperma* f. sp. *glycinea* and stimulated glyceollin production in inoculated soybean hypocotyls. No evidence was obtained for in vitro suppression of metalaxyl toxicity. The similarity of results obtained with metalaxyl application at the inoculated site and through the base of the hypocotyl indicates that glyphosate did not simply interfere with metalaxyl uptake. It appears, therefore, that some mechanism of host resistance that can by inhibited by glyphosate contributes to the action of metalaxyl in restricting tissue colonization by *P. megasperma* f. sp. *glycinea*. Glyceollin production is the most obvious candidate for this role. However, although glyphosate clearly suppressed metalaxyl-stimulated glyceollin production, concentrations in all but two instances remained above ED_{50} values (200 μg mℓ$^{-}$)[157] and, apart from the general trend, there was no precise correlation between the extent to which lesions spread and the amount of glyceollin that accumulated. Similar results were obtained previously in lesions only partially restricted by metalaxyl.[57] Thus, it would appear either that glyceollin is much less active in vivo than in vitro or that it does not accumulate uniformly throughout the lesion and the periphery in sufficient quantity to completely prevent the growth of the pathogen. The results are the same[58] for the suppression of both natural resistance to *P. megasperma* f. sp. *glycinea* and glyceollin production following treatments with glyphosate. The results were interpreted to indicate that glyceollin accumulation accounts for resistance to *P. megasperma* f. sp. *glycinea,* although glyceollin levels did not always correspond with incompatibility as frequently as had been observed.[59] The concentrations of glyceollin exceeded ED_{50} levels in the glyphosate-induced compatible interactions. In fact, the utilization of resources for glyceollin production, despite the inhibition of the shikimate pathway, is remarkable and may occur at the expense of other depleted and possibly essential activities. In addition, the effects of glyphosate on both natural and metalaxyl-induced resistance must be interpreted with caution, because the inhibition of chorismate formation from shikimate influences many aspects of metabolism besides phenylpropanoid biosynthesis. Aromatic amino acids are depleted and indole acetic acid levels are suppressed by shikimate or contributory pathways. The possibility remains, therefore, that undetermined changes in host metabolism permit the pathogen to spread more freely in host tissue treated with glyphosate and that this is the cause and not the result of decreased glyceollin accumulation.[59]

A complete suppression of fungal sporulation was obtained after a 24-hr application of metalaxyl directly to the roots of pea plants infected with downy mildew. No ultrastructural changes were evident in either fungus or host at this stage. After 5 days, encasements were evident around many fungal haustoria. By 15 days, many haustoria were encased and their cytoplasm was usually highly vacuolated. Similar encasements have been described in rust-infected bean plants 14 days after soil application of oxycarboxin.[60,61] Pring and Richmond[60] interpreted the formation of encasement in these plants as a host reaction to fungicide-induced necrosis of the fungal haustoria. In ultrastructural studies of natural resistant reactions in rust-infected plants, encasement of haustoria is interpreted as a nonspecific response by the host.

Crute[52] has described a different response to metalaxyl in which lettuce cotyledons were inoculated with sporangia of *Bermia lactucae* 14 days after germination of the seedlings on a filter paper already impregnated with the fungicide. Within a short period of fungal penetration of the host cell, a hypersensitive reaction was observed similar to that described in resistance cultivars in lettuce.[52]

Earlier Ward et al.[55] suggested that stimulation of glyceollin production in *P. megasperma* f. sp. *glycinea*-inoculated hypocotyls treated with metalaxyl could be due to the release of elicitors from damaged hyphae. However, recent evidence indicates that metalaxyl, which is only fungistatic, does not inhibit spore germination, respiration or hyphal wall synthesis and does not affect membrane permeability.[50,51] It is unlikely therefore that the stimulation of putative host defense reponses in metalaxyl-treated hypocotyls is due directly to the loss of hyphal integrity and the consequent release of molecules with elicitor activity. The primary action of metalaxyl was shown to be the inhibition of RNA and protein synthesis. Should this be the main effect of metalaxyl or *P. megasperma* f. sp. *glycinea* in treated tissues, the development of symptoms of incompatibility, including glyceollin production, must also result from the inhibition of protein synthesis in the pathogen. This strongly indicates that in untreated tissues RNA and protein synthesis by *P. megasperma* f. sp. *glycinea* is essential for compatibility or for the suppression of resistance mechanisms. It is tempting to speculate that natural disease resistance could proceed either via its interference with synthesis or through the activity of specific pathogen proteins that constrain the incompatibility that must exist between such widely different organisms as *P. megasperma* f. sp. *glycinea* and soybean. Evidence that *P. megasperma* f. sp. *glycinea* produces glycoproteins that suppress glyceollin formation in soybean has been reported by Ziegler and Pontzen.[61] Because of its specificity for pathogen-protein synthesis as opposed to host-protein synthesis, metalaxyl should be a useful tool in the study of host/pathogen interactions involving members of Peronosporales.

Similar host defense reactions would also provide a rational explanation for various reports of unusual or unexpected host resistance to fungal invasion following treatment with various fungicides. Nevertheless, the biocides may not necessarily have caused the observed host responses either directly or indirectly, because the surfactants used in their formulation may also have been responsible. Triton, a surfactant, induced the accumulation of phytoalexins in French beans; interestingly, the quantity accumulated was related to the phytotoxicity of the surfactant, which was dependent upon the length of the ethanol side chain.[55] This suggests the crucial role of the inert material incorporated in the formulation of a fungicide, and this necessitates caution in evaluating the physiological and biochemical effects of fungicides on plants after their application. Whether or not prolonged use of such nonfungicidal compounds will lead to a failure of disease control due to the development of resistance, as has been observed in systemic fungicides, has yet to be determined. When a compound induces several changes in the host plant that lead to resistance or to the formation of a nonspecific fungicide, adaptation of the pathogen, the potency of which is inhibited by a chemical, can restore its pathogenicity. Uesugi[62] reported that successive tetrachlorophthalide treatments of rice plants, when inoculated in succession with *Pyricularia oryzae,* did not lead to reduced disease control.

VIII. FUTURE OUTLOOK

The possibility that plant diseases can be controlled by manipulating the defense mechanism of the plant itself has been suggested several times, but few chemicals which are able to do this have been named. It is noteworthy that the compounds themselves do not stimulate phytoalexin production, but rather increase the capacity of rice to synthesize phytoalexins in response to infection. Such chemicals that specifically interfere with pathogenesis open up the possibility of a more sophisticated method of plant disease control. The ecological advantages are obvious since saprophytic fungi and other organisms are not affected. It should be realized that highly selective agents may be economically less attractive, especially when different pathogens on the same host plant have to be eradicated. It is important to study their fundamental aspects of interference with pathogenesis, however, since increased knowledge in this area may allow for new perspectives in the design of fungicides. The only limitation of specific compounds is their narrow disease-control spectrum. This limitation and the fungal resistance problems encountered with certain systemic fungicides will necessitate the continued use of older, less specific surface protectants.

REFERENCES

1. **Rathmell, W. G.,** The discovery of new methods of chemical disease control: current development, future prospects and the role of biochemical and physiological research, *Adv. Plant Pathol.,* 2, 259, 1984.
2. **Sisler, H. D. and Ragsdale, N. N.,** Mode of action and selectivity of fungicides, in *Agricultural Chemicals of the Future,* Hilton, J. L., Ed., Rowman and Allanheld, Beltsville, Md., 1983, 176.
3. **Langcake, P.,** Alternative chemical agents for controlling plant disease, *Philos. Trans. R. Soc. London Ser. B,* 295, 83, 1981.
4. **Wade, M.,** Antifungal agents with an indirect mode of action, in *Proc. Symp. British Mycological Society,* Trinci, A.P.J. and Ryley, J. F., Eds., Cambridge University Press, Cambridge, 1984, 283.
5. **Dekker, J.,** The development of resistance to fungicides, *Progress in Pesticide Biochemistry and Toxicology,* Vol. 4, Hutson, D. H. and Roberts, T. R., Eds., John Wiley & Sons, New York, 1985, 1965.
6. **Gramlich, L. V., Schwer, J. F., and Brown, I. F.,** Characteristics and field performance of a new broad spectrum fungicide, *Proc. 5th Br. Insectic. Fungic. Conf.,* 2, 576, 1969.
7. **Smith, P. M. and Spencer, D. M.,** Evaluation of systemic fungicides for use in the glasshouse industry, *Pestic. Sci.,* 2, 201, 1971.
8. **Langcake, P. and Wickins, S. G. A.,** Studies on the action of the dichlorocyclopropanes on the host-parasite relationship in the rice blast disease, *Physiol. Plant Pathol.,* 7, 113, 1975.
9. **Langcake, P., Cartwright, D., and Ride, J. P.,** The dichlorocyclopropanes and other fungicides with indirect mode of action, in *Systemic Fungicides and Antifungal Compounds,* Lyr, H. and Potter, C., Eds., Academie-Verlag, Berlin, 1981, 630.
10. **Cartwright, D., Langcake, P., Pryce, R. J., Lewerthy, D. P., and Ride, J. P.,** Isolation and characterization of phytoalexins from rice as momilactones A & B, *Phytochemistry,* 20, 535, 1980.
11. **Cartwright, D., Langcake, P., Pryce, R. J., Lewerthy, D. P., and Ride, J. P.,** Chemical activation of host defense mechanisms as a basis for crop protection, *Nature (London),* 267, 511, 1977.
12. **Cartwright, D. W., Langcake, P., and Ride, J. P.,** Phytoalexin production in rice and its enhancement by a dichloropropane fungicide, *Physiol. Plant Pathol.,* 17, 259, 1980.
13. **Oku, H., Shiraishi, T., and Ouchi, S.,** Suppression of phytoalexin pisatin, *Naturwissenschaften,* 64, 643, 1977.
14. **Varns, J. L. and Kuc, J.,** Suppression of rishitin and phytotuberin accumulation and hypersensitive response in potato by compatible races of *Phytophthora infestans, Phytopathology,* 61, 178, 1971.
15. **Keen, N. T.,** Specific elicitors of phyotalexin production: determinants of race specificity in pathogens, *Science,* 187, 74, 1975.
16. **Froyd, J. D., Page, S. J., Guse, L. R., Dreikorn, B. A., and Pafford, J. L.,** Tricyclazole — a new systemic fungicide for the control of *Pyricularia oryzae* in rice, *Phytopathology,* 66, 1135, 1976.
17. **Tokousbalides, M. C. and Sisler, H. D.,** Effect of tricyclazole on growth and secondary metabolism in *Pyricularia oryzae, Pestic. Biochem. Physiol.* 8, 26, 1978.
18. **Tokousbalides, M. C. and Sisler, H. D.,** Site of inhibition of tricyclazole in the melanin biosynthetic pathway of *Verticillium dahliae, Pestic. Biochem. Physiol.,* 11, 64, 1979.
19. **Woloshuk, C. P. and Sisler, H. D.,** Tricyclazole, pyroquilon, tetrachlorophthalide, PCBA, coumarin and related compounds inhibit melanization and epidermal penetration by *Pyricularia oryzae, J. Pestic. Sci.,* 7, 161, 1982.
20. **Yamaguchi, I., Sekido, S., and Misato, T.,** The effect of nonfungicidal anti-blast chemicals on the melanin biosynthesis and infection by *Pyricularia oryzae, J. Pestic. Sci.,* 7, 523, 1982.
21. **Wheeler, M. H.,** Melanin biosynthesis in *Vertiicillium dahlia:* dehydration and reduction reactions in cell free homogenates, *Exp. Mycol.,* 6, 171, 1982.
22. **Watanabe, T., Igarashi, H., Matasumoto, K., Seki, S., Mase, S., and Sekizawa, Y.,** The characteristics of probenazole (Oryzemate) for the control of rice blast, *J. Pestic. Sci.,* 2, 291, 1977.
23. **Watanabe, T.,** Effects of probenazole (Oryzemate) on each stage of rice blast fungus *(Pyricularia oryzae* Cav.) in its life cycle, *J. Pestic. Sci.,* 2, 395, 1977.
24. **Uchiyama, M., Abe, H., Sato, R., Shimura, M., and Watanabe, T.,** Fate of 3-allyloxyl-1,2-benzisothiazole-1,1-dioxide (Oryzemate) in rice plants, *Agric. Biol. Chem.,* 37, 737, 1973.
25. **Sekizawa, Y. and Mase, S.,** Mode of controlling action of probenazole against rice blast disease with reference to induced resistance mechanism in rice plant, *J. Pestic. Sci.,* 6, 91, 1981.
26. **Iwata, M., Susuki, Y., Watanabe, T., Mase, S., and Sekizawa, Y.,** Effects of probenazole on the activities of enzymes related to the resistant reaction in rice plant, *Ann. Phytopathol. Soc. Jpn.,* 46, 297, 1980.
27. **Sekizawa, Y., Watanabe, T., Shimura, M., Matsumoto, K., and Iwata, M.,** Mode of action of rice blast protectant, probenazole, Proc. 5th Int. Cong. Pesticide Chemistry, Abstr. IVC-4, IUPAC, Kyoto, Japan, Aug. 29 to Sept. 1984.
28. **Langcake, P., Carwright, D., Lewerthy, D. P., Pryce R. J., and Ride, J. P.,** The dichlorocyclopropane fungicides with an indirect mode of action, *Neth. J. Plant Pathol.,* 83 (Suppl 1), 153, 1978.

29. **Sekizawa, Y. and Mase, S.,** Recent progress in studies on nonfungicidal controlling agent probenazole, with reference to the induced resistance mechanism of rice plants, *Rev. Plant Prot. Res.,* 13, 114, 1980.

30. **Watanabe, T., Sekizawa, Y., Shimura, M., Suzuki, Y., Matsumoto, K., Iwata, M., and Mase, S.,** Effect of probenazole (Oryzemate) on rice plants with reference to controlling the blast, *J. Pestic. Sci.,* 4, 53, 1979.

31. **Zentmeyer, G. A. and Ohr, H. D.,** Systemic soil fungicides for control of phytophthora root rot and stem canker of avocado, *Proc. Am. Phytopathol. Soc.,* 5, 124, 1978.

32. **Laville, E.,** Utilisation d'un nouveau fongicide systemique l'Aliette, dans la lutte contre la gommose a phytophthora des agrumes, *Fruits,* 34, 35, 1979.

33. **Bompeix, G., Ravise, A., Raynall, G., Fettouche, F., and Durand, M. C.,** Stimulation des reactions de defense et de synthese de phytoalexines chez la vigne et la tomato sous l'influence du tris-o-ethyl phosphonate d'aluminium, in Abstr. 18eme Colloq. Soc. Franc. Phytopathologie, Toulouse, April 1980, 130.

34. **Farih, H., Tsao, P. H., and Menge, J. A.,** Fungitoxic activity of fosethyl aluminium on growth, sporulation and germination of *Phytophthora parasitica* and *P. citrophthora, Phytopathology,* 71, 934, 1981.

35. **Vegh, I., Baillot, F., and Roy, J.,** Study of the activity of aluminium ethyl phosphite (LS 74-783) with respect of *Phytopthora cinnamomi* Rands causal agent of decline of ornamental shrubs, *Phytiatr. Phytopharm.,* 26, 85, 1977.

36. **Clerjeau, M. and Beyries, A.,** Comparative study of the preventive action and systemic capacity of some new fungicides (phosphites, prothiocarb and pyroxychlor) against *Phytophthora capsici, Phytiat. Phytopharm.,* 26, 73, 1977.

37. **Vo-thi, H., Bompeix, G., and Ravise, A.,** Role du tris-o-ethyl phosphonate d'aluminium dans la stimulation des reactions de defense des tissue de tomate contre le *Phytophthora capsici, C. R. Acad. Sci. Ser. D,* 288, 1171, 1979.

38. **Bompeix, G., Ravise, A., Raynal, G., Fettouche, F., and Durand, M. C.,** Modalites de l'obtention bloquantes sur feuilles detachées de tomate par l'action du tris-o-ethyl phosphonate d'aluminium (phosethyl d'aluminum), hypotheses sur son mode d'action in vivo, *Annu. Rev. Phytopathol.,* 12, 337, 1980.

39. **Bompeix, G., Fettouche, F., and Saindrenan, P.,** Mode d'action du phosethyl-Al, *Phytiat. Phytopharm.,* 30, 257, 1981.

40. **Hubele, A., Kunz, W., Eckaardt, W., and Sturm, E.,** The fungicidal activity of acylalanines, in *Pesticide Chemistry, Human Welfare and the Environment,* Vol. 1, Doyle, P. and Fujita, T., Eds., Pergamon Press, Oxford, 1983, 233.

41. **Schwinn, F. J., Staub, T., and Urech, P. A.,** A new type of fungicide against diseases caused by Oomycetes, *Meded. Fac. Landbouwet. Rijksuniv. Gent,* 42, 1181, 1977.

42. **Urech, P. A., Schwinn, F., and Staub, T.,** CGA 48988, a novel fungicide for the control of late blight, downy mildews and related soil borne diseases, *Proc. 1977 Br. Crop Prot. Conf. Pests and Diseases,* 1, 623, 1977.

43. **Moser, H., Riks, G., and Senter, H.,** Der Einfluß von Abopisomerie und chiralem Zenhum auf die biologische Aktivitat des Metolachlor, *Z. Naturforsch.,* 876, 451, 1982.

44. **Bergamaschi, P., Borsari, T., Cararaglia, C., and Mirenna, L.,** Methyl N-phenyl-N-2,6-xylyl-DL-alaninate (M 9834), a new systemic fungicide controlling downy mildew and other diseases caused by peronosporales, *Proc. 1981 Br. Crop Prot. Conf. Pests and Diseases,* 1, 11, 1981.

45. **Kaspers, H. and Reuff, J.,** Bekampfung von falschen Mehltaupilazs darch ein neues organisches Fungizic (RD 20615), *Mitt. Biol. Bundesanst. Land Forstwirtsch. Berlin-Dahlem,* 191, 240, 1979.

46. **Baumert, D. and Baschaus, H.,** Cyprofuram, a new fungicide for the control of Phycomycetes, *Meded. Fac. Landbouwet. Rijksuniv. Gent,* 47, 979, 1982.

47. **Gisi, U., Harr, J., Sandmeier, S., and Wiedmer, H.,** SAN 371F, a new systemic oxazolidinone fungicide against diseases caused by Peronosporales, *Meded. Fac. Landbouwet. Rijksuniv. Gent,* 48, 541, 1983.

48. **Kerkenaar, A. and Sijpesteijn, A. K.,** Antifungal activity of metalaxyl and furalaxyl, *Pestic. Biochem. Physiol.,* 15, 71, 1961.

49. **Bruin, G. C. A.,** Resistance in Peronosporales to Acylalanine-type Fungicides, Ph.D. thesis, University of Guelph, Ontario, Canada, 1980, 110.

50. **Kerkenaar, A.,** On the antifungal mode of action of metalaxyl, an inhibitor of nucleic acid synthesis in *Pythium splendens, Pestic. Biochem. Physiol.,* 16, 1, 1981.

51. **Fisher, D. J. and Hayes, A. L.,** Mode of action of the systemic fungicides furalaxyl, metalaxyl and Ofurace, *Pestic. Sci.,* 13, 330, 1982.

52. **Crute, I. R.,** Studies on new systemic fungicides active against *Bremia lactucae* in 3rd Int. Congr. Plant Pathology, Munich, Aug. 16 to 23, 1978.

53. **Staub, T., Dahmen, H., and Schwinn, F.,** Effects of Ridomil on the development of *Plasmopara viticola* and *Phytophthora infestans* on their host plants, *Z. Pflanzenkr. Pflanzenschutz,* 87, 83, 1978.

54. **Cohen, Y., Reivini, M., and Eyal, H.,** The systemic antifungal activity of Ridomil against *Phytophthora infestans* on tomato plants, *Phytopathology,* 69, 645, 1979.

55. **Ward, E. W. B., Lazarovits, G., Stossel, P., Barrie, S. D., and Unwin, D. H.**, Glyceollin production associated with control of *Phytophthora* rot of soybeans by the systemic fungicide metalaxyl, *Phytopathology*, 70, 738, 1980.
56. **Borner, H., Schatz, G., and Grisebach, H.**, Influence of the systemic fungicide metalaxyl on glyceollin accumulation in soybean infected with *Phytophthora megasperma* f. sp. *glycinea*, *Physiol. Plant Pathol.*, 23, 145, 1983.
57. **Lazarovits, G. and Ward, E. W. B.**, Relationship between localized glyceollin accumulation and metalaxyl treatment in the control of phytophthora rot in soybean hypocotyls, *Phytopathology*, 72, 1217, 1982.
58. **Keen, N. T., Holliday, M. J., and Yoshikawa, M.**, Effects of glyphosate on glyceollin production and the expression of resistance to *Phytophthora megasperma* f. sp. *glycinea*, *Phytopathology*, 72, 1467, 1982.
59. **Ward, E.W.B.**, Suppression of metalaxyl activity by glyphosate: evidence that host defence mechanisms contribute to metalaxyl inhibition of *Phytophthora megasperma* f. sp. *glycinea* in soybeans, *Physiol. Plant Pathol.*, 28, 318, 1984.
60. **Pring, R. J. and Richmond, D. V.**, An ultrastructural study of the effect of oxycarboxin on *Uromyces phaseoli* infecting leaves of *Phaseolus vulgaris*, *Physiol. Plant Pathol.*, 8, 155, 1976.
61. **Ziegler, E. and Pontzen, R.**, Specific inhibition of glucan-elicited glyceollin accumulation in soybeans by an extracellular mannan-glycoprotein of *Phytophthora megasperma* f. sp. *glycinea*, *Physiol. Plant Pathol.*, 20, 321, 1982.
62. **Uesugi, Y.**, *Pyricularia oryzae* in rice, in *Fungicide Resistance in Crop Protection*, Dekker, J. and Georgopoulus, S. G., Eds. Pudoc, Wageningen, 1982, 207.

Appendix I

FUNGICIDES WITH THEIR COINED, FORMULATED, AND CHEMICAL NAMES AND MANUFACTURER

Coined name	Formulated name	Chemical name	Manufacturer
		Nonsystemic Fungicides	
Anilazine	Dyrene	2,4-Dichloro-6(2-chloroanilino)-1,3,5-triazine	Bayer, Mobay
—	Aretan	2-Methoxyethyl mercury chloride	ICI
Binapicryl	Morocide	2-sec-Butyl-4,6-dinitrophenyl-3-methylbut-2-enoate	Hoechst
Bordeaux mixture	Several	Copper sulfate-hydrated lime	Several
Captafol	Difolatan	N-1,2,3-Tetrachloroethyl-thiocyclohex-4-ene-1,2-dicarboximide	Chevron
Captan	Several	N-(Trichloromethylthio)-4-cyclohex-4-en-1,2-dicarboximide	Shell
Ceresan	Several	N-Ethylmercuri-p-toluene sulfonanilide	Bayer
Chloraniformethane	Milfaron, Imvgan	N-(2,2,2-Trichloro-1-[3,4-dichloroanilino]-ethyl)-formamide	ICI
Chlorothalonil	Deconil	Tetrachloroisophthalonitrile (2,4,5,6-tetrachloro-1,3-dicyanobenzene)	Dimond Chemrone
Copper oxychloride	Several		Several
Cycloheximide	Actidione	3-(2-[3,5-Dimethyl-2-oxocyclohexyl]-2-hydroxyethyl) glutarimide	Upjohn
Dazomet	Mylone	Tetrahydro-3,5-dimethyl-1,3,5-thiadiazine-2-thione	Verchin
Dichlofluanid	Elvaron	N-Dichlorofluoromethylthio-N,N-dimethyl-N-phenylsulpamide	Bayer
Dichlone	Phygone	2,3-Dichloro-1,4-naphthoquinone	Uniroyal
Dicloran	Botran	2,6-Dichloro-4-nitroaniline	FMC
Dinocap	Dikar	2,6-Dinitro-4-octylphenyl crotonate	Rohm & Haas
Dithianon	Delan	5,10-Dihydro-5,10-dioxonaphtho-(2,3-b)-1,4-dithi-in-2,3-dicarbonitrile	Celamerck
Dodine	Melprex, cyprex	1-Dodecylguanidinium acetate	Cyanamid, Sipcam
Drazoxolon	Ganocide	4(2-Chlorophenylhydrazono)-3-methyl-5-isoxazolone	ICI
Fenaminosulf	Dexon	Sodium-4-dimethylaminobenze-diazosulfonate	Bayer
Fenpropimorph	Carbel	(±)-cis-4-(3-(4-tert-Butylphenyl)-2-methylpropyl-2,6-dimethylmorpholine	BASF

Appendix I (continued)

FUNGICIDES WITH THEIR COINED, FORMULATED, AND CHEMICAL NAMES AND MANUFACTURER

Coined name	Formulated name	Chemical name	Manufacturer
Nonsystemic Fungicides			
Ferbam	Hexaferb	Iron tri-(dimethyldithiocarbamate)	Du Pont
Folpet	Several	N-(Trichloromethanesulfenyl) pthalimide	Bayer
Guazatme	Panoctin Plus	1,1-Iminodi(octamethylene)diguanidine	Kenogard
Hexachlorophene	Isobac	—	Kalo
Iprodione	Rovral	3-(3,5-Dichlorophenyl)-N-isopropylcarbamoyl-2,4-dioxoimidazolidine	Rhone Poulenc
Mancozeb	Dithane M-45	Complex of zinc (2.5%) and manganese (20%) ethylenebisdithiocarbamate	Rohm & Haas
Maneb	Dithane M-22	Manganese ethylenebisdithiocarbamate	Rohm & Haas
Metiram	Polyram-Combi	Zineb and polyethylene thiuram disulfide complex	BASF, Celamerck
Nabam	Nabame	Disodium ethylenebisdithiocarbamate	Du Pont
Propineb	Antriocol	Zinc propylenebisdithiocarbamate (polymeric)	Bayer
Quintozene	Brassical, Terraclor	Pentachloronitrobenzene	Hoechst
SMDC	Vapam	Sodium methyldithiocarbamate	Stauffer
TCMTB	Busan	2-(Thiocyanonethylthio) benzothirazole	Buckman
Thiram	Thiride	Tetramethylthiuramdisulfide	Du Pont
Viclozolin	Ronilan	3–(3,5-Dichlorophenyl)-5-methyl-5-vinyl-1,3-oxazoli-dine-2,4-dione	Rhone Poulenc
Zineb	Dithane Z-78	Zinc ethylenebisdithiocarbamate	Rohm & Haas
Ziram	Cuman	Zinc bis(dimethyldithiocarbamate)	Du Pont, Ciba-Geigy
Systemic Fungicides			
Aluminium-tris	Aliette	Aluminium triethyl phosphonate	May & Baker
Benalaxyl	Galben	Methyl-N-phenylacetyl-N-2,6-xylyl-DL-alaninate	Farmoplant
Benodanil	Calirus	2-Iodobenzoic acid anilide (2-iodobenzianilide)	BASF
Benomyl	Benlate	Methyl-1-(butylcarbamoyl) benzimidazole-2-yl carbamate (MBC)	Du Pont
Biloxazol, Bitertanol	Baycor	1-(1,1-Biphenyl)-4-yl-oxyl)-a-(1,1-dimethyl-ethyl)-1H-1,2,4-triazole-1-ethanol	Bayer

Bupirimate	Nimrod	5-Butyl-2-ethylamino-6-methylpyrimidine-4-yl dimethylsulfamate	ICI
Buthiobate	Denmert	S-n-Butyl-4-tert-butyl-benzyl-N-3-pyridyldithiocarbonimidate	Sumitomo
Carbendazim	Bavistin, Derosal, Lignasin	Methyl benzimidazole-2-yl carbamate (MBC)	BASF
Carboxin	Vitavax	5,6-Dihydro-2-methyl-1,4-oxathiin-3-carboxanidide(2,3-dihydro-6-methyl-5-phenylcarbamoyl-1,4-oxathiin) (DMOC)	Uniroyal
Chlorobenthiazone	—	4-Chloro-3-methylbenzothiazole-2 (3H)-one	Summitomo
Chloroneb	Demosan	1,4-Dichloro-2,5-dimethoxybenzene	Du Pont
Cyclafuramid	—	N-Cyclohexyl-2,5-dimethylfuramid (2-5-dimethyl-furan-3-carboxylic acid cyclohexamide)	BASF
Cymoxanil	Curzate	2-Cyano-N-(ethylaminocarbamyl)-2-(methoxyimino)acetamide (1-(2-Cyano-2-methoxyimino-acetyl)-3-ethylurea	Du Pont
Cypendazole	Folcidin	1-(5-Cyanopentylcarbamoyl)-benzimidazole-yl carbamate	Chemagro
Cyprofuram	—	(±)-h-(N-(3-Chlorophenyl) cyclopropanecarboxamido)-y-butyrolactone	Du Pont
Dichlozolinate	Serinal	Ethyl-3-(3,5-dichlorophenyl)-5-methyl-2,4-dioxo-5-oxazolinecarboxylate	Farmoplant
Diclobutrazol	Vigil	(2R,3R) and (2S,3S)-1-(2,4-dichlorophenyl)4,4-dimethyl-2-(1,2,4-triazole-1-yl)pentan-3-0)	Plant Protection
Dimethirimol	Milcurb	5-Butyl-2-dimethylamino-4-hydroxy-6-methylpyrimidine	Plant Protection
Dodemorph	Mehltaumittel, Meltatox	4-Cyclododecyl-2,6-dimethylmorpholine	BASF
Edifenfos	Hinosan	O-Ethyl-SS-diphenylphosphorodithioate (EDP)	Bayer
Etaconazole	Somax	1-(2-(2,4-Dichlorophenyl)-4-ethyl-1,3-dioxolan-2-yl-methyl)-1H-1,2,4-triazole	Ciba-Geigy
Ethazole (etridiazole)	Terrazole Truban, Terrachlor, Super X	5-Ethoxy-trichloromethyl-1,2,4-thiadiazole (ETMT)	Ohi Mathieson
Ethirimol	Milgo, Milstem	5-Butyl-2-ethylamino-4-hydroxy-6-methyl-pyrimidine	Plant Protection
Fenapanil	Sisthane	a-Butyl-a-phenyl-1H-imidazole-1-propanenitrile	Rohm & Haas
Fenarimol	Rubigan, Rimidin	a-(2-Chlorophenyl)-a-(4-chlorophenyl)-5-pyrimidine methanol	Eli Lilly
Fenfuram	Paroram	2-Methyl-furan-3-carboxylic acid anilide (2-methyl-3-furanilide)	Shell

Appendix I (continued)

FUNGICIDES WITH THEIR COINED, FORMULATED, AND CHEMICAL NAMES AND MANUFACTURER

Coined name	Formulated name	Chemical name	Manufacturer
		Systemic Fungicides	
Fenpropimorph	Carbel	($\pm$)-*cis*-4(3-(4-*tert*-Butylphenyl)-2-methylpropyl-2,6-dimenthyl morpholine	BASF
Fluotrimazole	Perusulon	1-(3-Trifluoromethyltrityl)-1,2,4-triazole	Bayer
Fuberidazole	Voronit	2-(2-Furyl) benzimidazole	Bayer
Furalaxyl	Fongarid	Methyl-*N*-(2,6-dimethyl-phenyl)-N-furoyl-(2)alaninate	Ciba-Geigy
Furcarbanil	Campogran	2,5-Dimethyl-3-furanilide	BASF
Furmecylox		*N*-Cyclonexyl-*N*-methoxy-2,5-dimethyl-3-furanacarboximide	BASF
Hymexazole	Tachigaren	3-Hydroxy-5-methylisozazole	Sankyo
IBP	Kitazin, Kitazin P	*S*-Benzyl-*OO*-di-isopropyl phosphorothioate	Kumiai
Imazalil	Fungaflor	1-(2-(2,4-Dichlorophenyl-2(2-propenyl-oxyethyl) 1H-imidazol) (1-(allyloxy-2,4-dichlorophenethyl)imidazole	Janseen
Iprodione	Rovral	3-(3,5-Dichlorophenyl)-*N*-isopropyl carbamoly-2,4-dioxoimidazolidine-1-carboxamide	Rhone Poulenc
Isoprothiolane	Fuziwan	Di-isopropyl-1,3-dithiolan-2-yl-idenemalonate	Nihon Nohyaku
Mebenil	BAS 305	O-Methyl benzoic acid anilide (2-toluanilide) or (*O*-toluanilide)	BASF
Metalaxyl	Ridomil, Acylon, Apron SD 35	DL-*N*-(2,6-Dimethylphenyl)-*N*-(2-methoxyacetyl)-alaninate	Ciba-Geigy
Milforum	Palafol	(2-Chloro-*N*-(2,6-dimethylphenyl)-*N*-tetrahydro-2-oxo-3-furanyl) acetamide	Chevron
Oxycarboxin	Plantvax	5,6-Dihydro-2-methyl-1,4-oxathiin-3-carboxianilid-4,4-dioxide (2,3-Dihydro-6-methyl-5-phenylcarbamoyl-1,4-oxath-iin-4,4-dioxide)	Uniroyal
Probenazole	Oryzemate	3-Allyloxybenzol 1,2-thiazole-1,2-dioxide	Meiji Seika Kaisha
Prochloraz	Sportak	*N*-Propyl-*N*(2-2,4,6-trichlorophenoxylethyl) imidazole-1-carboxamide	Schering

Procymidone	Sumilex, Sumisclex	*N*-(3,5-Dichlorophenyl)-1,2-dimethylcyclo-propandicarboximide	Sumitomo
Propamocarb	Previcur N	Propyl-3-(dimethylamino)-propylcarbamate NOR-AM hydrochloric acid (PDAC)	
Propiconazole	Desmel Tilt	1-(2-(2,4-Dichlorophenyl)-4-propyl-1,3-di-oxolan-2-yl-methyl)1H-1,2,4-triazole	Ciba-Geigy
Prothiocarb	Dynone, Previcur	*S*-Ethyl-*N*-(3-dimethylaminopropyl)-thiocarbamate hydrochloride	Schering
Pyracarbolid	Sicarol	2-Methyl-5,6-dihydro-4 H-pyran-3-carboxylic acid anilide (2-H-3,4-Dihydro-6-methylpyran-5-carboxanilide)	Hoechst
Pyrazofos	Curamil, Afugan	*O*,*O*-Diethyl-*O*-(5-methyl-6-ethoxycarbonyl-pyra-zolo(1,5a)-pyrimidine-2-yl)-thionophosphate (0-6-Ethoxycarbonyl-5-methylpyra-zolo[1,5a]pyrimidine-2-yl-*OO*-diethyl-phosphorothioate)	Hoechst
Pyroquilon	Fongoren	1,2,5,6-Tetrahydropyrrolo(3,2,1-ij) quinolin-4-one	Ciba-Geigy
Pyroxychlor	Nurelle	2-Chloro-5-methoxy-4-trichloromethyl pyridine	
Thiabendazole	Mertect, Mycozol, Storite, Tec-tab, Tecto, Termaole, Tifume	2-(4-Thiazol-yl) benzimidazole (TBZ)	Merck
Thiophanate (thiopanate-ethyl)	Cercobin, Chipco spot clean, Cycosin, Enovit, Nemafex, Pelt Sol, Topsin 3334, Furf Fungicide	1-2-Di-(3-ethoxycarbonyl-2-thioureido) benzene (die-thyl-4,4-a-phenylenebis (3-thioallophenate)	Nippon Soda
Thiophanate-methyl	Cercobin, Cycosin, Enovit-Super, Fungo, Homai, Mil-dothane, Pelt-44, Topsin-Methyl, Trevin	1,2-Di-(3-methoxycarbonyl-2-thioreido)benzene (di-methyl 4,4-a-phenylenebis (3-thioallophanate)	Nippon Soda
Triadimefon	Amiral, Bayleton	1-(4-Chlorophenoxy)-3,3-dimethyl-1-(1,2,4-triazol-1-yl)butan-2-one	Bayer
Triadimenol	Bayton	1-(4-Chlorophenoxy)-3,3-dimethyl-1-(1,2,4-triazole-1-yl)butan-2-ol	Bayer
Triamifos	Wepsyn	*p*-5-Amino-l-(bis-(dimethylamino)-phosphoryl)-3-phenyl-1,2,4-triazole (*p*-5 Amino-3-phenyl-1,2,4-triazole-l-yl-*N*,*N*,*N*-tetra-methylphosphonic diamide)	Philip Duphar
Triarimol		a-(2,4-Dichlorophenyl)-phenyl-5-pyrimidinemethanol	Eli Lilly

Appendix I (continued)

FUNGICIDES WITH THEIR COINED, FORMULATED, AND CHEMICAL NAMES AND MANUFACTURER

Coined name	Formulated name	Chemical name	Manufacturer
		Systemic Fungicides	
Triazbutyl	Indar	4-Butyl-1,2,4-triazole	Rohm & Haas
Tricyclozole	Beam, Bim	5-Methyl-1,2,4-triazole-(3,4-b) benzothiazole	Eli Lilly
Tridemorph	Calixin, Beacon	N-Tridecyl-2,6-dimethylmorpholine (2,6-dimethyl-4-tridecylmorpholine)	BASF
Triforine	Fungidex, Saprol	N,N-1,4-Piperazindiyl-bis-(2,2,2-trichloro-ethyliden)-bis-formamide (1,4-Di-(2,2,2-trichloro-1-formamidoethyl)-piperazole)	Celamerck

Note: Acronyms in parentheses are the alternate names.

INDEX

D

copper and, 86
white, 92

S

Saccharase, 150
Saccharomyces cerevisiae, 150
Saprophytes, 7—8, 16, 21, 131—132, 135
  buffering capacity of, 16
  colonization of, 17
  growth of, 132
  inhibition of, 14
  in life of plant, 39
  phylloplane fungi, 18
  phyllosphere, 13
  reduction of, 17, 93
  rhizosphere microflora, 161—166
  sensitivity of, 22
Scabs, 15, 19, 65
Scleroderma bovista, 175
Sclerotial fungi, 5
Sclerotinia, 5, 87, 88, 90—91
Sclerotium, 131, 133, 220
Seed-borne disease, control of, 220—221
Seeds
  fungicide absorption by, 20
  injury of with organic mercurials, 119
  treatment of, 1, 22, 40
    endomycorrhizae and, 174
    growth and, 62
    phylloplane mycoflora and, 25, 27
    rhizobia and, 193—195, 203—207
    rhizosphere and, 161
    yield and, 66—67
Selenophoma denacis, 61
S-enantiomers, 231
Senescence, 79, see also Leaf, senescence of
  accelerated, 65, 81
  delayed, 37, 39, 78, 104
  early, 64
  filamentous fungi and, 223
  microorganisms and, 81
  premature, 109
  prevention of, 100—101
Septoria nodorum, 18
Serine concentration, 46
Side effects of agrochemicals, 28, see also Nontarget
    effects
Sitosterol, 45
β-Sitosterol, 113—114
Smuts, 39, 67, 97
Sodium azide, 176
Sodium diethyldithiocarbamate, 146
Sodium, uptake and efflux of, 35
Soil, 8—9
  bacteria of, 137—139
  chemical properties of, 151—154
  denitrification of, 146—147
  enzymatic activity of, 150—151
  fertility of, 131, 139—141, 152
    bacteria and, 137

improved of, 59—60
  rejuvenation of, 185
flooded, 154
fumigation of, 217—218
fungi of, 18, 90, 131—137
iatrogenic effects of fungicides on, 89
microorganisms of, 131—140, 154—155
  actinomycetes, 139—140
  bacteria, 137—139
  fungi, 131—137
nitrification of, 141—146
  nonsystemic, 141—143
  systemic, 143—146
oxygen uptake in, 149—150
pathogens of, 87, 91, 92
pH of, 36, 142, 154
respiration, 148—150
rhizosphere, 161—166
  action mechanism, 165—166
  nonsystemic fungicides, 161—163
  systemic fungicides, 163—165
root-free, 161
Soil biota, 8—10
Soil-borne diseases, control of, 217—220
Soil fumigant, 179, 181
Soil-root interface, 161
Soil treatment, 1, 20—21, 35, 59—60
  endomycorrhizae and, 175—179
  rhizobia and, 207—208
Southern stem blight, 90
Spergon, 204
Spore-forming bacteria, 137
Spore germination, 13
Sporobolomyces, 14—19
Sporobolomyces roseus, 13, 20—21
Stemphylium, 16, 18
Stereum purpureum, 14
Sterol, 42—45, 109, 199
Stigmasterol, 113—114
Storage fungi, 66
Strawberry ringspot virus, 102
Streptomyces, 139
Stunting and deformation of grapes, 125
Sugar content, 49—50
Sulfoxide, 111
Sulfuric acid formation, 36
Sulfur, 1, 120, 151
Sulfur dioxide, 8, 110, 111
Sulfur fungicides, 13, 15
  dusts, 36, 60
  fruit set and, 70, 71
  phytotoxicity, 117—118
  plant growth and yield with, 59—60
Surface fungicides, see Fungicides, surface
Surfactants, 209—210, 234
Symbionts, 169, 173, see also Endomycorrhizae;
    Mycorrhizae
Symbiosis, 169
Symplast, 5—6
Symplastic movement, 16
Systemic fungicides, see Fungicides, systemic